Elements of Complex Analysis

Jacob Sonnenschein

Université Libre de Bruxelles

Simon Green

California State Polytechnic University, Pomona

Dickenson Publishing Company, Inc.
Encino, California and Belmont, California

Copyright © 1977 by Dickenson Publishing Company, Inc.
All rights reserved. No part of this book may be reproduced,
stored in a retrieval system, or transcribed, in any form or
by any means—electronic, mechanical, photocopying, recording,
or otherwise—without the prior written permission of the
publisher, 16250 Ventura Boulevard, Encino, California 91436.

ISBN: 0-8221-0169-6
Library of Congress Catalog Card Number: 76-27549

Printed in the United States of America
Printing (last digit): 9 8 7 6 5 4 3 2 1

Contents

Preface   *vii*

chapter 0
Algebraic and Topological Preliminaries
1

§1. Some notions of set theory   *1*
    1. Logical symbols   *1*
    2. Sets   *1*
    3. Functions   *5*
    4. Sequences and countable sets   *6*

§2. Sets with an algebraic structure   *9*
    1. Groups   *9*
    2. Rings and fields   *10*
    3. Vector spaces   *11*

§3. Sets with a topological structure—topological spaces   *12*
    1. Definition of a topology   *12*
    2. Continuous functions   *15*
    3. Connectedness and the connectedness argument   *17*
    4. Compactness   *20*

§4. Metric spaces   *22*
    1. Definitions   *22*
    2. Normed vector spaces   *24*
    3. Sequences in metric spaces; completeness   *25*
    4. Continuity in metric spaces   *28*
    5. Compactness in metric spaces   *30*

chapter 1
Complex Differentiable Functions
37

§1. Summary of the course   37
§2. The complex numbers   38
    1. Algebraic aspects   38
    2. Geometric representation of complex numbers   39
    3. Arguments of products and quotients   44
    4. The Riemann sphere   47
§3. Topology of the complex plane   51
    1. Neighborhoods and discs in the complex plane; open and closed sets   51
    2. Sequences in the complex plane   52
    3. Compact sets in the complex plane   53
    4. Connected sets and domains in the complex plane   55
    5. Arcs and closed curves   58
§4. Differentiable functions   60
    1. Continuous functions   60
    2. Differentiability   63
    3. The Cauchy-Riemann equations   66
    4. Primitives   70
    5. Geometric interpretation of the derivative   71
    6. Isomorphisms   73

chapter 2
Holomorphic Functions
77

§1. Homotopy and line integrals   77
    1. Homotopy of closed curves with respect to a domain D 77
    2. Line integrals   81
    3. Properties of line integrals   86
§2. Cauchy's theorem   93
    1. Cauchy's theorem for starlike domains   93
    2. Cauchy's theorem for general domains; residues   99
§3. Cauchy's formula   106
    1. The multivalued function log z 106
    2. Index of closed curve   110

3. Linear integral transformations   *112*
4. Cauchy's formula   *116*

§4. Compactness of a bounded and closed subset of $H(D)$ *120*

§5. Harmonic functions   *126*

§6. The mean-value property and maximum modulus principle   *129*

chapter 3
Analytic Functions
133

§1. Power series   *133*

1. Numerical series   *133*
2. Series of functions and uniform convergence   *136*
3. Power series   *137*
4. A function holomorphic on D is analytic on D *142*
5. Cauchy's inequalities   *145*
6. Schwarz's lemma   *146*
7. Zeros of an analytic function   *146*
8. Analytic continuation   *149*
9. Permanency principle of functional relations   *153*

§2. Laurent series   *156*

1. Definition of Laurent series   *156*
2. Isolated singularities and their classification   *160*
3. Meromorphic functions   *164*
4. The point at infinity   *165*

§3. Residues and their applications   *169*

1. The residue theorem   *169*
2. Applications of the residue theorem   *172*
3. The logarithmic residue   *183*

chapter 4
Conformal Mappings
191

§1. Conformal mappings   *191*

1. Definitions and relations   *191*
2. Mapping at points where $f' = 0$ *194*
3. The open mapping theorem   *196*

§2. Homographies *199*
  1. Definitions and basic properties *199*
  2. Symmetry *204*

§3. Riemann's mapping theorem *208*

Solutions to the Odd-Numbered Problems *213*

Index *279*

Preface

This book is the result of lectures given by Prof. Dr. J. Sonnenschein at the Brussels University. It is an introductory course in complex analysis of one complex variable. It contains sufficient material for about thirty lectures.

The aim of the book is to introduce the principal notions and theorems of complex analysis and to make the reader acquainted as quickly as possible and with as much rigor as can be obtained in a short course, with the knowledge necessary to use the most important results of complex analysis in pure and applied mathematics.

An introductory course to complex analysis is not supposed to contain original results. The only original features of this book lie in the presentation of the material.

Chapter 0 introduces the reader to the basic notions of topology which are necessary to understand the following chapters. Those students who have followed a course of topology may skip Chapter 0.

The following four chapters present the essentials of complex analysis. It seems to us that four notions form the basis of complex analysis. In each of the chapters 1 to 4 we have treated one of these notions: In the first chapter the notion of complex differentiability of a function, in the second one the property of a function that its integral over any closed curve of a certain family of curves is zero. This property we called holomorphicity. In the third chapter we have introduced the notion "analytic" for functions which possess about any point of a domain an expansion in a Taylor series. The fourth property of functions, which is studied in the fourth chapter, is a geometric property, namely the conformality of the mapping.

These four notions which by their definition are quite different are shown to imply each other, that is, to be equivalent.

In our definitions and proofs we have been influenced by many books on the same subject, especially by the books of L. Ahlfors: *Complex*

Analysis; H. Cartan: *Théorie élémentaire des fonctions analytiques d'une ou plusieurs variables complexes;* Z. Nehari: *Conformal mapping;* and others.

It is a pleasure for us to express our gratitude to Prof. Dr. J. P. Gossez of the Brussels University for his suggestions and valuable remarks and to Profs. Dr. Emil Herzog and Dr. Ta Li of California State Polytechnic University who checked the manuscript and prepared the solutions for the extensive Problem Section and the Teacher's Manual which will form a valuable addition to the text.

Jacob Sonnenschein
Simon Green

chapter 0

Algebraic and Topological Preliminaries

§1. Some notions of set theory

1. Logical symbols

We introduce here some logical symbols which we shall use throughout this book.

Let S_1 and S_2 be two statements, then the symbol $S_1 \Rightarrow S_2$ means that statement S_1 implies statement S_2. The symbol $S_1 \Leftrightarrow S_2$ means $S_1 \Rightarrow S_2$ and $S_2 \Rightarrow S_1$, and we say that statement S_1 is equivalent to statement S_2.

Another way to express $S_1 \Leftrightarrow S_2$ is to say S_1 holds if and only if S_2 holds, or, briefly S_1 holds *iff* S_2 holds. [*Iff*, with double *f*, is an abbreviation for "if and only if."]

The symbol $\exists$ means "there exists," and the symbol $\ni$ stands for "such that."

EXAMPLE

"$\exists$ at least one integer $x \ni \frac{3}{4} < x < \frac{5}{2}$" means "there exists at least one integer x such that $\frac{3}{4} < x < \frac{5}{2}$."

The symbol $\forall$, an inverted A, means "for all."

2. Sets

We assume that the reader of this book is acquainted with elementary set theory; in any case, we remind him of the following notations:

$a \in A$, the element a belongs to the set A.

$a \notin A$, the element a does not belong to the set A.

$A \subset B$ or $B \supset A$, A is a subset of B; that is, $a \in A \Rightarrow a \in B$.

$A \not\subset B$, A is not a subset of B. We shall write $A = B$ if $A \subset B$ and $B \subset A$.

We shall use the symbol $\emptyset$ for the void or empty set, $\{a\}$ for the set containing only the element a, $\{a_1, a_2, \ldots, a_n\}$ for the set containing the elements $a_1, a_2, a_3, \ldots, a_n$, and $\{a|P\}$ for the set of all a having property P.

$A \cup B$ is the union of the sets A and B: $A \cup B = \{a | a \in A \text{ or } a \in B\}$. It is the set of all elements a which belong either to A or to B or to both

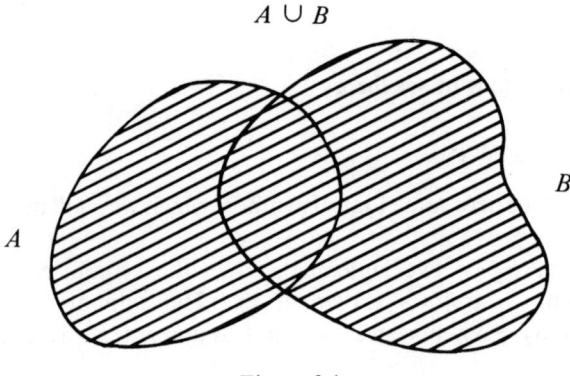

Figure 0.1

of them. $A \cap B$ is the intersection of the sets A and B: $A \cap B = \{a | a \in A \text{ and } a \in B\}$; $A \cap B$ is the set of all elements which belong to A and also to B.

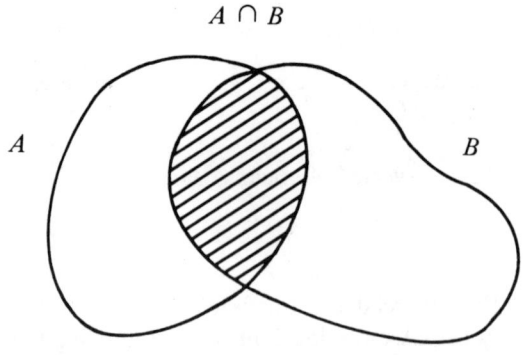

Figure 0.2

If $\mathscr{C}$ is a collection of sets C then the union or intersection of all sets C belonging to $\mathscr{C}$ will be written:

$$\bigcup_{C \in \mathscr{C}} C \quad \text{or} \quad \bigcap_{C \in \mathscr{C}} C$$

By $B\backslash A$ we mean the set of all elements belonging to B which do not belong to A; we do not require $B \supset A$: $B\backslash A = \{a | a \in B, a \notin A\}$. Clearly $B\backslash A = B\backslash(A \cap B)$.

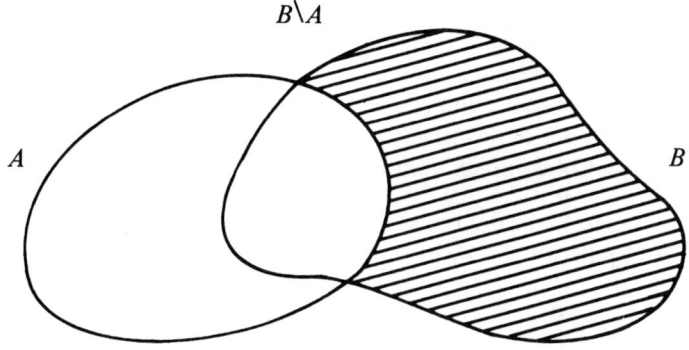

Figure 0.3

Let A, A_1, and A_2 be three sets. We verify easily de Morgan's duality relations:

$$A\backslash(A_1 \cup A_2) = (A\backslash A_1) \cap (A\backslash A_2)$$

and

$$A\backslash(A_1 \cap A_2) = (A\backslash A_1) \cup (A\backslash A_2)$$

Let us prove the first of the two relations: If $x \in A\backslash(A_1 \cup A_2)$, then $x \in A$, $x \notin A_1$ and $x \notin A_2$. Therefore, $x \in A\backslash A_1$ and $x \in A\backslash A_2$, which shows that $x \in (A\backslash A_1) \cap (A\backslash A_2)$. On the other hand, if $x \in (A\backslash A_1) \cap (A\backslash A_2)$, then $x \in A$, $x \notin A_1$ and $x \notin A_2$. It follows that $x \in A$ and $x \notin A_1 \cup A_2$, or $x \in A\backslash(A_1 \cup A_2)$.

The proof of the second duality relation is left to the reader.

DeMorgan's relations may be generalized for an infinite collection $\mathscr{C}$ of sets C in the following way:

$$A\backslash \bigcup_{C \in \mathscr{C}} C = \bigcap_{C \in \mathscr{C}} (A\backslash C)$$

and

$$A\backslash \bigcap_{C \in \mathscr{C}} C = \bigcup_{C \in \mathscr{C}} (A\backslash C)$$

Let A and B be two sets; we define the *Cartesian product* $A \times B$ to be the set of ordered pairs (a, b) with $a \in A$ and $b \in B$:

$$A \times B = \{(a, b) | a \in A, b \in B\}$$

We sometimes say that a is the first component and b the second component of the pair (a, b).

Further on, we shall denote by—

- **Z** the set of integers;
- **P** the set of positive integers, or natural numbers;
- **Q** the set of rational numbers;
- **R** the set of real numbers, $\mathbf{R}^+$ the set of positive real numbers, and $\mathbf{R}^-$ the set of negative real numbers;
- **C** the set of complex numbers.

It follows from the definition of the Cartesian product that $\mathbf{R} \times \mathbf{R} = \mathbf{R}^2$ is the set of pairs (x_1, x_2) with $x_1 \in \mathbf{R}$ and $x_2 \in \mathbf{R}$, and

$$\mathbf{R}^n = \{(x_1, x_2, \ldots, x_n) | x_1, x_2, \ldots, x_n \in \mathbf{R}\}$$

Let A be a set. A subset S of $A \times A$ is called a *relation* in A. For example, if $A = \mathbf{R}$, then $S = \{(a, b) | b = a^2, a \in \mathbf{R}\} \subset A \times A$ is a relation in A.

A relation in A is called an *equivalence relation* if it satisfies the following three conditions:

(i) $(a, a) \in S \quad \forall a \in A \quad$ (*reflexive*)

(ii) $(a, b) \in S \Rightarrow (b, a) \in S \quad$ (*symmetric*)

(iii) (a, b) and $(b, c) \in S \Rightarrow (a, c) \in S \quad$ (*transitive*)

EXAMPLE

The set of pairs $(a, b) \in \mathbf{R} \times \mathbf{R}$ with $b > a$ defines the relation "greater than," but is not an equivalence relation since it is neither reflexive nor symmetric.

EXAMPLE

Let A be the set of all intervals $I = [a, b] \subset \mathbf{R}$, and denote by $L(I)$ the length of I. The set of pairs $(I_1, I_2) \in A \times A$ for which $L(I_1) = L(I_2)$ defines an equivalence relation.

3. Functions

Let A, B be two sets. We define a *function*, or *mapping*, $f: A \to B$ (A into B) by associating with each element $a \in A$ one and only one element $b \in B$, which we denote by $b = f(a)$.

A function or mapping f from A into B is thus a set of ordered pairs $(a, b) \in A \times B \ni \forall a \in A$ ∃ exactly one ordered pair (a, b) in the set with first component a. We call $b = f(a)$ the value of a by f, or we say that a is transformed into b by f. We call the set A of all first components the *domain of definition* of f, and the set of the second components the *image* of A under f, denoted by $f(A)$; clearly $f(A) \subset B$. If the image of A under f is B ($f(A) = B$)—that is, if for each $b \in B$ ∃ an element a in A with $f(a) = b$—then f is said to be a mapping of A *onto* B.

We say that f is *univalent* if $f(a_1) = f(a_2) \Rightarrow a_1 = a_2$. If f is onto and univalent, then we say f is a *one-to-one correspondence* between A and B, and that A and B are *equivalent* (in the sense of cardinality or set theory). A trivial one-to-one mapping is the identity mapping $A \to A$ with $f(a) = a \,\forall\, a \in A$.

Let f be a one-to-one mapping from A onto B; then to a given $b \in B$ there corresponds one and only one element $a \in A$. This means b is mapped by a certain function into A. We denote this function by f^{-1} and call it the *inverse function*, or *inverse mapping*, of f. Hence, if f is one-to-one, then $f(a) = b \Rightarrow f^{-1}(b) = a$.

If $A_1 \subset A$ and $B_1 \subset B$, then we write $f(A_1) = \{f(a) | a \in A_1\}$ and $f^{-1}(B_1) = \{a | f(a) \in B_1\}$. If f is a mapping $A \to B$ and g is a mapping $B \to C$, then we call the mapping of $A \to C$ which contains all pairs $(a, g(f(a)))$, with $a \in A$, $f(a) \in B$, and $g(f(a)) \in C$, the composition of g and f, denoted by $g \circ f$. Evidently, $\{a, g(f(a))\}$ is a subset of $A \times C$. If f and g are one-to-one, then $g \circ f$ is also one-to-one, and its inverse function is $(g \circ f)^{-1} = f^{-1} \circ g^{-1}$.

If f maps $A \to B$ and A_1 is a subset of A, then we call the set of pairs $(a, f(a))$ with $a \in A_1$ the *restriction of f to A_1*, which is denoted by f/A_1 and which maps $A_1 \to B$. Clearly, if f is univalent, so is f/A_1.

EXAMPLE

Let $x \in \mathbf{R}$; then $f = x^2$ is a mapping $\mathbf{R} \to \mathbf{R}$ and contains the ordered pairs (a, b) with $a \in \mathbf{R}$, $b \in \mathbf{R}$ and $b = a^2$, for example, $(1, 1)$, $(2, 4)$, $(0, 0)$, $(-1, 1)$. This mapping is not from $\mathbf{R}$ onto $\mathbf{R}$ because $x^2 \geq 0$, so that $f(\mathbf{R})$ does not cover $\mathbf{R}$ but only $\mathbf{R}^+ \cup \{0\}$, where $\mathbf{R}^+$ is the set of real positive numbers. The function f is not univalent since we have $f(-a) = f(a)$, so there is no inverse function f^{-1}.

Let I be the interval $\{x\,|\,0 \le x \le 2\}$; then $f(I)$ is the interval $I_1 = \{y\,|\,0 \le y \le 4\}$, and $f^{-1}(I_1)$ is the interval $\{x\,|-2 \le x \le 2\}$. The restriction $f/\mathbf{R}^+ = x^2/\mathbf{R}^+$ has an inverse function, namely, the positive square root, and $f/\mathbf{R}^-$ also has an inverse function, namely, the negative square root.

EXAMPLE

Let z be a complex number; then $f = 1/z$ is a mapping which is defined throughout $\mathbf{C}\backslash\{0\}$. $f\colon \mathbf{C}\backslash\{0\} \to \mathbf{C}\backslash\{0\}$ is a mapping onto which is univalent since $1/z_1 = 1/z_2 \Rightarrow z_1 = z_2$; thus it is one-to-one. This mapping is called the *inversion mapping*; it contains the ordered pairs $(z, 1/z)$ with $z \in \mathbf{C}\backslash\{0\}$.

The inverse function f^{-1} exists since f is univalent and onto; f^{-1} contains the ordered pairs $(1/z, z)$ with $1/z \in \mathbf{C}\backslash\{0\}$. It is clear that $f^{-1} \circ f$ contains the pairs $(z, f^{-1}(f(z))) = (z, f^{-1}(1/z)) = (z, z)$ which defines the identity mapping. Set $z' = 1/z$; then f^{-1} contains the pairs $(z', 1/z')$, the same pairs as f. So f^{-1} is the same function as f.

Problems

0.1.: Show that equivalence in the sense of set theory is an equivalence relation.
0.2.: Show that the distributive law holds for the two operations $\cup$ and $\cap$:

$$A \cup (B \cap C) = (A \cup B) \cap (A \cup C)$$

and

$$A \cap (B \cup C) = (A \cap B) \cup (A \cap C)$$

---

4. Sequences and countable sets

DEFINITION A *sequence* is a set whose element are indexed by the positive integers:

$$a_1, a_2, a_3, \ldots, a_n, \ldots = \{a_n\}$$

EXAMPLE

$\{a_n = (-1)^n\}$ defines the sequence $-1, 1, -1, 1, \ldots$.

DEFINITION A set which is equivalent to the set of positive integers is said to be *countable*.

Clearly every countable set can be indexed in the form of a sequence; conversely, every sequence is countable since we have the one-to-one correspondence $n \leftrightarrow a_n$.

EXAMPLE

1 The set of all positive even integers is countable and forms the sequence

$$a_1 = 2, \quad a_2 = 4, \quad \ldots, \quad a_n = 2n, \quad \ldots$$

The set of all positive even integers is equivalent to the set of natural numbers. The one-to-one correspondence is easily seen to be given by:

$$\begin{array}{cccccc} 1, & 2, & 3, & 4, & \ldots & n, & \ldots \\ \updownarrow & \updownarrow & \updownarrow & \updownarrow & & \updownarrow & \\ 2, & 4, & 6, & 8, & \ldots & 2n, & \ldots \end{array}$$

The function $f(n) = 2n$ which describes this one-to-one correspondence is univalent, since $2n = 2m \Leftrightarrow n = m$. Note that in this example a set is equivalent to one of its subsets. Clearly this is possible only for infinite sets.

2 The set of all integers is countable. Indeed, if we write the integers in the following order, $0, 1, -1, 2, -2, \ldots$, then $a_1 = 0$, $a_2 = 1$, $a_3 = -1$, $a_4 = 2, \ldots$, and in general $a_{2n} = n$ and $a_{2n+1} = -n$.

Let us establish two important properties of countable sets.

1. *Every infinite subset S' of a countable set S is countable.*

If $S = \{a_1, a_2, \ldots, a_n, \ldots\}$, then denote by a_1' the first element of S' you encounter in the sequence S, by a_2' the second, and so on. Then $S' = \{a_1', a_2', \ldots, a_n', \ldots\}$ is countable.

EXAMPLE

The odd integers are countable since they are an infinite subset of all the integers, which are countable.

2. *The union of a countable family of countable sets is countable.*

Let $\{S_1, S_2, \ldots, S_n, \ldots\}$ be the countable family of countable sets, and a_{ij} the jth element of S_i. All the elements a_{ij} may then be written in the following order:

$$a_{11}, a_{12}, a_{13}, \ldots$$
$$a_{21}, a_{22}, a_{23}, \ldots$$
$$a_{31}, a_{32}, a_{33}, \ldots$$
$$\ldots, \ldots, \ldots, \ldots$$

We have listed the elements of S_1 in the first row, the elements of S_2 in the second row, and so on. Now we can write all these elements in the form of a sequence in the following way: $a_{11}, a_{12}, a_{21}, a_{13}, a_{22}, a_{31}, \ldots$, such that a_{mn} preceeds a_{rs} if $m + n < r + s$, or, in case $m + n = r + s$, if $m < r$. The order used is illustrated as follows:

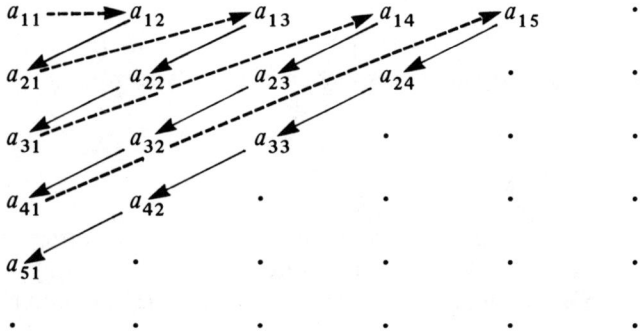

This method of ordering elements a_{ij} into a sequence is called the diagonal method.

COROLLARY *The Cartesian products $\mathbf{P} \times \mathbf{P}$ and $\mathbf{Z} \times \mathbf{Z}$ are countable.*

EXAMPLE

The set of rational numbers is countable. In fact, any rational number is the ratio of two relatively prime integers p/q with $q > 0$. Hence, the set of rational numbers is equivalent to an infinite subset of $\mathbf{Z} \times \mathbf{Z}$, where $\mathbf{Z}$ denotes the set of integers.

Problems

0.3.: Show that the set of real numbers is not countable.
0.4.: Show that the set of points of $\mathbf{R}^2$ with rational coordinates is countable. How could you generalize this result?

§2. Sets with an algebraic structure

1. Groups

DEFINITION A set G with mapping $G \times G \to G$ which associates with each pair (a, b), a and $b \in G$, an element of G denoted by $a * b$ is called a *group* $(G, *)$ if the following three conditions are satisfied:

1. The "group operation" is associative; that is,
$$(a * b) * c = a * (b * c), \quad \forall a, b, c \in G$$

2. $\exists$ a "neutral element" $e \in G \ni$
$$a * e = e * a = a, \quad \forall a \in G$$

3. Every element $a \in G$ has an "inverse element" $a^{-1} \in G \ni$
$$a * a^{-1} = a^{-1} * a = e$$

The group is said to be *Abelian* if $a * b = b * a$, $\forall a, b \in G$.

EXAMPLE

The one-to-one mappings of a set A onto itself form a group $G(A)$ if we take as the group operation the composition of mappings. Indeed, if f, g and $h \in G(A)$, then obviously $f \circ g$ and $g \circ f \in G(A)$, and we have $(f \circ g) \circ h = f \circ (g \circ h)$. The neutral element is the identity mapping and the inverse element to f is f^{-1}. The proof is easy and left to the reader.

REMARK

If A and B are equivalent sets, that is, if $\exists$ a one-to-one correspondence $g : A$ onto B, then we find all one-to-one correspondences A onto B by composing the elements of $G(A)$ with g or g with the elements of $G(B)$.

The element $a * b$ is sometimes called the product of a and b and denoted by $a \cdot b$, or the sum of a and b and denoted by $a + b$. If we use addition $(+)$ as group operation, then we call the neutral element the zero element, denoted by 0.

EXAMPLES

1 The rational numbers without zero, $\mathbf{Q}\backslash\{0\}$, with operation $\cdot$, $(\mathbf{Q}\backslash\{0\}, \cdot)$, form a group.

2 The integers $\mathbf{Z}$ with the operation $+$, $(\mathbf{Z}, +)$, form a group.

A subset G' of a group $(G, *)$ is called a *subgroup* if $(G', *)$ is a group.

Two groups $(G, *)$ and $(H, *)$ are called *isomorphic* if there exists a one-to-one mapping $f : G$ onto H which preserves the algebraic structure; that is, $f(g_1 * g_2) = f(g_1) * f(g_2)$, where g_1 and g_2 are two arbitrary elements of G. Clearly isomorphicity is an equivalence relation; indeed:

(i) Every group $(G, *)$ is isomorphic to itself (take the identity mapping for f).

(ii) If $(G, *)$ is isomorphic to $(H, *)$, then $(H, *)$ is isomorphic to $(G, *)$, since, if f is the one-to-one mapping $G \to H$ which preserves the algebraic structure, then f^{-1} is a one-to-one mapping $H \to G$ and preserves the algebraic structure.

(iii) If $(G, *)$ is isomorphic to $(H, *)$ and $(H, *)$ is isomorphic to $(J, *)$, then we find easily that $(G, *)$ is isomorphic to $(J, *)$.

2. Rings and fields

DEFINITION A set R with two operations $+$ and $\cdot$, $(R, +, \cdot)$, is said to be a *ring* if $(R, +)$ is an Abelian group, and if the operation $\cdot$ is associative and distributive; that is, $(a \cdot b) \cdot c = a \cdot (b \cdot c)$ and $a \cdot (b + c) = a \cdot b + a \cdot c$, $(b + c) \cdot a = b \cdot a + c \cdot a$.

A ring R is said to be *commutative* if $a \cdot b = b \cdot a \ \forall a, b \in R$.

EXAMPLE

The integers with the two operations $+$ and $\cdot$, $(\mathbf{Z}, +, \cdot)$, form a commutative ring. Note that $(\mathbf{Z}, \cdot)$ is not a group since $a \cdot x = b$ with $a, b \in \mathbf{Z}$ does not always have a solution in $\mathbf{Z}$.

DEFINITION A commutative ring F is called a *field* if the equation $a \cdot x = b$ has a solution for any pair $a, b \in F$, except $a = 0$, or in other words if $(F \backslash \{0\}, \cdot)$ is an Abelian group.

Two fields $(G, +, \cdot)$ and $(H, +, \cdot)$ are said to be *isomorphic* if there exists a one-to-one mapping $f : G$ onto H which preserves the algebraic structure; that is,

$$f(g_1 + g_2) = f(g_1) + f(g_2)$$

and

$$f(g_1 \cdot g_2) = f(g_1) \cdot f(g_2)$$

Problems

0.5.: Show that if $(R, +, \cdot)$ is a ring, then $(R, \cdot)$ need not be a group.
0.6.: Show that the polynomials in one variable with real coefficients form a commutative ring, but not a field.

3. Vector spaces

DEFINITION A real (complex) *vector space* V is a set of elements called vectors, denoted by $x, y, z, \ldots$, and two operations, addition $(+)$ and scalar multiplication, which satisfy the following conditions:
1. $(V, +)$ is an abelian group.
2. The multiplication of a vector by a real (complex) number, called a scalar, is defined, and is distributive and associative: if a and b are two real (complex) numbers, then
$$(a + b)x = ax + bx, \quad a(x + y) = ax + ay$$
and
$$a(bx) = (ab)x$$
Moreover, we require that $1x = x$.

EXAMPLE

The set $\mathbf{R}^n$ consisting of the elements $(x_1, x_2, \ldots, x_n)$ with $x_i \in \mathbf{R}$, $i = 1, 2, \ldots, n$, is a vector space if we define addition by
$$x + y = (x_1 + y_1, x_2 + y_2, \ldots, x_n + y_n)$$
and scalar multiplication by
$$ax = (ax_1, ax_2, \ldots, ax_n)$$

A commutative ring with a scalar multiplication (real or complex) satisfying the conditions:
$$(a + b)x = ax + bx, \quad a(x + y) = ax + ay$$
and
$$a(bx) = (ab)x, \quad a(xy) = (ax)y = x(ay)$$
is called an *algebra*.

EXAMPLE

The set of all polynomials with real coefficients forms an algebra if we take as scalars the real numbers.

Problem

0.7.: Show that the polynomials of degree $\leq n$ in one variable, with real coefficients, form a real vector space.

---

§3. Sets with a topological structure—topological spaces

1. Definition of a topology

In order to define connectedness, compactness, continuity, and other concepts, it is necessary to endow with a *topology* the sets we are considering—such as **C**, the set of complex numbers, **R**, the set of real numbers, and sets whose elements are functions. Quite frequently we shall use the word "space" for "set" and the term "points" for "elements."

DEFINITION   Let E be a space and $\mathscr{T}$ a collection of subsets of E. We shall call $\mathscr{T}$ a *topology* for E if it satisfies the following conditions:

O_1: $\varnothing \in \mathscr{T}$ and $E \in \mathscr{T}$.

O_2: The union of an arbitrary family $\mathscr{O}$ of sets belonging to $\mathscr{T}$ belongs to $\mathscr{T}$.

O_3: The intersection of a finite family $\mathscr{O}$ of sets belonging to $\mathscr{T}$ belongs to $\mathscr{T}$.

The subsets $\mathscr{T}$ are called the *open sets* of E.

According to this definition, a given space may have different topologies.

EXAMPLE

Consider the space of the real line **R**. Let S be a subset of **R**. We say that $c \in S$ is an interior point of S if $\exists$ an open interval $]a, b[\subset S$ $\ni c \in]a, b[$; that is, $a < c < b$. We say that S is open if every point of S

is an interior point. It is easy to verify that the interval $]a, b[$ is open and that the interval $[a, b]$ is not open because a and b which belong to $[a, b]$ are not interior points of $[a, b]$. The collection of open sets of $\mathbf{R}$, we have defined in this way is called the *standard topology* of $\mathbf{R}$. We shall prove later on (see Theorem 0.9) for the more general case of metric spaces that the collection of sets of the standard topology of $\mathbf{R}$ satisfies conditions O_1, O_2 and O_3.

It is also easy to see that, if we declare all subsets of $\mathbf{R}$ to be open, then the family of all subsets satisfies also the conditions O_1, O_2, O_3; it defines another topology for $\mathbf{R}$ called the *discrete topology*. We get still another topology for $\mathbf{R}$ if we take as the only open sets $\emptyset$ and $\mathbf{R}$.

Let us consider a space E and assume that we have defined a family $\mathcal{T}$ of subsets of E, which satisfy conditions O_1, O_2, O_3. Then the pair $(E, \mathcal{T})$ is called a *topological space*.

DEFINITION A set $S \subset E$ is called *closed* if its complement $E \backslash S$ is open.

$\emptyset$ and E are closed because they are complements of E and $\emptyset$, respectively. Therefore, $\emptyset$ and E are both open and closed.

REMARK

Using the duality relation of de Morgan, we find that the intersection of an arbitrary family $\mathscr{F}$ of closed sets is closed, and that the union of a finite family of closed sets is again closed. Indeed, if $\mathscr{F}$ is an arbitrary family of closed sets, then for any $\mathbf{C} \in \mathscr{F}$ the set $E \backslash \mathbf{C}$ is open and so is $\bigcup_{\mathbf{C} \in \mathscr{F}} (E \backslash \mathbf{C})$. Since

$$\bigcup_{\mathbf{C} \in \mathscr{F}} (E \backslash \mathbf{C}) = E \backslash \bigcap_{\mathbf{C} \in \mathscr{F}} \mathbf{C}$$

we see that $\bigcap_{\mathbf{C} \in \mathscr{F}} \mathbf{C}$ is closed. In the same way, we see that if $\mathscr{F}$ is a finite family of closed sets $\mathbf{C}$, then $E \backslash \mathbf{C}$ is open as well as $\bigcap_{\mathbf{C} \in \mathscr{F}} (E \backslash \mathbf{C})$, and since $\bigcap_{\mathbf{C} \in \mathscr{F}} (E \backslash \mathbf{C}) = E \backslash \bigcup_{\mathbf{C} \in \mathscr{F}} \mathbf{C}$, we see that $\bigcup_{\mathbf{C} \in \mathscr{F}} \mathbf{C}$ is closed. One can define a topology for E by selecting a family of subsets of E to be the closed sets of the topology.

DEFINITION We call a set $N(p)$ a *neighborhood* of point p if $\exists$ an open set $O \ni p \in O \subset N(p)$.

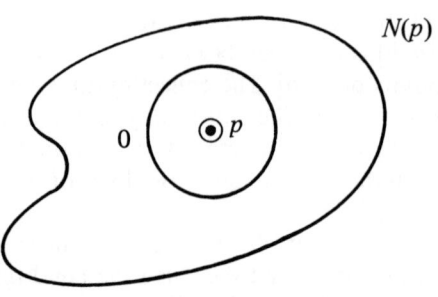

Figure 0.4

$N(p)$ need be neither open nor closed. Clearly, if $N(p) \subset M$, then M is also a neighborhood of p. Furthermore, the intersection of a finite number of neighborhoods of p is again a neighborhood of p. It is possible to introduce a topology for a space E by defining neighborhoods for its points.

DEFINITION The point x is called an *interior point* of a set S if $\exists$ a neighborhood $N(x)$ of x *contained in* S; in short, *if* S *is a neighborhood of* x.

THEOREM 0.1 *A set S is open iff all its points are interior.*

PROOF

Indeed, suppose S is open. Then S is a neighborhood of all its points. Hence all its points are interior points.

On the other hand, suppose all points $x \in S$ are interior; then each $x \in S$ is contained in some open set $O \subset N(x) \subset S$. The union of these open sets is S; hence S is open.

DEFINITION If S is any subset of E, then the *interior* of S, denoted by S^0, is the union of all open subsets of S. The *closure* of S, denoted by $\bar{S}$, is the intersection of all closed sets containing S.

Obviously S^0, as a union of open sets, is open, and $\bar{S}$, as an intersection of closed sets, is closed.

DEFINITION The *boundary* of S, denoted by $b(S)$, is the set $b(S) = \bar{S} \cap \overline{E \backslash S}$.

Obviously, $b(S)$ is closed, since it is intersection of two closed sets.

2. Continuous functions

DEFINITION Given two topological spaces $(E_1, \mathcal{T}_1)$ and $(E_2, \mathcal{T}_2)$, and the mapping $f: E_1 \to E_2$, we say that f is *continuous at* $x \in E_1$ if, for each neighborhood $N_2 \subset E_2$ of $f(x)$, $\exists$ neighborhood $N_1 \subset E_1$ of $x \ni f(N_1) \subset N_2$. We say that f is a *continuous mapping or function*, on E_1 provided that f is continuous at each point of E_1.

THEOREM 0.2 *The following statements are equivalent.*

(i) $f: E_1 \to E_2$ is continuous on E_1.

(ii) Given any open set $O \subset E_2$, the set $f^{-1}(O) \subset E_1$ is open.

(iii) Given any closed set $F \subset E_2$, the set $f^{-1}(F) \subset E_1$ is closed.

PROOF

$(i) \Rightarrow (ii)$: If $f(x) \in O$, or $x \in f^{-1}(O)$, and if $N_2 \subset O$ is a neighborhood of $f(x)$ (N_2 may be equal to O), then $\exists$ a neighborhood $N_1 \subset E_1$ of $x \ni f(N_1) \subset N_2 \subset O$. Therefore, $N_1 \subset f^{-1}(N_2) \subset f^{-1}(O)$ and so x is an interior point of $f^{-1}(O)$. Since x is arbitrary in $f^{-1}(O)$, all points of $f^{-1}(O)$ are interior points, and $f^{-1}(O)$ is open.

$(ii) \Rightarrow (iii)$: Given a closed set $F \subset E_2$, the set $E_2 \backslash F$ is open and $f^{-1}(E_2 \backslash F)$ is open; now $f^{-1}(F) = E_1 \backslash f^{-1}(E_2 \backslash F)$, so $f^{-1}(F)$ is closed.

$(iii) \Rightarrow (i)$: Given a neighborhood $N_2(f(x)) \subset E$ of $f(x)$, then $N_2(f(x))$ contains an open set O_2 which itself contains $f(x)$. $E_2 \backslash O_2$ is closed and so is $f^{-1}(E_2 \backslash O_2)$ because of (iii). Now $f^{-1}(O_2) = E_1 \backslash f^{-1}(E_2 \backslash O_2)$. It follows that $f^{-1}(O_2)$ is a neighborhood $N_1(x)$ of x and $f(N_1(x)) \subset O_2 \subset N_2(f(x))$.

If E_1, E_2, E_3 are three topological spaces, f is a continuous mapping $E_1 \to E_2$, and g is a continuous mapping $E_2 \to E_3$, then $g \circ f$ is a continuous mapping $E_1 \to E_3$. Indeed, if O_3 is any open set in E_3, then $(g \circ f)^{-1}(O_3) = f^{-1}(g^{-1}(O_3))$. Now $O_2 = g^{-1}(O_3)$ is an open set of E_2 because of the continuity of g, and $f^{-1}(O_2)$ is an open set of E_1 because of the continuity of f. Thus for any open set $O_3 \in E_3$, the set $(g \circ f)^{-1}(O_3)$ is open in E_1; hence $g \circ f$ is continuous.

The function f is called an *open map* $E_1 \to E_2$ if for any open set $O \subset E_1$ the image $f(O)$ is open. If f is an open mapping and one-to-one, then the inverse function f^{-1} is continuous since $(f^{-1})^{-1} = f$ transforms open sets into open sets. A one-to-one correspondence f is called a *homeomorphism* if f and f^{-1} are both continuous.

EXAMPLE

Let $a \in \mathbf{R}$ with $a \neq 0$; then the mapping $f: x \to ax$ ($\mathbf{R} \to \mathbf{R}$) is a homeomorphism. Indeed, the mapping f is one-to-one, since $ax_1 = ax_2 \Rightarrow x_1 = x_2$. Moreover, f and f^{-1} are both continuous on $\mathbf{R}$, for if $O \subset \mathbf{R}$ is open and $x \in O$, then $\exists$ an open interval $]x - \varepsilon, x + \varepsilon[\subset O$ such that ax is contained in $f(O)$. It follows that $f(O)$ is open, since all its points are interior; hence f is an open mapping and f^{-1} is continuous. Similarly, we can show that f^{-1} is an open mapping and hence that f is continuous.

If $\exists$ a homeomorphism of S_1 onto S_2, then we say that S_1 and S_2 are *homeomorphic sets*.

EXAMPLE

The open interval $]0, 1[$ and $\mathbf{R}^+$ are homeomorphic. Indeed, if

$$f(x) = \frac{1}{1+x} : \mathbf{R}^+ \to]0, 1[$$

f is continuous on $\mathbf{R}^+$. Its inverse

$$f^{-1}(y) = \frac{1}{y} - 1 :]0, 1[\to \mathbf{R}^+$$

is continuous on $]0, 1[$.

It can easily be seen that homeomorphicity is an equivalence relation (see 0.1.2). In fact:

(i) *(reflexive)* Any set is homeomorphic to itself (take for f the identity mapping).
(ii) *(symmetric)* If f is a homeomorphism $S_1 \to S_2$, then f^{-1} exists and is a homeomorphism $S_2 \to S_1$.
(iii) *(transitive)* If f is a homeomorphism $S_1 \to S_2$ and g is a homeomorphism $S_2 \to S_3$, then $g \circ f$ is a homeomorphism $S_1 \to S_3$.

DEFINITION Let $(E, \mathscr{T})$ be a topological space and $S \subset E$. The topology $\mathscr{T}$ induces a topology $\mathscr{T}_S$ for S in the following way: Denote by $\mathscr{T}_S$ the set $\mathscr{T}_S = \{O \cap S \mid O \in \mathscr{T}\}$. Then it is easy to see that the collection $\mathscr{T}_S$ of the subsets $O_S = O \cap S$ of S satisfy the conditions O_1, O_2, O_3 (see 0.3.1) for open sets. So the pair $(S, \mathscr{T}_S)$ is a topological space. The topology $\mathscr{T}_S$ is called the *relative topology* of $S \in E$.

Note that the subsets $O_S = O \cap S$ which are the open sets of S may not belong to $\mathcal{T}$, which means that they are not necessarily open sets in the topology $\mathcal{T}$ of E.

EXAMPLE

Let $\mathcal{T}$ be the standard topology of **R**, and S the closed interval $[0, 1]$; then the interval $O = \,]0, 2[$ is open in $(\mathbf{R}, \mathcal{T})$, so that $[0, 1] \cap \,]0, 2[\, = \,]0, 1]$ is open in $(S, \mathcal{T}_S)$ but not in $(\mathbf{R}, \mathcal{T})$.

3. Connectedness and the connectedness argument

Intuitively, to be connected means to be of one piece. To be precise, we introduce a more sophisticated definition:

DEFINITION A topological space E is *connected* if it is not the union of two disjoint nonempty open sets. If S is a subset of E, then we say that S is connected if it is connected in the relative topology, that is, if S is not the union of two disjoint nonempty open sets in the relative topology $\mathcal{T}_S$.

THEOREM 0.3 E is connected iff E and $\emptyset$ are the only sets which are both open and closed in E.

PROOF

Assume E is connected and suppose $S \subset E$ is both open and closed. Then $E \backslash S$ is also both open and closed. Since $E = S \cup (E \backslash S)$ and E is connected, either $E \backslash S = \emptyset$ and $S = E$, or $S = \emptyset$ and $E \backslash S = E$. Conversely, if E and $\emptyset$ are the only sets which are both open and closed, and if $O \subset E$ is open with $O \neq \emptyset$ and $O \neq E$, then $E \backslash O$ is necessarily closed but not open, and E is not the union of two disjoint nonempty open sets. Hence E is connected.

We shall use this property of connectedness of sets in the following way: If we can prove that a nonempty subset S of a connected topological space E is both open and closed, then $S = E$. This method of proof will be called the connectedness argument.

EXAMPLE

Any interval I on $\mathbf{R}$ is connected. Let $O \subset I$, $O \neq \emptyset$ and $O \neq I$, be an open subset of I in the relative topology of I. Let a be an element of O and b an element of $I \setminus O$; then either $a < b$ or $a > b$. Assume $a < b$, and let $s = \text{l.u.b.}\{x \mid x \in O, x < b\}$. Then $a < s \leq b$ and $s \in I$. It is easy to show that s does not belong to O: If $s = b$ then $s \in I \setminus O$. If $s < b$ and $s \in O$, then, since O is open, $\exists$ an $\varepsilon > 0 \ni s \in \,]s - \varepsilon, s + \varepsilon[\subset O$ and hence $s + \varepsilon < b$, which contradicts the fact that $s = \text{l.u.b.}\{x \mid x \in O, x < b\}$. It is clear that s cannot be an interior point of $I \setminus O$; hence $I \setminus O$ is not open. It follows that I is not the union of two disjoint nonempty open sets; that is, I is connected. The case $a > b$ can be treated in a similar manner.

REMARK

An *upper bound* u of a set $S \subset \mathbf{R}$ is a real number which satisfies $u \geq x \; \forall \; x \in S$. The *least upper bound* of S is denoted by l.u.b.S and exists whenever an upper bound exists.

A *lower bound* l of a set $S \subset \mathbf{R}$ is a real number which satisfies $l \leq x \; \forall \; x \in S$. The *greatest lower bound* of S is denoted by g.l.b.S and exists whenever a lower bound exists.

To realize the power of the connectedness argument, we will use it to prove an important theorem. We first introduce the definition of local constancy:

DEFINITION A function f is said to be *locally constant* if for each $c \in E \; \exists$ a neighborhood $N(c)$ of $c \ni x \in N(c) \Rightarrow f(x) = f(c)$.

THEOREM 0.4 *If E_1 is connected and if $f : E_1 \to E_2$ is continuous and locally constant on E_1, then f is a constant, that is, $f(x) = f(c)$ for all $x \in E_1$.*

PROOF

Assuming $c \in E_1$ and $f(c) = \gamma$, let us denote by $S = \{x \mid f(x) = \gamma\} = f^{-1}(\gamma)$. It follows that S is closed, since $\{\gamma\}$ is closed (one point) and f is continuous. On the other hand, S is open because f is locally constant. Since S is not empty ($c \in S$), it follows by the connectedness argument that $S = E_1$.

THEOREM 0.5 *Let E_1 be connected and f a continuous mapping $E_1 \to E_2$. Then $f(E_1)$ is connected.*

PROOF

If there exist two disjoint nonempty sets $O_1, O_2 \ni O_1$ and O_2 are open in the relative topology of $f(E_1)$ and $f(E_1) = O_1 \cup O_2$, then $f^{-1}(O_1)$ and $f^{-1}(O_2)$ are open, since f is continuous. Furthermore, $f^{-1}(O_1) \cap f^{-1}(O_2) = \emptyset$. On the other hand, we have $E_1 = f^{-1}(O_1) \cup f^{-1}(O_2)$, so that E_1 cannot be connected. This contradicts our hypothesis.

In the above context, if $S \subset E_2$ is connected and f is continuous, it may well happen that $f^{-1}(S) \subset E_1$ is not connected. Indeed, if we take disjoint segments in the complex plane and project them on the real axis, the projection may be connected.

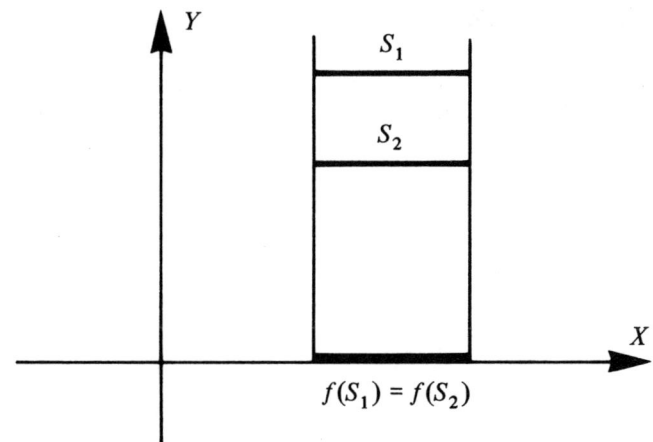

Figure 0.5

Problems

0.8.: Show that in the discrete topology of **R** (see 0.3.1) there are no connected sets except points.

0.9.: Let S_1 and S_2 be two connected sets with $S_1 \cap S_2 \neq \emptyset$. Show that $S_1 \cup S_2$ is connected. How can one generalize this result?

4. Compactness

Given a topological space $(E, \mathcal{T})$ and $S \subset E$. The family (set) $\mathcal{F}$ of open sets is said to be an *open covering* of S if $S \subset \bigcup_{O \in \mathcal{F}} O$. Given the open covering $\mathcal{F}$, an *open subcovering* of S is a subset of $\mathcal{F}$ which is also an open covering of S. An open covering is called finite if it contains a finite number of open sets.

DEFINITION We say that S is *compact* if every open covering of S contains a finite subcovering.

EXAMPLE

$\mathbf{R}$ is not compact. Indeed, consider the open covering of $\mathbf{R}$ by the sets $\{]n - 1, n + 1[, n \in \mathbf{Z}\}$. No finite subfamily can cover $\mathbf{R}$. We shall see later that a closed interval $[a, b] \subset \mathbf{R}$ is compact, and that any bounded and closed set of $\mathbf{R}^n$ is compact. The notion of compactness is important because compact spaces behave very much like bounded, closed subsets of $\mathbf{R}^n$.

We say that a topological space is separated, or is a *Hausdorff space*, if, given any two points a, b in the space, $a \neq b$, there exist two disjoint open sets O_a and $O_b \ni a \in O_a$ and $b \in O_b$.

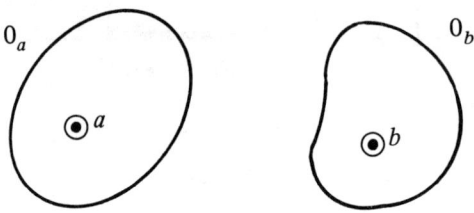

Figure 0.6

THEOREM 0.6 *A compact set K in a Hausdorff space E is closed.*

PROOF

We shall show $E \backslash K$ is open by showing that any point of $E \backslash K$ is an interior point of $E \backslash K$. Take any $a \in E \backslash K$ and any $x \in K$. Then $\exists$ two disjoint open sets $A(x)$ and $O(x)$ with $a \in A(x)$ and $x \in O(x)$. The family $\{O(x) | x \in K\}$ is an open covering of K; K is compact, so there exists a finite subcovering $O(x_1), O(x_2), \ldots, O(x_n)$ of K. Let $A(x_1), A(x_2), \ldots, A(x_n)$

be the corresponding open sets containing a; then $A(x_i) \cap O(x_i) = \emptyset$ for $i = 1, 2, \ldots, n$. Let A be the intersection of the $A(x_i)$, $i = 1, 2, \ldots, n$; that is, $A = \bigcap_{i=1}^{n} A(x_i)$. Then A is open and $A \cap O(x_i) = \emptyset$, $i = 1, 2, \ldots, n$, so $A \cap K = \emptyset$. Thus $A \subset E \backslash K$ and, since $a \in A$, a is an interior point of $E \backslash K$. Since a is arbitrary, $E \backslash K$ is open and K is closed.

A closed set need not be compact; in fact, **R** is closed without being compact. But a closed subset K^* of a compact set K is compact. Indeed, let $\mathscr{F}$ be an open covering of K^*; then $K \subset (K \backslash K^*) \cup \left(\bigcup_{O \in \mathscr{F}} O \right)$ and $K \backslash K^*$ is open in the relative topology of K, so $\{K \backslash K^*\} \cup \mathscr{F}$ is an open covering of K. Since K is compact, $\exists$ an finite open subcovering of K consisting of $K \backslash K^*$ and a finite number of elements $O_1, O_2, \ldots, O_n$ belonging to $\mathscr{F}$. The finite number of elements $O_1, O_2, \ldots, O_n$ belonging to $\mathscr{F}$ obviously covers K^*; hence K^* is compact.

THEOREM 0.7 *Let f be a continuous mapping $K \to E_2$ where K is a compact and E_2 is a Hausdorff space. Then $f(K)$ is compact.*

PROOF

Let $\mathscr{F}$ be an open covering of $f(K)$. Since f is continuous, the set $f^{-1}(O)$ is open for every $O \in \mathscr{F}$. The family $\mathscr{F}' = \{f^{-1}(O) | O \in \mathscr{F}\}$ covers K. Since K is compact, a finite number of sets in $\mathscr{F}'$ cover K, say $f^{-1}(O_1), f^{-1}(O_2), \ldots, f^{-1}(O_n)$. Obviously, the union $\bigcup_{i=1}^{n} O_i$ contains $f(K)$. Hence $f(K)$ is compact.

THEOREM 0.8 *A continuous one-to-one mapping f of a compact set K into a Hausdorff space E is a homeomorphism.*

PROOF

Since f is continuous, it is sufficient to show that f^{-1} is continuous. This may be done by proving that for any closed subset $\mathbf{C} \subset K$, $(f^{-1})^{-1}(\mathbf{C})$ is closed. Since $(f^{-1})^{-1} = f$, and any closed subset $\mathbf{C}$ of a compact set K is compact, it follows that $f(\mathbf{C})$ is compact. But compactness implies closure in a Hausdorff space, so f is a homeomorphism and K and $f(K)$ are homeomorphic sets.

Problems

0.10.: Show that the intersection and union of two compact sets are again compact.
0.11.: Show that in a Hausdorff space the boundary of a compact set is compact.

§4. Metric spaces

1. Definitions

In the following chapters, we shall be interested in special topological spaces, namely, certain metric spaces such as $\mathbf{R}^2$ and some sequence and function spaces. We will therefore study certain properties of metric spaces which we will need later on.

DEFINITION A *metric space* (E, d) is a space E with a distance d, that is, a mapping $d : E \times E \to \mathbf{R}^+ \cup \{0\}$ which has the following properties:

(i) $d(x, y) \geq 0$ for every pair $x, y \in E$
     $d(x, y) = 0 \Leftrightarrow x = y$
(ii) $d(x, y) = d(y, x)$ for every pair $x, y \in E$
(iii) $d(x, z) \leq d(x, y) + d(y, z)$ for every $x, y, z \in E$
     (triangle inequality)

DEFINITION If $a \in E$ and if $r > 0$, then the set $\{x \mid x \in E, d(a, x) < r\}$ denoted by $B(a, r)$ is called the *open ball* with center a and radius r. The set $\{x \mid x \in E, d(a, x) \leq r\}$ is called the *closed ball* with center a and radius r and is denoted by $\bar{B}(a, r)$.

DEFINITION If S is a subset of the metric space (E, d) and $a \in S$, then we say that a is an *interior point* of S if $\exists B(a, r) \subset S$.

DEFINITION A set S in the metric space (E, d) is called *open* if all its points are interior points.

It is clear that an open set contains at least one ball.

THEOREM 0.9 *The collection $\mathcal{T}$ of all open sets in a metric space (E, d) is a topology; that is, $\mathcal{T}$ satisfies conditions O_1, O_2, O_3.*

PROOF

O_1: The empty set $\emptyset$ is open since it does not contain any points, and the space E is open, since it contains any ball $B(a, r)$.

O_2: If $\mathscr{F}$ is any family of open sets O, then their union $\bigcup_{O \in \mathscr{F}} O$ is also open: Take an arbitrary point a belonging to at least one open set of the union, say O_0, and, since O_0 is open, $\exists$ an open ball $B(a, r)$ contained in O_0. $B(a, r)$ is also contained in the union; hence $\bigcup_{O \in \mathscr{F}} O$ is open.

O_3: Let $\{O_1, O_2, \ldots, O_n\}$ be a finite number of open sets with $\bigcap_{i=1}^{n} O_i \neq \emptyset$ and $a \in \bigcap_{i=1}^{n} O_i$. Then, since $a \in O_i$ and O_i is open, $\exists r_i > 0 \ni B(a, r_i) \subset O_i$. Let $r_{\min}$ be the minimum of the r_i ($i = 1, 2, \ldots, n$). Then $B(a, r_{\min})$ is contained in every O_i ($i = 1, 2, \ldots, n$). Therefore, $B(a, r_{\min})$ is contained in $\bigcap_{i=1}^{n} O_i$, which proves that a is an interior point of $\bigcap_{i=1}^{n} O_i$. Since a is an arbitrary point of $\bigcap_{i=1}^{n} O_i$, it follows that $\bigcap_{i=1}^{n} O_i$ is open.

The argument used to prove Theorem 0.9 does not hold for an infinite number of open sets O_i because the minimum, or g.l.b., of the r_i, may be 0. To illustrate this, consider the space $\mathbf{R}$ with $d(x, y) = |x - y|$. Then $(\mathbf{R}, d)$ is a metric space. In $\mathbf{R}$ the open ball $B(a, r)$ is the open interval $]a - r, a + r[$. If $S \subset \mathbf{R}$, then $a \in S$ is an interior point of S if $\exists$ an $r > 0 \ni]a - r, a + r[\subset S$. Obviously, the open interval $]a - r, a + r[$ is an open set (that is, all its points are interior points). Consider now the sequence of open intervals $\{I_n\}$, $I_n =]-1/n, +1/n[$ ($n = 1, 2, \ldots$). It is clear that $\bigcap_{n=1}^{\infty} I_n = \{0\}$. But $\{0\}$, the set containing only the origin, is a closed set, since $\mathbf{R}\backslash\{0\}$ is open. Thus the intersection of an infinite number of open sets may be closed.

As another example of a metric space, consider $\mathbf{R}^2$ with the Euclidean metric: If $x = (x_1, x_2)$ and $y = (y_1, y_2)$ are two points in $\mathbf{R}^2$, we define

$$d(x, y) = \sqrt{(x_1 - y_1)^2 + (x_2 - y_2)^2}.$$

It is easy to show that $d(x, y)$ satisfies all the properties of a distance. In $\mathbf{R}^2$ the open ball $B(a, r) = \{x \mid d(a, x) < r\}$ will be called the *open disc*.

THEOREM 0.10 *A metric space is a Hausdorff space.*

PROOF

Indeed, if a and b are two points of a metric space and $a \neq b$, then $d(a, b) = r > 0$. Consider then the two open balls $B(a, r/3)$ and $B(b, r/3)$.

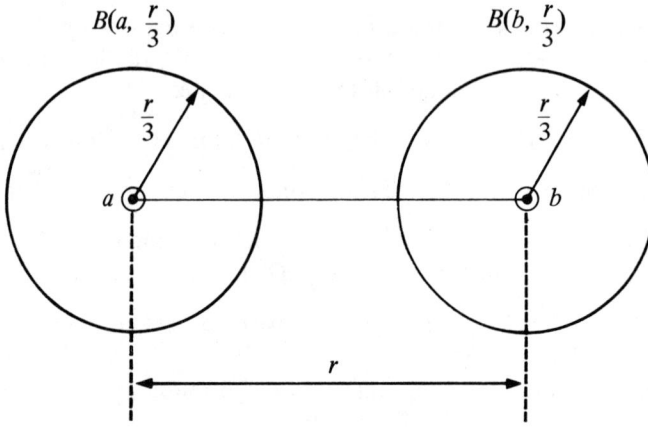

Figure 0.7

It is easy to show that these two open sets are disjoint; for if $x \in B(a, r/3)$, then $d(x, a) < r/3$ and $d(a, b) \leq d(a, x) + d(x, b) \Rightarrow d(x, b) \geq d(a, b) - d(a, x) \geq r - r/3 = 2r/3$, and $x \notin B(b, r/3)$. It follows that $B(a, r/3) \cap B(b, r/3) = \emptyset$.

2. Normed vector spaces

A very important subclass of metric spaces are the normed vector spaces, which have both a topological and an algebraic structure.

DEFINITION A *normed* real (complex) *vector space* $(V, \|\cdot\|)$ is a vector space where a norm $\|\cdot\|$ is defined. The norm $\|\cdot\|$ is a mapping $V \to \mathbf{R}^+ \cup \{0\}$ which has the following properties:

(i) $\|x\| \geq 0$ and $\|x\| = 0 \Leftrightarrow x = 0$
(ii) $\|ax\| = |a|\|x\|$ for $x \in V$ and a real (complex)
(iii) $\|x + y\| \leq \|x\| + \|y\|$

A normed real or complex vector space is a metric space when we define $d(x, y) = \|x - y\|$. It is easy to show that $\|x - y\|$ has the three properties of a distance.

EXAMPLE

1 Obviously $\mathbf{R}^n$ is a normed real vector space if we define $\|x\|^2 = x_1^2 + x_2^2 + \cdots + x_n^2$ and is a metric space with the Euclidean distance

$$d(x, y) = \|x - y\| = \sqrt{(x_1 - y_1)^2 + (x_2 - y_2)^2 + \cdots + (x_n - y_n)^2}.$$

2 Another important example of a normed vector space is the set of all real-valued functions which are continuous on the closed interval $[a, b]$. This set will be denoted by $C[a, b]$. It is a real vector space since the sum of two continuous functions and the product of a continuous function and a real number are both continuous. We define the norm $\|\cdot\|$ in $C[a, b]$ by $\|f\| = \text{l.u.b.}\{|f(x)| \, | \, x \in [a, b]\}$. It is left to the reader to show that $\|\cdot\|$ satisfies all the conditions of a norm for $C[a, b]$. Note that it is possible to define other norms in $C[a, b]$, for example, the norm used in the Hilbert space $L^2[a, b]$:

$$\|f\|_{L^2}^2 = \int_a^b f^2(x)\,dx.$$

The norm $\|f\| = \text{l.u.b.}\{|f(x)| \, | \, x \in [a, b]\}$ is called the *norm of uniform convergence*, since $\|f_n - f\| \to 0$ as $n \to \infty$ implies that the sequence f_n converges uniformly to f on $[a, b]$. (For reference, we recall the definition of uniform convergence: Let (E_1, d_1) and (E_2, d_2) be two metric spaces and let $\{f_n\}$ be a sequence of functions $E_1 \to E_2$. We say that the sequence f_n converges uniformly to f in E_1 if $\forall \, \varepsilon \, \exists \, N(\varepsilon) \ni n > N(\varepsilon) \Rightarrow d(f_n(x), f(x)) < \varepsilon \, \forall \, x \in E_1$.)

Problem

0.12.: Consider the set B of bounded sequences of real numbers $a = \{a_n\}$. Define $\|a\| = \text{l.u.b.}\{|a_n| \, | \, n \in \mathbf{P}\}$. Show that B is a normed vector space.

---

3. Sequences in metric spaces; completeness

DEFINITION Let $\{x_n\}$ be a sequence in a metric space E. We say that $\{x_n\}$ is a *Cauchy sequence* if for any $\varepsilon > 0$ there is a number $N(\varepsilon) \ni$ if m and n are any pair of positive integers for which $m > N(\varepsilon)$ and $n > N(\varepsilon)$, then $d(x_m, x_n) < \varepsilon$.

DEFINITION We say that the sequence $\{x_n\}$ in E *converges* to $x \in E$ if $\exists$ an $x \in E \ni \lim_{n \to \infty} d(x_n, x) = 0$.

Note that this definition of convergence depends on the space E. It may happen that a given sequence does not converge in E, but does converge in some extension of E. As an example, consider the set **Q** of all rational numbers, and let $r_1, r_2, \ldots, r_n, \ldots$ be a sequence of rational numbers which converges to an irrational number, for example, the sequence 3, 3.1, 3.14, 3.141, ... which converges to the irrational number π. Since $\pi \notin \mathbf{Q}$, the sequence does not converge in **Q**, but it does converge in the space **R**, which is an extension of **Q**.

THEOREM 0.11 *A convergent sequence on E is a Cauchy sequence on E.*

PROOF

If $\lim_{n \to \infty} d(x_n, x) = 0$, and $\varepsilon > 0$, then, for $m > N(\varepsilon/2)$ and $n > N(\varepsilon/2)$, we have $d(x_m, x) < \varepsilon/2$ and $d(x_n, x) < \varepsilon/2$. Now, using the triangle inequality, we have

$$d(x_m, x_n) \leq d(x_m, x) + d(x, x_n) < \frac{\varepsilon}{2} + \frac{\varepsilon}{2} = \varepsilon$$

Thus $\{x_n\}$ is a Cauchy sequence.

The converse of Theorem 0.11 does not hold, since a Cauchy sequence in E does not necessarily converge in E, as seen in the example above. In order to distinguish spaces in which every Cauchy sequence is convergent from those in which this is not the case, we introduce the concept of completeness.

DEFINITION We say that the metric space E is *complete* if every Cauchy sequence in E converges in E.

EXAMPLE

The set of all rational numbers **Q** is not complete, but the set of all real numbers **R** is complete. Using the completeness of **R**, we can easily show that $\mathbf{R}^k$, $k = 2, 3, \ldots$, is complete. Indeed, if $x^1, x^2, \ldots, x^n, \ldots$ is a Cauchy sequence in $\mathbf{R}^k$ with $x^n = (x_1^n, x_2^n, \ldots, x_k^n)$ and $d(x^m, x^n)^2 = (x_1^m - x_1^n)^2 + (x_2^m - x_2^n)^2 + \cdots + (x_k^m - x_k^n)^2$, then for any given $\varepsilon > 0$ we can find $N(\varepsilon) \ni$ for $m > N(\varepsilon)$ and $n > N(\varepsilon)$, $d(x^m, x^n) < \varepsilon$. Since

$|x_i^m - x_i^n| \leq d(x^m, x^n) < \varepsilon$ for $i = 1, 2, \ldots, k$, the sequence $x_i^1, x_i^2, \ldots, x_i^n, \ldots$ is a Cauchy sequence in **R** and converges, say to x_i, for $i = 1, 2, \ldots, k$. Since $x = (x_1, x_2, \ldots, x_k)$ is a point in $\mathbf{R}^k$, the sequence $\{x^n\}$ converges to x in $\mathbf{R}^k$. Thus $\mathbf{R}^k$ is complete.

DEFINITION A complete normed vector space is called a *Banach space*.

EXAMPLE

1 It is easy to show that $C[a, b]$, with the norm $\|f\| = $ l.u.b.$\{|f(x)| \, | \, x \in [a, b]\}$, is a Banach space.

2 $\mathbf{R}^n$ is a Banach space.

THEOREM 0.12 *A subset of a complete metric space E is closed iff $x_n \in S$ $(n = 1, 2, \ldots)$ and $\lim_{n \to \infty} x_n = x \Rightarrow x \in S$.*

PROOF

(i) Let S be closed, $x_n \in S$, and $\lim_{n \to \infty} x_n = x$. We will show that $x \in S$. If $x \in E\backslash S$, then, since $E\backslash S$ is open, $\exists$ a ball $B(x, r) \subset E\backslash S$, containing all of the points x_n for $n >$ some N. Hence $x_n \notin S$ for all sufficiently large n, contrary to the assumption. Therefore $x \in S$.

(ii) If S is not closed, then $E\backslash S$ is not open and contains at least one point x_o which is not an interior point. So, if we consider the balls $B(x_o, 1/n)$, $n = 1, 2, \ldots$, then each ball contains at least one point $x_n \in S$. Obviously, the sequence $x_1, x_2, \ldots$ converges to $x_o \notin S$, in contradiction to $\lim_{n \to \infty} x_n = x_o \Rightarrow x_o \in S$.

COROLLARY *A Cauchy sequence of a closed subset S of a complete metric space E is convergent in S; that is, the sequence has its limit point in S.*

Problems

0.13.: Show that the function space $C[a, b]$ is complete with the norm $\|f\| = $ l.u.b.$\{|f(x)| \, | \, x \in [a, b]\}$, but is not complete with the Hilbert norm $\| \, \|_{L^2}$.

Hint: First, show that if $\{f_n\}$ is a sequence of functions belonging to $C[a, b]$ which converges uniformly to f, then $f \in C[a, b]$. Secondly, consider the function φ defined by

$$\varphi(x) = \begin{cases} 0 & \text{for } a \leq x < c \\ 1 & \text{for } c \leq x \leq b \end{cases}$$

$\varphi \notin C[a, b]$, but it is easy to construct a sequence $\{\varphi_n\}$, with $\varphi_n \in C[a, b]$, $\ni \lim_{n \to \infty} \|\varphi_n - \varphi\|_{L^2} = 0$.

0.14.: Show that the set B in problem 0.12 is a Banach space.

0.15.: Let $S \subset \mathbf{R}$ be bounded. Does l.u.b.S or g.l.b.S belong to S? Give sufficient conditions.

0.16.: Show that if S is connected, then $\bar{S}$ is connected. (Recall that $\bar{S}$ is the closure of S.)

0.17.: Show that if (E, d) is a metric space, then so is $\left(E, \dfrac{d}{1+d}\right)$.

---

4. Continuity in metric spaces

THEOREM 0.13 *Let (E_1, d_1), (E_2, d_2) be two metric spaces. Then the mapping $f : E_1 \to E_2$ is continuous at x iff to each ball $B(f(x), \varepsilon) \subset E_2$ we can associate a ball $B(x, \delta(\varepsilon)) \subset E_1 \ni f(B(x, \delta(\varepsilon))) \subset B(f(x), \varepsilon)$.*

PROOF

(i) Suppose f is continuous at x; then since $B(f(x), \varepsilon)$ is a neighborhood of $f(x)$, $\exists$ a neighborhood $N(x) \ni f(N(x)) \subset B(f(x), \varepsilon)$. Since $N(x)$ contains an open set $O(x)$ with x as one of its interior points, $\exists$ a ball $B(x, \delta) \subset O(x) \subset N(x)$. Clearly, $f(B(x, \delta)) \subset B(f(x), \varepsilon)$.

(ii) Suppose with each ball $B(f(x), \varepsilon) \subset E$ we can associate a ball $B(x, \delta(\varepsilon)) \subset E \ni f(B(x, \delta(\varepsilon))) \subset B(f(x), \varepsilon)$. Let $N(f(x))$ be a neighborhood of $f(x)$; then this neighborhood contains a ball $B(f(x), \varepsilon)$; hence $\exists$ a ball $B(x, \delta(\varepsilon))$ which is a neighborhood of x with $f(B(x, \delta(\varepsilon))) \subset B(f(x), \varepsilon) \subset N(f(x))$. Thus the mapping is continuous.

DEFINITION We say that the mapping $f : E_1 \to E_2$, where (E_1, d_1) and (E_2, d_2) are metric spaces, is *sequentially continuous* if $x_n \to x$ implies $f(x_n) \to f(x)$.

THEOREM 0.14 *f is continuous $\Leftrightarrow$ f is sequentially continuous.*

PROOF

(i) *f is continuous $\Rightarrow$ f is sequentially continuous*: Suppose f is continuous in x and the sequence $\{x_n\}$ converges to x; then for $B(f(x), \varepsilon) \subset E_2$ $\exists$ a ball $B(x, \delta(\varepsilon)) \subset E_1 \ni f(B(x, \delta(\varepsilon))) \subset B(f(x), \varepsilon)$. Now, for $n >$ some N, all x_n lie in $B(x, \delta(\varepsilon))$, hence the corresponding $f(x_n)$ lie in $B(f(x), \varepsilon)$; thus $d(f(x_n), f(x)) < \varepsilon$ for all $n > N$, indicating $f(x_n) \to f(x)$.

(ii) *f is sequentially continuous $\Rightarrow$ f is continuous*: Suppose f is not continuous in x. Then there is a ball $B(f(x), \varepsilon) \ni$, whatever δ we may choose, $f(B(x, \delta)) \not\subset B(f(x), \delta)$. Let us choose $\delta = 1/n$. Then $\exists$ an $x_n \in B(x, 1/n) \ni f(x_n) \notin B(f(x), \varepsilon)$; that is, $d(f(x_n), f(x)) \geq \varepsilon$. It is obvious that $x_n \to x$, while the assumptions $d(f(x_n), f(x)) \geq \varepsilon$ for $n = 1, 2, \ldots$ indicate that $f(x_n)$ does not converge to $f(x)$, contrary to the hypothesis that f is sequentially continuous.

DEFINITION A subset S of a metric space (E, d) is said to be *bounded* if $\exists$ a ball $B(a, r)$ containing S.

It is left to the reader to show that, if $S_1, S_2, \ldots, S_n$ is a finite collection of bounded subsets of E, then $\bigcup_{i=1}^{n} S_i$ is bounded. To show that this is not true for an infinite collection of bounded sets, we consider the following example: Take for (E, d) the space $(\mathbf{R}, |x - y|)$ and the infinite collection of bounded sets I_n, $n = 1, 2, \ldots$, with $I_n = \left[n - \frac{1}{n}, n + \frac{1}{n}\right]$. Then $\bigcup_{n=1}^{\infty} I_n$ is not bounded.

Problems

0.18.: Let f and g be real-valued and continuous functions defined on the metric space (E, d). Show that $f + g$, fg and $|f|$ are continuous.

0.19.: Let (E, d) be a metric space. Show that the identity is an homeomorphism between (E, d) and $\left(E, \frac{d}{1+d}\right)$. $\left(\text{Note that } 0 \leq \frac{d}{1+d} < 1.\right)$

5. Compactness in metric spaces

DEFINITION A subset S of a metric space (E, d) is said to be *totally bounded* if to an arbitrary $\varepsilon > 0$ we can associate a finite number of balls $B_i = B(a_i, \varepsilon)$, $i = 1, 2, \ldots, n$, which cover S.

Obviously, the notions "bounded" and "totally bounded" are equivalent in $\mathbf{R}^n$, but they are not in infinite dimensional spaces.

DEFINITION Let S be a subset of a metric space (E, d) and $S' \subset S$. The set S' is said to be *dense* in S if with any $x \in S$ and any $\varepsilon > 0$ we can associate an $x' \in S' \ni d(x', x) < \varepsilon$, or $x \in B(x', \varepsilon)$.

EXAMPLE

The set of rational numbers $\mathbf{Q}$ is dense in the set of real numbers $\mathbf{R}$.

THEOREM 0.15 *A totally bounded set S in a metric space possesses a countable or finite dense subset S'.*

PROOF

Since S is totally bounded, $\exists$ a finite collection C_1 of balls $B(c_{11}, 1)$, $B(c_{12}, 1)$, $\ldots$, $B(c_{1k_1}, 1)$ with radius 1 which cover S, and another finite collection C_2 of balls $B(c_{21}, \frac{1}{2})$, $B(c_{22}, \frac{1}{2})$, $\ldots$, $B(c_{2k_2}, \frac{1}{2})$ with radius $\frac{1}{2}$ which cover S, and, for each $m = 3, 4, \ldots$, a collection C_m of balls $B(c_{m1}, 1/m), \ldots$, $B(c_{mk_m}, 1/m)$ with radius $1/m$ which cover S. Obviously, the set S' of the centers of these balls c_{ij} with $i = 1, 2, \ldots$ and $j = 1, 2, \ldots, k_i$ is countable and dense in S. (S' is finite if S is finite.)

DEFINITION A subset S of a metric space (E, d) is said to be *precompact* if any sequence $\{x_n\}$ of points of S contains a subsequence which is a Cauchy sequence.

(A subsequence is a subset of the sequence which is itself a sequence.)

THEOREM 0.16 *A subset S of (E, d) is totally bounded $\Leftrightarrow S$ is precompact.*

PROOF

(i) *Totally bounded $\Rightarrow$ precompact*: Let S be totally bounded and $\{x_n\}$ an infinite sequence belonging to S. Since S is totally bounded, with each integer $m > 0$ we can associate a finite collection C_m of balls with radius

$1/m$ which cover S, where $C_m = \{B(c_{m1}, 1/m), B(c_{m2}, 1/m), \ldots, B(c_{mk_m}, 1/m)\}$. Obviously, in at least one of the balls of C_1 there is an infinite subsequence $\{x_{1n}\}$ of $\{x_n\}$, and in at least one of the balls of C_2 there is an infinite subsequence $\{x_{2n}\}$ of $\{x_{1n}\}$. Similarly, for each integer $m > 0$, in at least one of the balls of C_m there is an infinite subsequence $\{x_{mn}\}$ of $\{x_{m-1,n}\}$. The "diagonal" sequence $x_{11}, x_{22}, \ldots, x_{nn}, \ldots$ is a Cauchy sequence: for if $n > m$, the points x_{nn} belong to one of the balls of C_m, and we have $d(x_{nn}, x_{n'n'}) < 2/m$ for $n, n' > m$. When m is sufficiently large, we have $2/m < \varepsilon$ for any preassigned $\varepsilon > 0$. This shows that total boundedness implies precompactness.

(ii) *Precompact* $\Rightarrow$ *totally bounded.* It suffices to show that if S is not totally bounded, it cannot be precompact. If S is not totally bounded, then for some $\varepsilon > 0$ we cannot find a finite number of balls with radius ε which cover S. Choose an arbitrary $x_1 \in S$; then the ball $B(x_1, \varepsilon)$ does not cover S. Then choose $x_2 \notin B(x_1, \varepsilon)$; the union $B(x_1, \varepsilon) \cup B(x_2, \varepsilon)$ does not cover S. Similarly, we choose $x_3 \in S \ni x_3 \notin B(x_1, \varepsilon) \cup B(x_2, \varepsilon)$. Again, we know that the union $\bigcup_{i=1}^{3} B(x_1, \varepsilon)$ does not cover S. We repeat this process and find $x_n \in S$ with $x_n \notin \bigcup_{i=1}^{n-1} B(x_i, \varepsilon)$. The sequence $\{x_n\}$, $n = 1, 2, \ldots$, has the property $d(x_m, x_n) > \varepsilon$ for any pair m, n. It follows that $\{x_n\}$ does not contain a subsequence which is a Cauchy sequence. This means S is not precompact. Thus it follows that for S to be precompact it is necessary that S be totally bounded.

DEFINITION A subset S of a complete metric space (E, d) is said to be *sequentially compact* if it is precompact and closed; that is, if any sequence $\{x_n\}$ of points of S contains a convergent subsequence. (S closed implies that a Cauchy sequence is convergent).

Since S is precompact $\Leftrightarrow S$ is totally bounded, we may say that "S sequentially compact" is equivalent to "S totally bounded and closed."

DEFINITION Let S be a subset of the complete metric space (E, d) and $\{O_\alpha\}$ an open covering of S. The real number $\delta > 0$ is called a *Lebesgue number* of the covering $\{O_\alpha\}$ if every open ball B with diameter less than δ for which $B \cap S \neq \varnothing$ is contained in at least one of the open sets O_α of the covering.

THEOREM 0.17 *If S is sequentially compact, then any open covering $\{O_\alpha\}$ of S possesses a Lebesgue number $\delta > 0$.*

PROOF

Suppose a Lebesgue number $\delta > 0$ of $\{O_\alpha\}$ does not exist; then for each $n > 0$ we can find a ball B_n with diameter less than $1/n \ni B_n \cap S \neq \emptyset$ and B_n is not contained in any of the open sets of the covering $\{O_\alpha\}$. Let $x_n \in B_n \cap S$, $n = 1, 2, \ldots$. Since S is sequentially compact, a subsequence $\{x_{n'}\}$ of $\{x_n\}$ is convergent: $x_{n'} \to x_o$. Since $x_o \in S$, $\exists$ an open set O_o in the covering $\{O_\alpha\}$ with $x_o \in O_o$. Since O_o is open, $\exists$ a ball $B(x_o, r) \subset O_o$. By virtue of $d(x_{n'}, x_o) \to 0$, for sufficiently large n', say $n' > N(r/2)$, all the $x_{n'} \in B(x_o, r/2) \subset O_o$. If we take $1/n' < r/2$, then we have $B_{n'} \subset B(x_o, r) \subset O_o$, in contradiction to the choice of the balls $B_{n'}$.

THEOREM 0.18 *A complete metric space (E, d) is compact $\Leftrightarrow$ sequentially compact $\Leftrightarrow$ totally bounded and closed.*

PROOF

(i) *Compact $\Rightarrow$ totally bounded and closed*: Let S be compact. Consider the open balls $B(x, \varepsilon)$, where $\varepsilon > 0$ and $x \in S$. The collection of open balls $B(x, \varepsilon)$ covers S and, since S is compact, a finite number of those open balls cover S. Hence S is totally bounded, and, since compactness in a Hausdorff space implies closure, S is closed.

(ii) *Sequentially compact $\Rightarrow$ compact*: Suppose $\{O_\alpha\}$ is an open covering of S. Let us show that it possesses a finite subcovering. Since sequential compactness implies the existence of a Lebesgue number $\delta > 0$ corresponding to the covering $\{O_\alpha\}$, then, because of the total boundedness of S, $\exists$ a finite number of balls $B_1, B_2, \ldots, B_k$ with diameter $\delta' < \delta$ covering S, such that each of the B_j is contained in some open set O_j of the covering $\{O_\alpha\}$. The collection $\{O_j | j = 1, \ldots, k\}$ obviously covers S and thus forms a finite subcovering, indicating that S is compact.

In the definition of continuity (see 1.4.4) the quantity $\delta = \delta(\varepsilon)$ depends not only on ε, but also on x. In the case of uniform continuity, we can find a $\delta(\varepsilon)$ which is suitable for all x:

DEFINITION Let (E_1, d_1), (E_2, d_2) be two metric spaces and f a continuous mapping $E_1 \to E_2$. We say that f is *uniformly continuous* on E_1 if for any $\varepsilon > 0$ we can find a $\delta(\varepsilon) > 0 \ni \forall \, x \in E_1$, $f[B(x, \delta(\varepsilon))] \subset B(f(x), \varepsilon)$.

The following theorem shows that compact spaces behave much like bounded and closed sets in $\mathbf{R}^n$. From real analysis, we know that if $K \subset \mathbf{R}^n$ is bounded and closed and f is a continuous mapping $\mathbf{R}^n \to \mathbf{R}$, then f is uniformly continuous on K. This result can be generalized as follows:

THEOREM 0.19 Let (E_1, d_1), (E_2, d_2) be two metric spaces and let f be a continuous mapping $E_1 \to E_2$. If E_1 is compact, then f is uniformly continuous on E_1.

PROOF

For a given $\varepsilon > 0$, there corresponds to each ball $B(f(x), \varepsilon/2) \subset E$, and $\forall\, x \in E_1$, a ball $B(x, \delta(\varepsilon, x)) \ni f[B(x, \delta(\varepsilon, x))] \subset B(f(x), \varepsilon/2)$.

The balls $B\left(x, \dfrac{\delta(\varepsilon, x)}{2}\right)$ cover E_1. Since E_1 is compact, a finite number of them cover E_1, say

$$B\left(x_1, \frac{\delta(\varepsilon, x_1)}{2}\right),\ B\left(x_2, \frac{\delta(\varepsilon, x_2)}{2}\right),\ \ldots,\ B\left(x_n, \frac{\delta(\varepsilon, x_n)}{2}\right).$$

Set

$$\delta(\varepsilon) = \min\left(\frac{\delta(\varepsilon, x_1)}{2}, \frac{\delta(\varepsilon, x_2)}{2}, \ldots, \frac{\delta(\varepsilon, x_n)}{2}\right).$$

Suppose x and x' are arbitrarily given with $d(x, x') < \delta(\varepsilon)$; then x and x' both belong to one of the balls $B(x_i, \delta(\varepsilon, x_i))$. Indeed, x belongs to one of the balls $B\left(x_i, \dfrac{\delta(\varepsilon, x_i)}{2}\right)$, and, since $d(x, x') < \delta(\varepsilon)$, it follows that

$$d(x_i, x') \leq d(x_i, x) + d(x, x') \leq \frac{\delta(\varepsilon, x_i)}{2} + \delta(\varepsilon) \leq \delta(\varepsilon, x_i)$$

We have selected $B(x_i, \delta(\varepsilon, x_i)) \ni f[B(x_i, \delta(\varepsilon, x_i))] \subset B(f(x_i), \varepsilon/2)$. It follows that $f(x)$ and $f(x')$ both belong to the ball $B(f(x_i), \varepsilon/2)$, or that

$$d(f(x), f(x')) \leq d(f(x), f(x_i)) + d(f(x_i), f(x')) \leq \frac{\varepsilon}{2} + \frac{\varepsilon}{2} = \varepsilon$$

In other words, $f[B(x, \delta(\varepsilon))] \subset B(f(x), \varepsilon)$ with $\delta(\varepsilon)$ independent of x.

Another illustration of the similarity between bounded and closed subsets of $\mathbf{R}^n$ and compact subsets of metric spaces is given by the following theorem:

THEOREM 0.20 Let (E, d) be a compact metric space and f a continuous mapping $E \to \mathbf{R}$. Then f is bounded and attains its maximum and minimum on E.

PROOF

f is bounded since, by Theorem 0.7, $f(E)$ is compact. $f(E)$, being compact in **R**, is closed, and contains thus its maximum and minimum.

Let K be a compact metric space and let $C[K]$ be the set of all real- or complex-valued functions f which are continuous on K. It is clear that $C[K]$ is a vector space. We introduce now the norm of uniform convergence in $C[K]$ (see 0.4.2):

$$\|f\| = \text{l.u.b.} \{|f(x)| \,|\, x \in K\}$$

By an argument similar to that used to solve problem 0.13, we see that $(C[K], \|\cdot\|)$ is a complete normed vector space; that is, a Banach space.

DEFINITION   We say that the functions f of the subset S of $C[K]$ are *equicontinuous*, or, in other words, that S is equicontinuous, if to an arbitrary $\varepsilon > 0$ we can associate a $\delta(\varepsilon) > 0 \ni d(x, x') < \delta(\varepsilon) \Rightarrow |f(x) - f(x')| < \varepsilon\ \forall$ functions $f \in S$ and all $x, x' \in K$.

EXAMPLE

The functions $1, x, x^2, \ldots, x^n, \ldots$ are all continuous in $[0, 1]$ without being equicontinuous, since for any δ, $0 < \delta < 1$ we may choose first n sufficiently large $\ni |(1 - \delta)^n| < \varepsilon$, and then choose an x with $1 - \delta < x < 1$, $\ni x^n > 1 - \varepsilon$. Then we have $|x - (1 - \delta)| < \delta$, but $|x^n - (1 - \delta)^n| > (1 - \varepsilon) - \varepsilon = 1 - 2\varepsilon > \varepsilon$ for suitably chosen $\varepsilon > 0$.

DEFINITION   We say that the functions f of the subset S of $C[K]$ are *equibounded*, or that S is equibounded, if $\exists$ a real number $M > 0 \ni \|f\| \leq M\ \forall\ f \in S$.

A criterion for compactness of a subset S of $C[K]$ is given by the Ascoli-Arzela theorem:

THEOREM 0.21   *A subset S of $C[K]$ is compact if it is closed, equibounded, and equicontinuous.*

PROOF

We shall show that S is sequentially compact; that is, that any sequence $\{f_n\}$ of functions belonging to S contains a convergent subsequence $\{f_{n'}\}$. Since the functions f_n, $n = 1, 2, \ldots$, are equicontinuous, then with any $x \in K$ and arbitrary $\varepsilon > 0$ we can associate a $\delta = \delta(\varepsilon) > 0 \ni x, x' \in K$ and $d(x, x') < \delta(\varepsilon) \Rightarrow |f_n(x) - f_n(x')| < \varepsilon$ for $n = 1, 2, \ldots$. Using the compactness of K, we cover K with a finite number of balls $B(c_1, \delta)$, $B(c_2, \delta), \ldots, B(c_k, \delta)$ with radius $\delta = \delta(\varepsilon)$.

Consider now the sequence of values $\{f_n(c_1)\}$. Because of the equi-boundedness of the functions of S, the sequence is a bounded subset of $\mathbf{R}$, if the f_n are real valued, and a bounded subset of $\mathbf{R}^2$, if the f_n are complex valued. A subsequence $f_n^{(1)}(c_1)$ is convergent in $\mathbf{R}$ or $\mathbf{R}^2$, since boundedness in $\mathbf{R}^n$ is equivalent to total boundedness and closure in $\mathbf{R}^n$.

Consider then the sequence $\{f_n^{(1)}(c_2)\}$. This sequence is also bounded, and so $\exists$ a convergent subsequence $\{f_n^{(2)}(c_2)\}$ of $\{f_n^{(1)}(c_2)\}$. We repeat this process and finally arrive at the convergent sequence $\{f_n^{(k)}(c_k)\}$. The functions $f_n^{(k)}$ converge at the k points $c_1, c_2, \ldots, c_k$. It follows that for n, $m > N(\varepsilon)$ we have $|f_n^{(k)}(c_i) - f_m^{(k)}(c_i)| < \varepsilon$ for $i = 1, 2, \ldots, k$. Then, if x is an arbitrary point of K belonging to $B(c_j, \delta)$, we have

$$|f_n^{(k)}(x) - f_m^{(k)}(x)| \leq |f_n^{(k)}(x) - f_n^{(k)}(c_j)|$$
$$+ |f_n^{(k)}(c_j) - f_m^{(k)}(c_j)|$$
$$+ |f_m^{(k)}(c_j) - f_m^{(k)}(x)|$$

Since $x \in B(c_j, \delta)$, we have $|f_n^{(k)}(x) - f_n^{(k)}(c_j)| < \varepsilon$, as well as $|f_m^{(k)}(c_j) - f_m^{(k)}(x)| < \varepsilon$. Then the inequality $|f_n^{(k)}(c_j) - f_m^{(k)}(c_j)| < \varepsilon$ implies that $|f_n^{(k)}(x) - f_m^{(k)}(x)| < 3\varepsilon$ for any $x \in K$. This means that $\|f_n^{(k)} - f_m^{(k)}\| < 3\varepsilon$, which implies that the sequence $f_1^{(k)}, f_2^{(k)}, \ldots, f_n^{(k)}, \ldots$ is a Cauchy sequence in S. Since S is closed in the complete metric space $C[K]$, this sequence is convergent in S.

Problems

0.20.: Let K be a compact subset of the metric space (E, d) and $x \in E$. Define the distance

$$\text{dist}(x, K) = \text{g.l.b.}\{d(x, y) | y \in K\}$$

Show that if $x \notin K$, then dist $(x, K) > 0$.

0.21.: Let K and L be two compact subsets of the metric space (E, d). Define the distance dist (K, L) by

$$\text{dist}(K, L) = \text{g.l.b.}\{d(x, y) | x \in K, y \in L\}$$

Show that if $K \cap L = \varnothing$, then $\text{dist}(K, L) > 0$.

0.22.: Show that a closed, bounded set in $\mathbf{R}^n$ is compact.

0.23.: Show that the mapping $x \to 1/x$, $x \in \,]0, 1[$, is continuous but not uniformly continuous.

chapter 1
Complex Differentiable Functions

§1. Summary of the course

In this course we shall study functions of one complex variable. We shall extend concepts developed in real analysis, such as derivative, integral, and Taylor series, to the field of complex-valued functions of one complex variable.

Let us first give a summary of the course:

An open connected set in the xy-plane will be called a domain D. We shall introduce four properties of complex-valued functions defined on a domain D:

1) Complex differentiability (a notion which was introduced by Riemann): A function f is said to be complex differentiable at a point $a \in D$ if

$$\lim_{z \to a} \frac{f(z) - f(a)}{z - a} = f'(a)$$

exists. We say that f is complex differentiable on D if it is complex differentiable at each point of D.

2) Holomorphicity (this notion was introduced by Cauchy): A continuous complex-valued function f is said to be holomorphic on D if

$$\int_\gamma f(z)\,dz = 0$$

for each closed curve γ which can be shrunk to a point in D by continuous deformation in D.

3) Analyticity (a notion introduced by Weierstrass): A complex-valued function f is said to be analytic on D if with each point $c \in D$ one can

associate a power series $\sum_{n=0}^{\infty} a_n(z-c)^n$, which is convergent in an open disc $d(c, r) \subset D$ with $r > 0$, $\ni$

$$f(z) = \sum_{n=0}^{\infty} a_n(z-c)^n \quad \forall\, z \in d(c, r)$$

(*Note*: $d(c, r)$ is an open disc with center c and radius r.)

4) Conformality: The nonconstant function f is said to define a conformal mapping on D if it preserves the angle between two arcs at every point of D, except possibly at isolated points.

The aim of this course is to show the equivalence of these four notions.

§2. The complex numbers

1. Algebraic aspects

A complex number z is defined as an ordered pair of real numbers:

$$z = (a, b) \quad \text{where} \quad a \in \mathbf{R}, \quad b \in \mathbf{R}$$

Thus it may be considered to be an element of $\mathbf{R}^2$. The real numbers a and b will be called the *real part* and the *imaginary part* of z, respectively. In symbols,

$$a = \operatorname{Re} z, \quad b = \operatorname{Im} z$$

Two complex numbers $z_1 = (a_1, b_1)$ and $z_2 = (a_2, b_2)$ shall be considered to be equal iff $a_1 = a_2$ and $b_1 = b_2$. The set of all complex numbers will be denoted by $\mathbf{C}$.

The elementary operations of addition and multiplication in $\mathbf{C}$ are defined as follows:

$$z_1 + z_2 = (a_1 + a_2, b_1 + b_2)$$
$$z_1 z_2 = (a_1 a_2 - b_1 b_2, a_1 b_2 + a_2 b_1)$$

It is easy to verify that these operations obey all the rules of algebra, that is, the commutative, associative, and distributive laws as established for the field of real numbers.

The complex numbers $(0, 0)$ and $(1, 0)$ obviously assume the roles of the additive and multiplicative identities. The inverse operations to addition and multiplication—namely, subtraction and division—can always be performed within **C**, except for division by (0, 0). This is evidenced by the formulas

$$z_1 - z_2 = (a_1 - a_2, b_1 - b_2)$$

$$\frac{z_1}{z_2} = \left(\frac{a_1 a_2 + b_1 b_2}{a_2^2 + b_2^2}, \frac{a_2 b_1 - a_1 b_2}{a_2^2 + b_2^2}\right)$$

which are easily verified. We conclude, therefore, that **C** is closed with respect to the four elementary operations as defined above, except for the division by (0, 0); in other words, **C** is a field.

C contains a subfield $\mathbf{C}_0 = \{(a, 0) | a \in \mathbf{R}\}$ which is isomorphic to **R** itself. Indeed, we find

$$(a, 0) \pm (b, 0) = (a \pm b, 0)$$
$$(a, 0)(b, 0) = (ab, 0)$$
$$\frac{(a, 0)}{(b, 0)} = \left(\frac{a}{b}, 0\right)$$

so that $(a, 0) \leftrightarrow a$ for all $a \in \mathbf{R}$ establishes the isomorphism.

It is readily seen that

$$(a, b) = (a, 0) + (b, 0)(0, 1)$$

If we therefore introduce the abbreviation

$$(0, 1) = i$$

and identify $(a, 0)$ and $(b, 0)$ with a and b, respectively, we obtain

$$(a, b) = a + bi \quad \text{where} \quad i^2 = (-1, 0) = -1$$

Thus, the equation $x^2 + 1 = 0$, which has no solution in **R**, becomes solvable in **C**, where it has the roots $\pm i$. Furthermore, it has been established that operations involving the "imaginary" unit $i = \sqrt{-1}$ can safely be performed without fear of contradiction, and that $a + bi$ can be treated as a polynomial in i, with the added understanding that $i^2 = -1$.

2. Geometric representation of complex numbers

The one-to-one correspondence between **C** and $\mathbf{R}^2$ suggests a geometric interpretation of the complex numbers. All one has to do is to let the point $P(x, y)$ in a rectangular coordinate system represent the complex number $z = x + iy$. In that way, every complex number is mapped uniquely onto a specified point of the xy-plane and, vice versa, every point of that plane corresponds to one and only one complex number. The plane involved is called the complex, or Gaussian, plane and is denoted by **C**. Thus, no distinction is made between the plane **C** and the set **C** of the complex numbers.

In the **C**-plane, the x-axis of the coordinate system is called the *real axis*, whereas the y-axis is called the *imaginary axis*.

It is convenient to have a representation of z in polar coordinates. Introducing (r, θ) as indicated in Figure 1.1, it is easily seen that

$$x = r \cos \theta, \qquad y = r \sin \theta$$

from which we obtain

$$r = \sqrt{x^2 + y^2} > 0, \qquad \cos \theta = \frac{x}{r}, \qquad \sin \theta = \frac{y}{r}$$

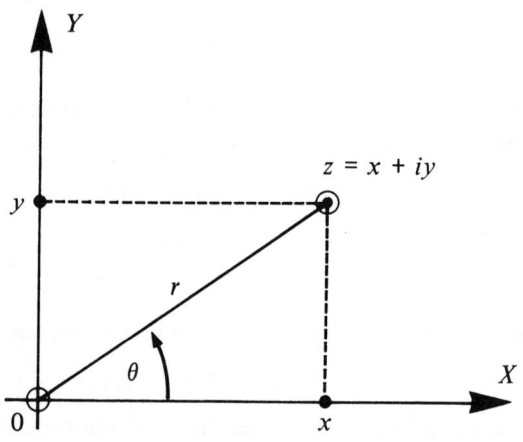

Figure 1.1

The positive quantity r is called the *modulus*, or *absolute value*, of z and is denoted by $r = |z|$, while θ is called the *argument* of z and is denoted by $\theta = \arg z$.

The argument or polar angle of z remains constant on any ray issuing from the origin. The ray corresponding to the argument θ will be denoted by $\mathbf{R}^\theta$. In particular, we have $\mathbf{R}^0 = \mathbf{R}^+$ and $\mathbf{R}^\pi = \mathbf{R}^-$.

Argument of z is not a single-valued function, since to any $z \neq 0$ there corresponds an infinity of values of $\arg z$. We shall say that $\arg z$ is a *multivalued* function. Indeed, if θ is one of the values of $\arg z$, we have

$$\arg z = \theta + 2k\pi, \qquad k = 0, \pm 1, \pm 2, \ldots$$

We may say that arg z is determined only mod 2π. For a given θ, the set of values $\theta + 2k\pi$, $k = 0, \pm 1, \pm 2, \ldots$, is an equivalence class mod 2π, and obviously $\exists$ a one-to-one correspondence between the equivalence classes mod 2π and the rays $\mathbf{R}^\theta$ issuing from the origin.

We can choose from each equivalence class mod 2π one representative θ which satisfies

$$\alpha - \pi < \theta \leq \alpha + \pi$$

where α is an arbitrary angle. We shall denote this representative by $\arg_\alpha z$. Obviously, $\arg_\alpha z$ is not defined at the origin and is discontinuous on the ray $\mathbf{R}^{\alpha-\pi}$ where it jumps from $\alpha - \pi$ to $\alpha + \pi$.

For $\alpha = 0$, we get a representative θ which satisfies

$$-\pi < \theta \leq \pi$$

We shall call this representative the principal value of arg z and denote it by Arg z:

$$\arg_0 z = \text{Arg } z$$

Clearly Arg z is discontinuous at $\mathbf{R}^\pi$. We shall show later on that Arg z is continuous throughout $\mathbf{C}\backslash\mathbf{R}_0^{\,-}$.

We shall denote by $\mathbf{R}_0^{\,\theta}$ the union of $\mathbf{R}^\theta$ and the origin:

$$\mathbf{R}_0^{\,\theta} = \mathbf{R}^\theta \cup \{0\}$$

Occasionally, we shall use the abbreviation

$$\text{cis } \theta = \cos \theta + i \sin \theta$$

in order to write the complex number z in the form

$$z = r \text{ cis } \theta$$

This implies, of course, that $|z| = r$, arg $z = \theta$, and $|\text{cis } \theta| = 1$.

The complex number $x - iy$ is called the *conjugate* of $z = x + iy$. It will be denoted by $\bar{z}$; that is, $\bar{z} = x - iy$. $\bar{z}$ is the reflection of z in the real axis. Obviously,

$$|z| = \sqrt{z\bar{z}} = \sqrt{x^2 + y^2}$$

Similarly, one obtains easily

$$\text{Re } z = \frac{1}{2}(z + \bar{z}), \qquad \text{Im } z = \frac{1}{2i}(z - \bar{z})$$

Let $z_1, z_2 \in \mathbf{C}$. Then the following statements hold:

(i) $\overline{z_1 + z_2} = \overline{z_1} + \overline{z_2}$

(ii) $\overline{z_1 z_2} = \overline{z_1}\,\overline{z_2}$

(iii) $\overline{\left(\dfrac{z_1}{z_2}\right)} = \dfrac{\overline{z_1}}{\overline{z_2}}, \quad |z_2| > 0$

(iv) $(\overline{\overline{z_1}}) = z_1$

We leave these to the reader to demonstrate. In general, if $z_1, z_2, z_3, \ldots, z_n \in \mathbf{C}$, and $\mathbf{R}(z_1, z_2, \ldots, z_n)$ is a rational function in $z_1, z_2, \ldots, z_n$ with real coefficients, then

$$\overline{\mathbf{R}(z_1, z_2, \ldots, z_n)} = \mathbf{R}(\overline{z_1}, \overline{z_2}, \ldots, \overline{z_n})$$

We note in passing that the modulus of a complex number behaves like a norm. In fact, if $z, z_1, z_2 \in \mathbf{C}$, then

1) $|z| \geq 0 \quad \forall z \in \mathbf{C}$ and $|z| = 0 \Leftrightarrow z = 0$
2) $|z_1 z_2| = |z_1| |z_2|$
3) $|z_1 + z_2| \leq |z_1| + |z_2|$ \quad (triangle inequality)

and, in addition to these, we have

4) $|z_1 - z_2| \geq \bigl||z_1| - |z_2|\bigr|$

Statement 1) holds without proof. The truth of statement 2) is revealed by

$$|z_1 z_2| = \sqrt{(z_1 z_2)(\overline{z_1 z_2})} = \sqrt{(z_1 \overline{z_1})(z_2 \overline{z_2})} = |z_1| |z_2|$$

To prove statements 3) and 4), we consider Figure 1.2. Let

$$\mathbf{P}_1 = (x_1, y_1) = z_1 = x_1 + iy_1$$
$$\mathbf{P}_2 = (x_2, y_2) = z_2 = x_2 + iy_2$$

and complete the parallelogram by introducing the point $\mathbf{P}_3$. Then we have

$$\mathbf{P}_3 = (x_1 + x_2, y_1 + y_2) = (x_1 + x_2) + i(y_1 + y_2) = z_1 + z_2$$

Now, in the triangle $O\mathbf{P}_1\mathbf{P}_3$ with the sides $\overline{O\mathbf{P}_1} = |z_1|$, $\overline{\mathbf{P}_1\mathbf{P}_3} = \overline{O\mathbf{P}_2} = |z_2|$ and $\overline{O\mathbf{P}_3} = |z_1 + z_2|$, any side must be smaller than or equal to the sum of the other two sides. Hence,

$$|z_1 + z_2| \leq |z_1| + |z_2|$$

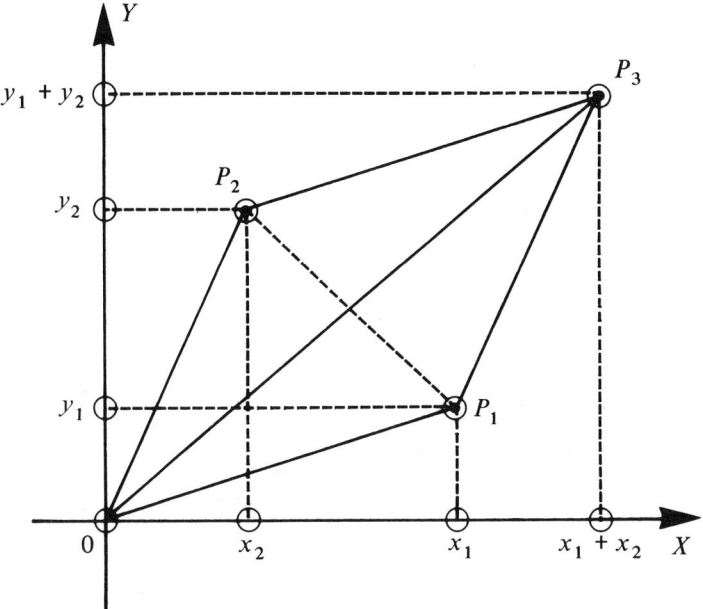

Figure 1.2

Similarly, in the triangle OP_1P_2, the sides are $\overline{OP_1} = |z_1|$, $\overline{OP_2} = |z_2|$ and $P_1P_2 = \sqrt{(x_1 - x_2)^2 + (y_1 - y_2)^2} = |z_1 - z_2|$, and any side must be larger than or equal to the difference of the other two, so that, in particular,

$$|z_1 - z_2| \geq ||z_1| - |z_2||$$

The triangle inequality can easily be generalized to encompass sums of more than two complex numbers:

$$\left|\sum_{k=1}^{n} z_k\right| \leq \sum_{k=1}^{n} |z_k|$$

with equality holding if and only if

$$\arg z_1 = \arg z_2 = \cdots = \arg z_n$$

The proof of this statement is left to the reader.

3. Arguments of products and quotients

The product of two complex numbers, $z_1 = r_1 \operatorname{cis} \phi_1$ and $z_2 = r_2 \operatorname{cis} \phi_2$, can be evaluated as follows:

$$\begin{aligned} z_1 z_2 &= r_1(\cos \phi_1 + i \sin \phi_1) r_2(\cos \phi_2 + i \sin \phi_2) \\ &= r_1 r_2 [(\cos \phi_1 \cos \phi_2 - \sin \phi_1 \sin \phi_2) \\ &\quad + i(\cos \phi_1 \sin \phi_2 + \sin \phi_1 \cos \phi_2)] \\ &= r_1 r_2 [\cos(\phi_1 + \phi_2) + i \sin(\phi_1 + \phi_2)] \\ &= r_1 r_2 \operatorname{cis}(\phi_1 + \phi_2) \end{aligned}$$

Thus, we have $|z_1 z_2| = r_1 r_2 = |z_1||z_2|$, as established previously. In addition, it follows that

$$\arg(z_1 z_2) = \phi_1 + \phi_2 = \arg z_1 + \arg z_2$$

It is easy to generalize these results to products of more than two factors: If $z_k = r_k \operatorname{cis} \phi_k$ ($k = 1, 2, \ldots, n$) then

$$\left| \prod_{k=1}^{n} z_k \right| = \prod_{k=1}^{n} r_k = \prod_{k=1}^{n} |z_k|, \qquad \arg\left(\prod_{k=1}^{n} z_k \right) = \sum_{k=1}^{n} \phi_k = \sum_{k=1}^{n} \arg z_k$$

The quotient of two complex numbers can be handled in a similar way. Considering that $|1/z| = 1/|z|$ and $\arg(1/z) = -\arg z$, we obtain the relations

$$\left| \frac{z_1}{z_2} \right| = \frac{|z_1|}{|z_2|}, \qquad \arg\left(\frac{z_1}{z_2} \right) = \arg z_1 - \arg z_2$$

A special case of particular importance involves products with identical factors, that is, powers of complex numbers. It follows immediately from the preceding discussion that

$$[r \operatorname{cis} \phi]^n = r^n \operatorname{cis} n\phi \qquad (n = 1, 2, \ldots)$$

and, in the special case of $r = 1$,

$$(\cos \phi + i \sin \phi)^n = \cos n\phi + i \sin n\phi$$

This relation is known as *DeMoivre's rule*, after Abraham DeMoivre, who first recognized its importance. Utilizing the rules for the quotients of complex numbers, one can easily verify that DeMoivre's rule also holds for negative integer values of the exponent.

We can use DeMoivre's rule to find the nth roots of a complex number $c = r \operatorname{cis} \theta$. More precisely, we will find all the roots of the equation $z^n = c$. Let $z = \rho \operatorname{cis} \phi$; then $z^n = \rho^n \operatorname{cis} n\phi$. It follows immediately that $\rho^n = r$, $\rho = \sqrt[n]{r}$, and, with due regard to the multivalued nature of the argument, $n\phi = \theta \pm 2k\pi$ ($k = 1, 2, 3, \ldots$), or, equivalently, $\phi = \theta/n \pm k \cdot 2\pi/n$. This last formula yields exactly n different values for z, namely,

$$z = \sqrt[n]{r}\operatorname{cis}\left(\frac{\theta}{n} + k \cdot \frac{2\pi}{n}\right), \quad k = 0, 1, 2, \ldots, n - 1$$

with all the other values of k producing the same n values of z in endless repetition.

Problems

1.1.: Reduce to the form $a + bi$.
  a) $(4 + i) - (3 - 2i)$
  b) $(4 + i)(2 - 3i)$
  c) $\dfrac{1}{1 + i}$

1.2.: Compute.
  a) $\left[\dfrac{-1 + i\sqrt{3}}{2}\right]^3$, $\left[\dfrac{-1 - i\sqrt{3}}{2}\right]^3$, $\left[\dfrac{1 + i\sqrt{3}}{2}\right]^3$,

  $\left[\dfrac{-1 + i\sqrt{3}}{2}\right]^6$, $\left[\dfrac{1 + i\sqrt{3}}{2}\right]^6$

  b) $(1 + i)^{10}$, $\dfrac{(-1 + i\sqrt{3})^{60}}{(1 - i)^{16}} + \dfrac{(-1 - i\sqrt{3})^{60}}{(1 + i)^{16}}$

1.3.: Prove that
  a) $(1 - i)^m = 2^{m/2}\left(\cos\dfrac{m\pi}{4} - i\sin\dfrac{m\pi}{4}\right)$

  b) $(1 + i\sqrt{3})^m = 2^m\left(\cos\dfrac{m\pi}{3} + i\sin\dfrac{m\pi}{3}\right)$

  c) $(1 - \cos\alpha + i\sin\alpha)^n = 2^n(i)^n\sin^n(\alpha/2)(\cos n\alpha/2 - i\sin n\alpha/2)$

1.4.: Show that $|z| = |\bar{z}|$.

1.5.: Show that $|z| \geq \max(|x|, |y|)$.

1.6.: a) For which values of z is $|z| = \pm z$?
      b) For which values of z is $|z| = 2\operatorname{Re} z$?

1.7.: Show that $|z_1 - z_2| \geq \||z_1| - |z_2|\|$.

1.8.: Prove algebraically that $|z_1| + |z_2| \geq |z_1 + z_2|$.

1.9.: a) Show by induction that $\left|\sum_{i=1}^{n} c_i\right| \leq \sum_{i=1}^{n} |c_i|$ and that the equal sign holds only if $\arg c_1 = \arg c_2 = \cdots = \arg c_n$.
b) Show that
$$\left|\prod_{i=1}^{n} c_i\right| = \prod_{i=1}^{n} |c_i| \quad \text{and} \quad \arg\left(\prod_{i=1}^{n} c_i\right) = \sum_{i=1}^{n} \arg c_i$$
c) Show for integral n that $|c|^n = |c^n|$ and $\arg c^n = n \arg c$.
d) Show also that $|\sqrt[n]{c}| = \sqrt[n]{|c|}$ and
$$\arg \sqrt[n]{c} = \frac{1}{n} \arg c + \frac{2k\pi}{n} \quad (k = 0, 1, 2, \ldots, n-1)$$
e) Show that if the segment joining z_1 to z_2 does not cut $\mathbf{R}_0^-$, where $\mathbf{R}_0^- = \mathbf{R}^- \cup \{0\}$, then
$$\text{Arg}\left(\frac{z_1}{z_2}\right) = \text{Arg } z_1 - \text{Arg } z_2$$

1.10.: Which part of the complex plane is defined by:
a) $\text{Re}(z) > 0$, $\quad \text{Im}(z) = 0$, $\quad \text{Re}(z) + \text{Im}(z) = 0$?
b) $|z - c| \leq R$. $\quad a \leq |z - z_0| \leq b$?
c) $|z - a| = R$, $\quad z = a + r \text{ cis } \theta$, $\quad 0 < \theta \leq 2\pi$?

1.11.: Determine the set of points satisfying each of the following:
a) $|z - a| + |z - b| = c$
b) $|z - a||z - b| = c$
c) $\left|\dfrac{z-a}{z-b}\right| = c$

1.12.: a) Show that the complex number z which divides the segment joining z_1 and z_2 in the real ratio λ is given by $z = \dfrac{z_1 + \lambda z_2}{1 + \lambda}$.
b) Show that the center of gravity z of the triangle with the vertices z_1, z_2, z_3 is given by $z = (z_1 + z_2 + z_3)/3$.
c) Determine the center of gravity of a triangle whose vertices divide each of the sides of the triangle z_1, z_2, z_3 in the ratio λ.

1.13.: a) If $w = \sqrt{z_1 z_2}$, show that
$$|z_1| + |z_2| = \left|\frac{z_1 + z_2}{2} + w\right| + \left|\frac{z_1 + z_2}{2} - w\right|$$
b) Prove the identity
$$|z_1 + z_2| + |z_1 - z_2| = \left|z_1 + \sqrt{z_1^2 - z_2^2}\right| + \left|z_1 - \sqrt{z_1^2 - z_2^2}\right|$$

1.14.: Show that the field of 2×2 matrices

$$\begin{pmatrix} a & b \\ -b & a \end{pmatrix}$$

is isomorphic to the field of complex numbers $a + bi$.

1.15.: Show that the mapping $z \to \bar{z}$ is an automorphism of **C**.

1.16.: a) Determine the locus of the points z which satisfy $\arg\left(\dfrac{z-a}{z-b}\right) = \dfrac{\pi}{2}$.

b) Determine the conditions for the points z_1, z_2, z_3 to be the vertices of an equilateral triangle.

---

4. The Riemann sphere

The Gaussian plane establishes a one-to-one correspondence between the complex numbers and the points of the plane. Frequently, another representation of the complex numbers is useful: the one provided by the stereographic projection of the Gaussian plane onto a sphere.

Let ξ, η, ζ be a right-handed Cartesian coordinate system in $\mathbf{R}^3$, such that the ξ and η axes coincide with the x and y axes of the Gaussian plane. Let K be a sphere with radius $\frac{1}{2}$ and the center $M(0, 0, \frac{1}{2})$, tangent to the Gaussian plane at the origin. Denote by $P'(x, y)$ the projection of an arbitrary point $P(\xi, \eta, \zeta)$ of the sphere onto the Gaussian plane, with the north pole $N(0, 0, 1)$ of the sphere as the center of projection. Furthermore, introduce polar coordinates ϕ, θ on the sphere, as indicated in Figure 1.3. Then the following relations are easily verified:

$$\left. \begin{array}{l} \xi = \tfrac{1}{2} \sin \theta \cos \phi \\ \eta = \tfrac{1}{2} \sin \theta \sin \phi \\ \zeta = \tfrac{1}{2} + \tfrac{1}{2} \cos \theta \end{array} \right| \quad \begin{array}{l} 0 \le \phi \le 2\pi \\ 0 \le \theta \le \pi \end{array}$$

$$x = \frac{\xi}{1-\zeta}, \quad y = \frac{\eta}{1-\zeta}$$

$$\xi = \frac{x}{1+x^2+y^2}, \quad \eta = \frac{y}{1+x^2+y^2}, \quad \zeta = \frac{x^2+y^2}{1+x^2+y^2}$$

If P is confined to a circle on the sphere—that is, to a plane—then ξ, η, ζ satisfy a linear equation, which can be written in the form

$$A\xi + B\eta + C(1-\zeta) + D = 0$$

Expressed in x and y, this is equivalent to

$$Ax + By + (C + D) + D(x^2 + y^2) = 0$$

thus indicating that P' is located on a circle or a straight line, depending on whether $D \neq 0$ or $D = 0$; that is, whether the circle on the sphere misses or crosses the north pole.

The stereographic projection maps the circles of the sphere onto the circles and straight lines of the plane, and vice versa.

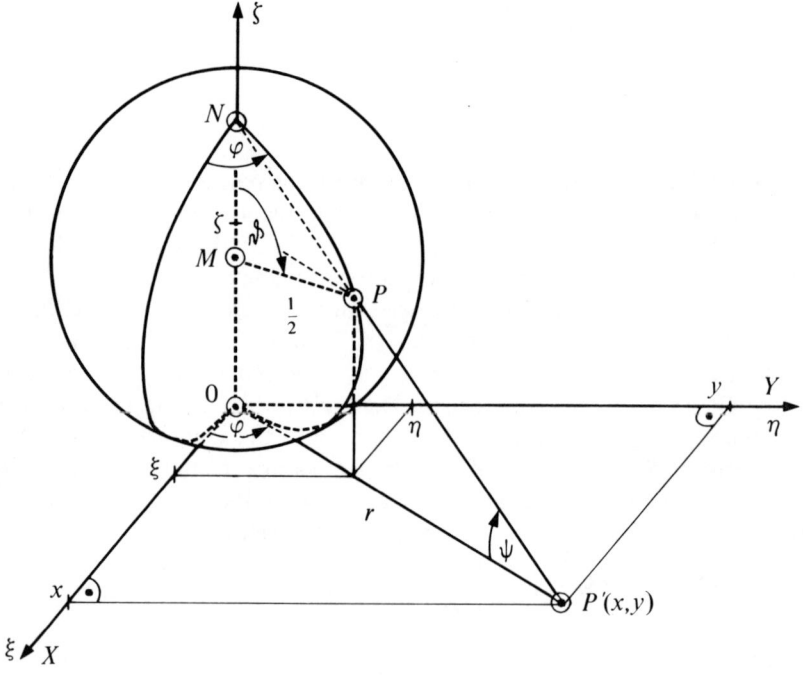

Figure 1.3

The stereographic projection is also isogonal, as can be ascertained in the following way: Two straight lines of the plane, crossing at P' and forming the angle α, are mapped onto two circles of the sphere, crossing at P and passing through N, where their tangents are parallel to the original lines. The intersecting circles form the same angle at P as at N; however, the sense of the angle is reversed.

The Riemann sphere S^2 is now obtained by associating with each point $P \neq N$ on the sphere the complex number which corresponds to

P' in the Gaussian plane. Thus, with $P(\xi, \eta, \zeta) = P(\phi, \theta)$ is associated the number

$$z = x + iy = \frac{\xi}{1-\zeta} + i\frac{\eta}{1-\zeta} = \frac{\xi + i\eta}{1-\zeta}$$

$$= \frac{\frac{1}{2}\sin\theta\cos\phi + \frac{i}{2}\sin\theta\sin\phi}{\frac{1}{2} - \frac{1}{2}\cos\theta}$$

$$= \frac{\sin\theta}{1 - \cos\theta}\operatorname{cis}\phi = \cot\frac{\theta}{2}\operatorname{cis}\phi$$

In particular, if two points $P_1(\theta, \phi)$ and $P_2(\pi - \theta, \phi)$ are symmetric with respect to the equatorial plane, we have

$$z_1 = \cot\frac{\theta}{2}\operatorname{cis}\phi$$

$$z_2 = \cot\left(\frac{\pi}{2} - \frac{\theta}{2}\right)\operatorname{cis}\phi = \tan\frac{\theta}{2}\operatorname{cis}\phi = \frac{1}{\overline{z_1}}$$

that is,

$$\overline{z_1}z_2 = z_1\overline{z_2} = 1$$

On the other hand, if $P_1(\theta, \phi)$ and $P_2(\theta + \pi, \phi)$ are diametrically opposed on the sphere, we have

$$z_1 = \cot\frac{\theta}{2}\operatorname{cis}\phi$$

and

$$z_2 = \cot\left(\frac{\theta}{2} + \frac{\pi}{2}\right)\operatorname{cis}\phi = -\tan\frac{\theta}{2}\operatorname{cis}\phi = -\frac{1}{\overline{z_1}}$$

that is,

$$\overline{z_1}z_2 = \overline{z_2}z_1 = -1$$

It should be pointed out at this stage that the one-to-one correspondence between the complex numbers and the points of the Riemann sphere excludes the point N from consideration, as there exists no counterpart of it in the plane. Indeed, if P approaches N in any direction whatsoever, we have $\theta \to 0$ and consequently

$$|z| = \left|\cot \frac{\theta}{2} \operatorname{cis} \phi\right| = \cot \frac{\theta}{2} \to \infty$$

In order to remove this inconsistency, we introduce, therefore, an ideal "number" ∞ and associate it with the point N on the sphere. Similarly, we add an ideal "point" ∞ to the Gaussian plane and call it the stereographic projection of N onto the plane.

The union $\mathbf{C} \cup \{\infty\}$ is called the *extended complex plane* and is denoted by $\mathbf{C}^\infty$. We define the neighborhood of the point ∞ as the stereographic projection onto $\mathbf{C}$ of a neighborhood of N on the sphere. It follows that the complement $\mathbf{C} \setminus d(c, r)$ of any disc $d(c, r)$ with respect to the complex plane is a neighborhood of the point ∞. It is then obvious that $\mathbf{C}^\infty$ is a compact topological space which is homeomorphic to S^2.

Problem

1.17.:
   a) Describe the relative positions of z and the following points in the complex plane, relative to the Riemann sphere:

   1) $-z$ 2) $\bar{z}$,   3) $1/z$ 4) $\dfrac{z + \bar{z}}{2}$

   5) $\dfrac{z - \bar{z}}{2}$ 6) $\pm iz$ 7) $z/\bar{z}$ 8) $z/|z|$

   b) Show that the equation of the stereographic projection is

$$z = \frac{\xi + i\eta}{1 - \zeta}, \quad \zeta \neq 1$$

   and conversely

$$\xi = \frac{x}{1 + |z|^2}, \quad \eta = \frac{y}{1 + |z|^2}, \quad \zeta = \frac{|z|^2}{1 + |z|^2}$$

   c) Show that the stereographic projection is a homeomorphism between the complex plane and $S^2 \setminus \{N\}$.
   d) Prove that the stereographic projections of straight lines in $\mathbf{C}$ are circles on the Riemann sphere, passing through N.

§3. Topology of the complex plane

1. Neighborhoods and discs in the complex plane; open and closed sets

The set $\mathbf{C}$ of complex numbers is equivalent to $\mathbf{R}^2$, and we shall endow it with the standard topology of $\mathbf{R}^2$. As shown in 0.4.3 $\mathbf{R}^2$ is a Banach space; that is, a complete normed vector space, where the vectors are complex numbers, and the scalars may be real or complex numbers, as explained in 1.1.1. We take as the norm the Euclidean norm of $\mathbf{R}^2$: If $c = a + bi$, then $\|c\| = |c| = \sqrt{c\bar{c}} = \sqrt{a^2 + b^2}$. The distance between two points c, c' is then given by $|c - c'|$. Since $|c|$ has all properties of a norm, the function $|c - c'|$ has all properties of a distance.

The set of all points z whose distance from a given point c is smaller than r is called an *open disc* and is denoted by $d(c, r)$. In symbols, $d(c, r) = \{z \mid |z - c| < r\}$. The boundary of the open disc $d(c, r)$ is the circle $\partial(c, r)$ defined by $\partial(c, r) = \{z \mid |z - c| = r\}$. The open disc together with its boundary is called the *closed disc* and is denoted by $\bar{d}(c, r)$.

DEFINITION   Let c be a point of the set S in the complex plane. We say c is an *interior point* of S if $\exists$ an open disc $d(c, r)$ which is contained in S (see 0.4.1).

EXAMPLE

Each point of an open disc $d(c, r)$ is an interior point. Indeed, if $z_0 \in d(c, r)$, we can choose $r' = |z_0 - c| \ni$ the disc $d\left(z_0, \dfrac{r - r'}{2}\right)$ is completely contained in the disc $d(c, r)$. The boundary points of the closed disc $\bar{d}(c, r)$, obviously, are not interior points.

DEFINITION   A set is called *open* if all its points are interior points.

Thus, the open disc $d(c, r)$ is an open set, but the closed disc $\bar{d}(c, r)$ is not an open set, because the boundary points belong to the disc. Note that this definition is the same as that for general metric spaces (see 0.4.1).

We will denote by $\mathbf{C} \setminus S$ the complement of S relative to the complex plane, that is, the complex plane $\mathbf{C}$ minus the set S.

DEFINITION   A set S is called *closed* if its complement $\mathbf{C} \setminus S$ is open.

EXAMPLES

1 The set of all points outside the open disc $d(c, r)$, that is, the set $\{z \| z - c| \geq r\}$, is closed.

2 The closed disc $\overline{d}(c, r)$ is closed, while its complement is open.

3 The set consisting of one point is closed, because its complement is open.

The empty set $\varnothing$ is considered to be both open and closed. It follows that the entire complex plane **C**, being the complement of $\varnothing$ is also both open and closed. There are also sets which are neither open nor closed.

EXAMPLE

The open disc $d(0, 1)$ plus the point $z = 1$ is a set S which is neither open nor closed. Indeed, $z = 1$ belongs to S and is obviously not an interior point of S. On the other hand, the complement $\mathbf{C} \backslash S$ of S is not open either, because the boundary points of $d(c, r)$ different from $z = 1$— take, for instance, $z = i$—belong to the complement $\mathbf{C} \backslash S$ and are not interior points of the complement.

2. Sequences in the complex plane

Let $c_1, c_2, \ldots, c_n, \ldots$ be a sequence of complex numbers. Using the same definition as in general metric spaces, we say that the sequence $\{c_n\}$ is a Cauchy sequence if for any arbitrary $\varepsilon > 0 \; \exists$ an $N(\varepsilon) \ni |c_m - c_n| < \varepsilon$ for all $m, n > N(\varepsilon)$. We say this sequence converges to c if for any $\varepsilon > 0 \; \exists \; N(\varepsilon) \ni c_n \in d(c, \varepsilon)$ for $n > N(\varepsilon)$. We write in this case $\lim_{n \to \infty} c_n = c$. The preceding definition may also be formulated as follows: $\lim_{n \to \infty} c_n = c$ if the sequence of real numbers $|c_n - c|$ converges to zero. The second formulation shows us how we can reduce convergence in the complex plane **C** to convergence in **R**. The point c is called the *limit point* of the sequence.

The convergence of the sequence $\{c_n\} = \{a_n + ib_n\}$ to $c = a + ib$ implies the convergence of $\{a_n\}$ to a and of $\{b_n\}$ to b. Conversely, if the sequences $\{a_n\}$ and $\{b_n\}$ converge to a and b, respectively, then $\{c_n\} = \{a_n + ib_n\}$ converges to $c = a + ib$. This result is easily shown by using the inequalities

$$\max(|a_n - a|, |b_n - b|) \leq |c_n - c| \leq |a_n - a| + |b_n - b|$$

It is clear that the complex plane **C** is complete, since it has the same topology as $\mathbf{R}^2$ (see 0.4.3).

If $\lim_{n\to\infty} c_n = c$, then $\lim_{n\to\infty} |c_n| = |c|$. Indeed, $||c_n| - |c|| \leq |c_n - c|$.

The algebraic operations with sequences are continuous: If $\{c_n\}$ and $\{d_n\}$ are two sequences with $\lim_{n\to\infty} c_n = c$ and $\lim_{n\to\infty} d_n = d$, then $\{c_n + d_n\}$ and $\{c_n d_n\}$ are also convergent sequences, and

$$\lim_{n\to\infty} (c_n + d_n) = \lim_{n\to\infty} c_n + \lim_{n\to\infty} d_n = c + d$$

$$\lim_{n\to\infty} c_n d_n = \lim_{n\to\infty} c_n \cdot \lim_{n\to\infty} d_n = cd$$

This can be shown by means of the inequalities

$$|(c_n + d_n) - (c + d)| \leq |c_n - c| + |d_n - d|$$

and

$$|c_n d_n - cd| \leq |(c_n - c)d_n| + |c(d_n - d)|$$

If $\lim_{n\to\infty} c_n = c$ and $c \neq 0$, then $\lim_{n\to\infty} (1/c_n) = 1/c$. Indeed, if $\varepsilon > 0$ is arbitrarily small, but smaller than $|c/2|$, and $N(\varepsilon)$ so large that for $n > N(\varepsilon)$ we have $c_n \in d(c, \varepsilon)$, it follows that

$$\left| \frac{1}{c_n} - \frac{1}{c} \right| = \left| \frac{c_n - c}{c_n c} \right| \leq 2 \frac{|c_n - c|}{|c|^2} \leq \frac{2\varepsilon}{|c|^2}$$

because $|c_n| > c/2$. If $\lim_{n\to\infty} c_n = c \neq 0$ and $\lim_{n\to\infty} d_n = d$, then obviously $\lim_{n\to\infty} (d_n/c_n) = \lim_{n\to\infty} (1/c_n) \cdot \lim_{n\to\infty} d_n = d/c$.

3. Compact sets in the complex plane

Any bounded set S in the complex plane is totally bounded (see 0.4.5), since for any $\varepsilon > 0$ we can find a finite number of discs $d(c_i, \varepsilon)$ ($i = 1, 2, 3, \ldots, n$) covering S. Therefore, a bounded set S in the complex plane is relatively compact. If it is also closed, then it is compact. Conversely, if S is compact, it is bounded and closed. (See Theorem 0.18.)

EXAMPLE

The boundary S' of a bounded set S is compact since it is closed and bounded; it is bounded because it is a subset of $\bar{S}$, the closure of S, which is bounded.

THEOREM 1.1 Let $K_1 \supset K_2 \supset K_3 \supset \cdots K_n \supset K_{n+1} \supset \cdots$ be a sequence of nonempty compact subsets of $\mathbf{C}$. Then the intersection $\bigcap_{n=1}^{\infty} K_n$ is nonempty.

PROOF

Choose arbitrarily $z_1 \in K_1$, $z_2 \in K_2$, ... $z_n \in K_n$, ...; the sequence $\{z_n\}$ lies in K_1 which is compact. A subsequence $\{z_{n'}\}$ of $\{z_n\}$ is convergent. Let z_o be the limit of this subsequence. Then $z_o \in K_n$ ($n = 1, 2, \ldots$). Indeed, for $n' > n$ all the points $z_{n'} \in K_n$ and, since K_n is compact, $z_o \in K_n$ for any n; it follows that $z_o \in \bigcap_{n=1}^{\infty} K_n$. This theorem holds true for general metric spaces.

THEOREM 1.2 The Cartesian product of $K_1 \times K_2$ of two compact sets $K_1, K_2 \subset \mathbf{C}$ is again a compact set.

PROOF

Let $\{(c_{1n}, c_{2n})\}$ be a Cauchy sequence of elements belonging to $K_1 \times K_2$, with $c_{1n} = a_{1n} + ib_{1n} \in K_1$ and $c_{2n} = a_{2n} + ib_{2n} \in K_2$. Then, using the topology of $\mathbf{R}^4 = \mathbf{C}^2$, where

$$\|(c_{1n}, c_{2n})\|^2 = |c_{1n}|^2 + |c_{2n}|^2 = |a_{1n}|^2 + |b_{1n}|^2 + |a_{2n}|^2 + |b_{2n}|^2$$

we have

$$\max(|c_{1n} - c_{1, n+p}|, |c_{2n} - c_{2, n+p}|) \leq \|c_{1n} - c_{1, n+p}, c_{2n} - c_{2, n+p}\|$$

Since $\{(c_{1n}, c_{2n})\}$ is a Cauchy sequence, $\|c_{1n} - c_{1, n+p}, c_{2n} - c_{2, n+p}\| < \varepsilon$ for $n > N(\varepsilon)$ and all $p = 1, 2, \ldots$. Hence $|c_{1n} - c_{1, n+p}| < \varepsilon$ and $|c_{2n} - c_{2, n+p}| < \varepsilon$, and both sequences $\{c_{1n}\}$ and $\{c_{2n}\}$ are Cauchy sequences. Because K_1 and K_2 are compact, $\exists\, c_1 \in K_1$ and $c_2 \in K_2 \ni \lim_{n \to \infty} c_{1n} = c_1$ and $\lim_{n \to \infty} c_{2n} = c_2$. Consequently, $\{(c_{1n}, c_{2n})\}$ converges to $(c_1, c_2) \in K_1 \times K_2$, so $K_1 \times K_2$ is closed because every Cauchy sequence is convergent. Since $K_1 \times K_2$ is the product of two bounded subsets of finite dimensional spaces, it is bounded; hence $K_1 \times K_2$ is compact. (See Theorem 0.18.)

DEFINITION If K_1 and K_2 are two compact sets, we define the distance of K_1 to K_2 by

$$\text{dist}(K_1, K_2) = \text{g.l.b.}\{|x - y| \,|\, x \in K_1, y \in K_2\}$$

THEOREM 1.3 *If K_1 and K_2 are two compact sets in the complex plane, and $K_1 \cap K_2 = \emptyset$, then $\mathrm{dist}(K_1, K_2) > 0$.*

PROOF

Suppose $\mathrm{dist}(K_1, K_2) = 0$; then for any arbitrary $n \; \exists \; x_n \in K_1$ and $y_n \in K_2 \ni |x_n - y_n| < 1/n$. Since K_1 is compact, $\exists$ a subsequence $\{x_{n'}\}$ of $\{x_n\}$ which is convergent. Let $x = \lim_{n \to \infty} x_{n'}$. Clearly $x \in K_1$. Similarly, we find the subsequence $\{y_{n'}\}$ which converges to y, with $y \in K_2$. It follows that for a given $\varepsilon > 0 \; \exists$ an $N(\varepsilon) \ni n' > N(\varepsilon) \Rightarrow |x_{n'} - x| < \varepsilon$, as well as $|y_{n'} - y| < \varepsilon$. Hence

$$|x - y| \leq |x - x_{n'}| + |x_{n'} - y_{n'}| + |y_{n'} - y| \leq \varepsilon + \frac{1}{n'} + \varepsilon < 3\varepsilon$$

for $n' > 1/\varepsilon$, and the distance $|x - y|$ is less than any arbitrarily small number 3ε. It follows that $x = y$, or $K_1 \cap K_2 \neq \emptyset$, in contradiction with out hypothesis.

4. Connected sets and domains in the complex plane

DEFINITION We denote by $[a, b]$ the closed segment which joins point a to point b, $a, b \in \mathbf{C}$. The union $\bigcup_{i=0}^{n-1} [a_i, a_{i+1}]$ will be called a *polygonal line*.

DEFINITION We say that the open set S is *polygonally connected* if any two points $a, b \in S$ can be joined by a polygonal line $\bigcup_{i=0}^{n-1} [a_i, a_{i+1}]$ belonging to S where $a = a_0$ and $b = a_n$.

THEOREM 1.4 *An open set $S \subset \mathbf{C}$ is connected iff it is polygonally connected.*

PROOF

(i) *Polygonally connected $\Rightarrow$ connected*: If S is the union of two disjoint open sets S_1 and S_2 with $S_1 \neq \emptyset$, we will show that $S_1 = S$ and $S_2 = \emptyset$. Let $a \in S_1$ and denote by b a point of S_2, assuming $S_2 \neq \emptyset$. Then $\exists$ a polygonal line $\bigcup_{i=0}^{n-1} [a_i, a_{i+1}]$ which joins a and b. Obviously, $\exists$ one segment of that line, say $[a_k, a_{k+1}]$, with $a_k \in S_1$ and $a_{k+1} \in S_2$

and $[a_k, a_{k+1}] \subset S$. Since $S_1 \cap [a_k, a_{k+1}]$ and $S_2 \cap [a_k, a_{k+1}]$ are open in the relative topology of $[a_k, a_{k+1}]$, it follows that $[a_k, a_{k+1}]$ is the union of two open subsets in the relative topology; specifically, $[a_k, a_{k+1}] = S \cap [a_k, a_{k+1}] = (S_1 \cap [a_k, a_{k+1}]) \cup (S_2 \cap [a_k, a_{k+1}])$, in contradiction with the connectedness of a segment (see the first example in 0.3.3). Hence it is necessary that $S_2 = \emptyset$.

(ii) *Connected* $\Rightarrow$ *polygonally connected*: If S is not polygonally connected, take an arbitrary point $a \in S$, and denote by S_1 the set of all points b which can be joined by a polygonal line to a. Denote by S_2 all other points of S. Obviously, S_1 and S_2 are disjoint. Let us now show that both S_1 and S_2 are open: S_1 is open, for, if b is any point of S_1, $\exists$ a disc $d(b, r) \subset S$, because S is open. Any point b' of (b, r) belongs to S_1, since by adjoining the segment $[b, b']$ to the polygonal line joining a to b we obtain a polygonal line joining a to b'. This indicates that every point of S_1 is an interior point of S_1; thus S_1 is open. S_2 is also open, since, if $c \in S_2$, $\exists$ a disc $d(c, \rho) \subset S$ because S is open, and any point c' of $d(c, \rho)$ belongs to S_2. For, if there were a polygonal line joining a to c', then by adjoining the segment $[c', c]$ to this line we would obtain a line joining a to c, which is impossible. Thus $c \in S_2$ is an interior point of S_2. Therefore, S_2 is open. Hence S is the union of two nonempty, disjoint open sets S_1 and S_2, which is in contradiction to the connectedness of S.

DEFINITION   An open connected set $S \subset \mathbf{C}$ is called a *domain*.

EXAMPLES

1   An open set S is called *starlike* if $\exists \; c \in S \ni$ for any $z \in S$ the segment $[c, z]$ belongs to S. c is called the center of the star. The center is not necessarily unique. Clearly, an open starlike set S is poly-

Starlike domain

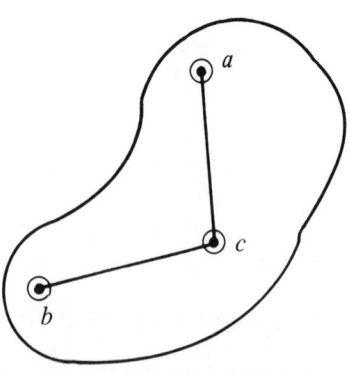

Figure 1.4

gonally connected, since for any two points a, $b \in S$ the polygonal line $[a, c] \cup [c, b]$ joining a to b belongs to S. Thus, an open starlike set is a domain.

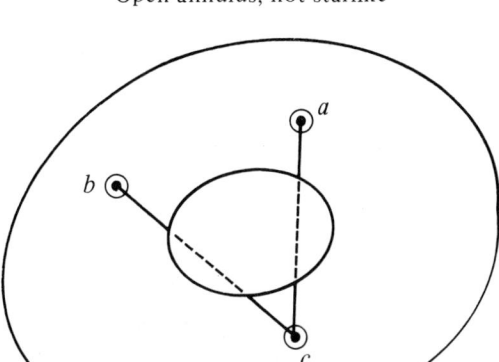

Figure 1.5

2 A set S is said to be *convex* if, for any pair of points a, $b \in S$, the segment $[a, b]$ belongs to S. Obviously, a convex set is polygonally connected since $[a, b]$ is a polygonal line. A convex set is starlike with any of its points as center. So an open convex set is a domain. An open disc, an open ellipse, an open triangle, and an open square are examples of convex domains.

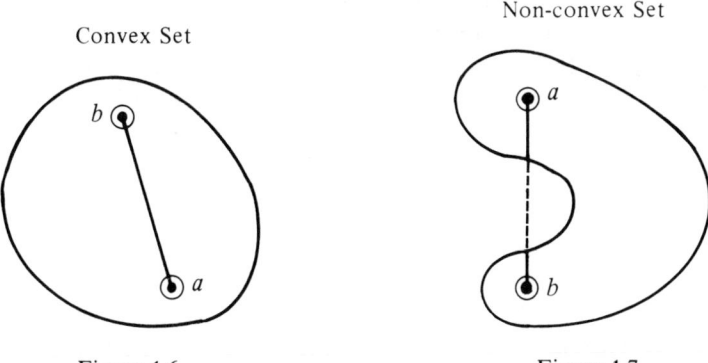

Figure 1.6 Figure 1.7

We call the set $d(c, r)\setminus\{c\}$ that is, the disc $d(c, r)$ without its center c, a *punctured disc*, and denote it by $\dot{d}(c, r)$. Obviously, a punctured disc is a domain which is neither convex nor starlike.

5. Arcs and closed curves

Our model of an arc is a closed interval on the real t-axis; for example, the closed interval $0 \le t \le 1$ or $a \le t \le b$.

DEFINITION An *arc* is the image of the interval $[0, 1]$ or $[a, b]$ under a continuous mapping γ in the complex plane, $\gamma: [0, 1] \to \mathbf{C}$.

Analytically, an arc is given in parametric form by $\gamma(t) = x(t) + iy(t)$, $0 \le t \le 1$, where $x(t)$ and $y(t)$ are continuous real-valued functions. Since the interval $[0, 1]$ is compact and γ is a continuous mapping, it follows that an arc is compact (see Theorem 0.7). If $\gamma(t)$ is a homeomorphism (see 0.3.2), then the arc is called a Jordan arc. Arcs will be denoted by Greek letters, for example, γ or Γ. We shall often denote the arc and the mapping by which it is defined by the same letter.

An arc is more than a set of points, since its points are ordered in some way. We say it has an *orientation*. If γ is a Jordan arc, we can define an order on γ which is the image of the order on $[0, 1]$ or $[a, b]$ by saying that the point $\gamma(t_2)$ *succeeds* $\gamma(t_1)$ if $t_2 > t_1$.

The arc γ starts at $\gamma(a)$ and ends at $\gamma(b)$ when $a \le t \le b$; that is, when t describes the closed interval $[a, b]$. Obviously, we get the same arc with the same orientation if we change the parameter by setting $t = t(\tau)$, where $t(\tau)$ is a continuous and strictly increasing function with $t(0) = 0$ and $t(1) = 1$, since $t_2 > t_1 \Rightarrow t(\tau_2) > t(\tau_1)$, for $0 \le \tau \le 1$. $\gamma(t)$, $0 \le t \le 1$, and $\gamma(t(\tau))$, $0 \le \tau \le 1$, define the same point set with the same orientation). We say that $\gamma(t(\tau))$ is another representation of the same arc.

If the arc γ is given by the mapping $\gamma(t)$, $0 \le t \le 1$, and we change the parameter setting $t = t(\tau)$, where $t(\tau)$ is a strictly decreasing continuous function with $t(0) = 1$ and $t(1) = 0$, then the point set is the same but the orientation is reversed. $\gamma(t(\tau))$ starts at $\gamma(t(0)) = \gamma(1)$ and ends at $\gamma(t(1)) = \gamma(0)$. We shall denote this arc by $-\gamma$.

EXAMPLE

$\gamma(t) = \cos \pi t + i\sin \pi t = \operatorname{cis} \pi t$, $0 \le t \le 1$, is a Jordan arc. When t varies from 0 to 1, $\gamma(t)$ covers the upper half of the circle $|z| = 1$, starting at $z = 1$ and ending at $z = -1$. If we change the parameter by using the strictly decreasing function $t = 1 - \tau$, $0 \le \tau \le 1$, then $\gamma(t(\tau)) = \operatorname{cis} \pi(1 - \tau) = -\operatorname{cis}(-\pi\tau)$ defines $-\gamma$.

Note that the image of the interval $[0, 1]$ on the t-axis may be a single point in the complex plane, since we may put $\gamma(t) = 0$, $0 \le t \le 1$.

Thus, a point is an arc, but certainly not a Jordan arc. One can show that certain arcs cover a square.

Even if γ is not a Jordan arc, if it contains more than one point, there is a partial order defined on it. Indeed, for every pair (t_1, t_2) with $0 \leq t_1 < t_2 \leq 1$ and $\gamma(t_1) \neq \gamma(t_2)$, we shall say that $\gamma(t_1)$ *precedes* $\gamma(t_2)$, or that $\gamma(t_2)$ *succeeds* $\gamma(t_1)$.

DEFINITION An arc $\gamma(t)$, $0 \leq t \leq 1$ or $a \leq t \leq b$, is called a *closed curve* or a *closed arc* if $\gamma(0) = \gamma(1)$.

One can take the circle $\partial(0, 1)$ instead of the interval $[0, 1]$ as model for a closed curve, and define γ as the image of $\partial(0, 1)$ by a continuous mapping in the complex plane. If the mapping is a homeomorphism, then γ is called a *closed Jordan curve*. Again, an orientation on $\partial(0, 1)$ will induce an orientation on γ.

DEFINITION We shall say that the arc γ is *smooth* if $\gamma(t)$ possesses a continuous derivative $\gamma'(t) = x'(t) + iy'(t) \neq 0$ for all values of t, $0 \leq t \leq 1$. We shall say that γ is *piecewise smooth* (p.w.s.) if it is smooth everywhere except at a finite number of points of $[0, 1]$, and if nonzero but unequal left and right derivatives exist for those points at which γ fails to be smooth.

We know from real analysis that a p.w.s. arc is rectifiable and that it possesses a continuous tangent except at the finite number of points where left and right derivatives are different. The direction of the tangent is determined by $\gamma'(t)$, and $\arg \gamma'(t)$ gives us the angle between the positive x-axis and the tangent.

EXAMPLE

It is easy to show that we can define an orientation on a polygonal line which makes it a p.w.s. curve.

Problems

1.18.: Show that an open interval $]a, b[$ on the real x-axis is not open in the topology of **C**.
1.19.: Give an example of two nonopen sets whose union is open.
1.20.: Give an example of two connected sets whose intersection is not connected.

1.21.: Give examples of two starlike domains D_1 and D_2 ∋
   a) $D_1 \cup D_2$ is not starlike.
   b) $D_1 \cap D_2$ is not starlike.
   c) $D_1 \cap D_2$ is convex.
1.22.: Construct a starlike domain which is not convex.
1.23.: Show that a ring (annulus) is not starlike.
1.24.: Show that the intersection of two convex sets is convex.
1.25.: Let S_1, S_2 be two open sets of **R**. Show that the Cartesian product $S_1 \times S_2$ is open in $\mathbf{R}^2$.
1.26.: a) Show that $\lim_{n \to \infty}(\sqrt{n+i} - \sqrt{n-i}) = 0$.

   b) Show that $\lim_{n \to \infty} n\left(\dfrac{a+bi}{|a|+|b|}\right)^n = 0$, for $a \neq 0$ and $b \neq 0$.

   c) For what values of z is $\sum_{n=0}^{\infty}\left(\dfrac{z-1}{z+1}\right)^n$ convergent?

   d) Show that $\lim_{n \to \infty} \dfrac{(n+1) \cdot i^n}{n^2 + 1} = 0$.

   e) Show that $a_n \to 0 \Leftrightarrow |a_n| \to 0$. What can you say about arg a_n?

1.27.: Show that if the sequence $\{a_n\}$ is convergent, then $\lim_{n \to \infty} \dfrac{1}{n}\sum_{i=1}^{n} a_i = \lim_{n \to \infty} a_n$.

1.28.: Let γ be the arc defined by $\gamma(t) = t + it^2$, $0 \leq t \leq 1$.
   a) Give another representation of γ.
   b) Represent $-\gamma$.

---

§4. Differentiable functions

1. Continuous functions

Let D be a domain in the complex plane **C** and f a function $D \to \mathbf{C}$. Since D and **C** are metric spaces (D with the relative topology), the following statements are equivalent (see 0.4.4):

(i) f is continuous on D, or continuous at each point $c \in D$; that is, with each disc $d(f(c), \varepsilon)$ we can associate a disc $d(c, \delta(\varepsilon)) \ni f[d(c, \delta(\varepsilon))] \subset d(f(c), \varepsilon)$.

(ii) f is sequentially continuous on D; that is, if $\{c_n\}$, $c_n \in D$, is a sequence converging to $c \in D$, then $\lim_{n \to \infty} f(c_n) = f(c)$.

(iii) Given any open set $S \subset \mathbf{C}$, the set $f^{-1}(S)$ is open in the relative topology of D.

EXAMPLES

1 The function which $\forall\, z \in \mathbf{C}$ is a complex constant c is continuous. The proof is trivial. Indeed, using definition (*iii*), we see that if $S \subset \mathbf{C}$ contains the point c, then $f^{-1}(S) = \mathbf{C}$, and if S does not contain c, then $f^{-1}(S) = \emptyset$. But $\mathbf{C}$ and $\emptyset$ are both open in the topology of $\mathbf{C}$.

2 The function $f(z) = z$, $z \in \mathbf{C}$, is continuous on $\mathbf{C}$. Obviously, $f^{-1}(S) = S$ for any set $S \subset \mathbf{C}$. Thus, $f^{-1}(S)$ is open whenever S is open.

3 The functions Re z (the real part of z) and Im z (the imaginary part of z) are both continuous throughout $\mathbf{C}$. Indeed, using definition (*ii*) of continuity, a result shown in 1.3.2 states that if $\{c_n\} = \{a_n + ib_n\} = \{\operatorname{Re} c_n + i \operatorname{Im} c_n\}$ is a sequence of complex numbers converging to $c = a + ib = \operatorname{Re} c + i \operatorname{Im} c$, then $\lim_{n \to \infty} a_n = a$ and $\lim_{n \to \infty} b_n = b$.

4 The function $|z| : \mathbf{C} \to \mathbf{R}^+$ is continuous. This follows from 1.3.2, since $\lim_{n \to \infty} c_n = c \Rightarrow \lim_{n \to \infty} |c_n| = |c|$.

5 The principal value Arg z of arg z is continuous on $\mathbf{C} \backslash \mathbf{R}_0^-$. Let us first show that Arg z is continuous at $z = 1$: Let $\{c_n\}$ be a sequence of complex numbers converging to 1. We must show that Arg c_n converges to Arg $1 = 0$. Let s_k be the length of the side of the regular k-polygon inscribed in the unit circle $\partial(0, 1)$. Then, for $n > N(s_k)$, c_n lies in the circle $\partial(1, s_k)$, and $|\operatorname{Arg} c_n| < 2\pi/k$. Clearly, $2\pi/k < \varepsilon$ for $k > K(\varepsilon)$.

Now let us show that Arg z is continuous at any point c which does not belong to $\mathbf{R}_0^-$. Let $\{c_n\}$ be a sequence converging to $c \notin \mathbf{R}_0^-$. Then the sequence $c_n/c \to 1$, and, because of the continuity at $z = 1$, we have $\lim_{n \to \infty} \operatorname{Arg}(c_n/c) = 0$. Moreover, since $c \notin \mathbf{R}_0^-$, $\exists\, d(c, \varepsilon) \ni d(c, \varepsilon) \cap \mathbf{R}_0^- = \emptyset$ and $\ni$ for $n > N(\varepsilon)$, $c_n \in d(c, \varepsilon)$. It follows that for $n > N(\varepsilon)$ the segment $[c_n, c]$ does not cut $\mathbf{R}_0^-$, so that $\operatorname{Arg}(c_n/c) = \operatorname{Arg} c_n - \operatorname{Arg} c$ (see problem 1.9(e)). Hence Arg c_n − Arg $c \to 0$, or Arg $c_n \to$ Arg c, and Arg z is continuous throughout the domain $\mathbf{C} \backslash \mathbf{R}_0^-$.

It is easy to show that at points of $\mathbf{R}_0^-$ the function Arg z is not continuous. It can also be shown that it is impossible to define arg z continuously in a domain containing the origin, or, more generally, in a domain which contains a closed curve enclosing the origin.

6 The function $f(z) = 1/z$ is continuous on $\mathbf{C} \backslash \{0\}$. The proof is left to the reader.

The set of all functions which are continuous on the domain D will be denoted by $C(D)$. We wish to show now that $C(D)$ is a complex algebra. Indeed, if f and $g \in C(D)$, and if a is any complex number, then

$af \in C(D), f + g \in C(D)$ and $fg \in C(D)$. To show this, we use the definition (ii) of a continuous function, and let $\{c_n\}$, with $c_n \in D$, be a sequence converging to $c \in D$. Then, because of the continuity of f and g, we have $\lim_{n \to \infty} f(c_n) = f(c_n)$ and $\lim_{n \to \infty} g(c_n) = g(c)$. It follows immediately that

$$\lim_{n \to \infty} af(c_n) = a \lim_{n \to \infty} f(c_n) = af(c)$$

$$\lim_{n \to \infty} (f(c_n) + g(c_n)) = \lim_{n \to \infty} f(c_n) + \lim_{n \to \infty} g(c_n) = f(c) + g(c)$$

and

$$\lim_{n \to \infty} f(c_n)g(c_n) = \lim_{n \to \infty} f(c_n) \lim_{n \to \infty} g(c_n) = f(c)g(c)$$

Hence $C(D)$ is a complex algebra.

COROLLARY *Any polynomial in z is continuous.*

If $f \in C(D)$ and if g is complex-valued and continuous on $f(D)$, then $g \circ f$ is continuous on D. Indeed, using definition (iii) of continuous functions in the same way as for general topological spaces (see 0.3.2), we find that if S is any open set in $\mathbf{C}$, then $g^{-1}(S)$ is open in the relative topology of $f(D)$, and $f^{-1}(g^{-1}(S))$ is open in the relative topology of D. But $f^{-1}(g^{-1}(S)) = (g \circ f)^{-1}(S)$, which shows that $g \circ f$ is continuous.

EXAMPLES

1 If $f \in C(D)$, then $|f| \in C(D)$, since $|\cdot|$ is a continuous operator and $|f| = |\cdot| \circ f$.

2 If $f \in C(D)$, then $\dfrac{1}{f}$ is continuous $\forall\, z$ for which $f(z) \neq 0$, since $\dfrac{1}{f} = \dfrac{1}{z} \circ f$. $\left(\dfrac{1}{z} \circ f \text{ means take } f(z) \text{ and then take its reciprocal.}\right)$

3 If $f \in C(D)$, then Re $f = U(x, y)$ and Im $f = V(x, y)$ are continuous functions on D. Indeed, $U(x, y) = \text{Re } f(z) = \text{Re} \circ f(z)$ and $V(x, y) = \text{Im } f(z) = \text{Im} \circ f(z)$, and, since Re z and Im z are continuous throughout $\mathbf{C}$ (see example 3 of continuous functions), it follows that $U(x, y)$ and $V(x, y)$ are continuous on D. Conversely, if Re $f = U(x, y)$ and Im $f = V(x, y)$ are continuous, then $f = U + iV$ is continuous as the sum of two continuous functions. $(V \in C(D) \Rightarrow iV \in C(D).)$

2. Differentiability

DEFINITION Let f be a complex-valued function defined on a domain D. We say f is *complex differentiable*, or, briefly, *differentiable*, at the point $c \in D$ if the limit

$$\lim_{\substack{z \to c \\ z \neq c}} \frac{f(z) - f(c)}{z - c} = \lim_{\substack{h \to 0 \\ h \neq 0}} \frac{f(c+h) - f(c)}{h}$$

exists.

This limit is denoted by $f'(c)$ and is called the derivative of f at c. It is understood, however, that the limit of $\dfrac{f(c+h) - f(c)}{h}$ must be independent of the path along which h approaches 0 in the complex plane. In other words, the limit must be the same, regardless whether h goes to 0 through real numbers, imaginary numbers, or in any other way.

In the future, we shall omit writing $z \neq c$ or $h \neq 0$ in the expression for the derivative.

We say that f is complex differentiable on D if it is complex differentiable at each point of D, and we denote the derivative by f' or $\dfrac{df}{dz}$.

Let f be differentiable on D, $z \in D$, and $h \neq 0$. Consider the expression

$$\Omega(z, h) = \frac{f(z+h) - f(z)}{h} - f'(z)$$

Then we have

$$f(z+h) = f(z) + hf'(z) + h\Omega(z, h)$$

with $\lim_{h \to 0} \Omega(z, h) = 0$. This relation is called the *Riemann relation* and will be referred to simply as *relation R*.

Conversely, if we can write

$$f(z+h) = f(z) + hf_1(z) + h\omega(z, h)$$

with $\lim_{h \to 0} \omega(z, h) = 0$, then obviously f is differentiable at z with $f'(z) = f_1(z)$. Using relation R, we see also that differentiability of f at z implies continuity at z, since $\lim_{h \to 0} f(z+h) = f(z)$.

In this book, we are only interested in functions which are differentiable in some domain. Thus, when we say f is differentiable at c, we always mean that f is differentiable in some domain containing c.

EXAMPLES

1 $f(z) = a$, where a is an arbitrary complex constant, is differentiable throughout the entire complex plane **C**. Indeed,

$$\lim_{z \to c} \frac{f(z) - f(c)}{z - c} = \lim_{z \to c} \frac{a - a}{z - c} = 0$$

That is, the derivative of a constant is zero, a result similar to that in real analysis.

2 $f(z) = z$ is differentiable in the entire plane, since

$$\lim_{z \to c} \frac{f(z) - f(c)}{z - c} = \lim_{z \to c} \frac{z - c}{z - c} = 1$$

for every complex number c.

3 $f(z) = az + b$ is differentiable, since

$$\lim_{z \to c} \frac{(az + b) - (ac + b)}{z - c} = a$$

for every complex number c.

4 If n is a positive integer, then $f(z) = z^n$ is differentiable in the entire plane, since

$$\lim_{z \to c} \frac{f(z) - f(c)}{z - c} = \lim_{z \to c} \frac{z^n - c^n}{z - c} = \lim_{z \to c} (z^{n-1} + z^{n-2}c + \cdots + c^{n-1}) = nc^{n-1}$$

5 $f(z) = \operatorname{Re} z$ is not differentiable. Indeed, if $z = x + iy$ and $c = a + ib$, then

$$\lim_{z \to c} \frac{\operatorname{Re} z - \operatorname{Re} c}{z - c} = \lim_{\substack{x \to a \\ y \to b}} \frac{x - a}{(x - a) + i(y - b)}$$

Therefore, if z converges to c on the line $y = b$ parallel to the x-axis, the limit is equal to $\lim_{x \to a} \frac{x - a}{x - a} = 1$; but, if z approaches c along the line $x = a$ parallel to the y-axis, the limit is equal to $\lim_{y \to b} \frac{0}{0 + i(y - b)} = 0$. Since the two values thus obtained are not the same, it follows that the limit is not independent of the path along which z tends toward c, as it should be if f is complex differentiable.

In the same manner, it can be shown that $f(z) = \operatorname{Im} z$ is not differentiable.

This example shows that $f(z) = U(x, y) + iV(x, y)$ may not be differentiable even though $U(x, y)$ and $V(x, y)$ both possess continuous partial derivatives.

6 $f(z) = 1/z$ is differentiable throughout $\mathbf{C}\setminus\{0\}$. Indeed, if $z \neq 0$, we have

$$\lim_{h \to 0} \frac{f(z+h) - f(z)}{h} = \lim_{h \to 0} \frac{1}{h}\left(\frac{1}{z+h} - \frac{1}{z}\right) = \lim_{h \to 0} \frac{-1}{z(z+h)} = -\frac{1}{z^2}$$

hence

$$\frac{d}{dz}\left(\frac{1}{z}\right) = -\frac{1}{z^2}$$

The set of all functions which are complex differentiable on the domain D will be denoted by $H(D)$. $H(D)$ is a complex algebra. Indeed, if f and $g \in H(D)$ and if a is any complex number, then: $af \in H(D)$ and $(af)' = af'$; $f + g \in H(D)$ and $(f+g)' = f' + g'$; and $f \cdot g \in H(D)$ with $(fg)' = f'g + fg'$. If $g \neq 0$, we have $\left(\dfrac{f}{g}\right)' = \dfrac{f'g - g'f}{g^2}$. It is left to the reader to prove these formulae of differentiation, which are well known in real analysis.

THEOREM 1.5 [*Chain Rule*] *If $f \in H(D)$ and $g \in H(D_1)$, where D_1 is a domain containing $f(D)$, then $g \circ f \in H(D)$ and $(g \circ f)' = (g' \circ f)f'$.*

PROOF

Let $d(z, r) \subset D$ and $|h| < r$. Then, since f is differentiable, we have $f(z+h) = f(z) + k$, where $k = hf'(z) + h\Omega_1(z, h)$ and $\lim_{h \to 0}\Omega_1(z, h) = 0$. Therefore, since g is differentiable, we obtain $g(f(z+h)) = g(f(z) + k) = g(f(z)) + kg'(f(z)) + k\Omega_2(f(z), k)$, with $\lim_{k \to 0}\Omega_2(f(z), k) = 0$. If we now substitute for k its expression in terms of h and introduce the expression

$$\Omega_3(z, h) = \Omega_1(z, h)g'(f(z)) + (f'(z) + \Omega_1(z, h))\Omega_2(f(z), k)$$

we obtain

$$g(f(z+h)) = g(f(z)) + hg'(f(z))f'(z) + h\Omega_3(z, h)$$

It is obvious that $\lim_{h \to 0}\Omega_3(z, h) = 0$. Therefore, we obtain

$$\frac{d}{dz}g(f(z)) = \lim_{h \to 0}\frac{g(f(z+h)) - g(f(z))}{h} = g'(f(z))f'(z)$$

or, in functional notation,

$$(g \circ f)' = (g' \circ f)f'$$

which completes the proof.

EXAMPLE

Let $g(z) = 1/z$ and $f \in H(D)$. Then, if $z \in D$ and $f(z) \neq 0$, we have

$$\left(\frac{1}{f(z)}\right)' = -\frac{1}{f^2(z)} \cdot f'(z) = -\frac{f'(z)}{f^2(z)}$$

We call a function *entire* if it is differentiable in the entire complex plane. A polynomial is an entire function, but the function $f(z) = 1/z$ is not entire because it is not differentiable at $z = 0$. A rational function $R(z) = P(z)/Q(z)$, where P and Q are polynomials with no common divisor, is differentiable everywhere except at points where $Q(z) = 0$. Thus, when Q is not a constant, then R is not an entire function.

The points where a function f is not differentiable will be called *singular points*. As said before, we are not interested here in functions which are singular in the entire plane. The function $f(z) = 1/z$ possesses a singular point at $z = 0$, but is differentiable elsewhere. While there are functions which are singular on a curve or on a domain, we shall confine ourselves to the study of isolated singular points. In a neighborhood of these singular points the function will be differentiable.

Problems

1.29.: Show that $f(z) = |z|$ is nowhere differentiable. (*Hint:* Take the derivative in the radial and tangential directions.)
1.30.: If $P(z)$ is a polynomial, show that $P(1/z)$ is singular at the origin and nowhere else.

---

3. The Cauchy-Riemann equations

Any complex-valued function f may be decomposed into its real part, Re $f(z) = U(x, y)$, and its imaginary part, Im $f(z) = V(x, y)$, $\ni f(z) = U(x, y) + iV(x, y)$. Both functions U and V are real-valued functions of the two real variables x and y. We shall sometimes write $f(z) = f(x, y)$ in order to show that f may be considered as a complex-valued function of two real variables x and y.

We now wish to find a criterion for differentiability of f in terms of the two functions U and V. We have seen (in example 5 of 1.4.2) that the existence of the partial derivatives U_x, U_y, V_x, V_y is not sufficient to guarantee the differentiability of f. Moreover, we shall now show that certain relations between the four partial derivatives are essential for the differentiability of f. These relations are called the *Cauchy-Riemann equations*.

DEFINITION The complex or real valued function f of the two real variables x and y is said to be *differentiable with respect to x and y* at the point (x, y) if, for sufficiently small real h_1 and h_2, it is possible to write

$$f(x + h_1, y + h_2) = f(x, y) + h_1 f_1(x, y)$$
$$+ h_2 f_2(x, y) + \sqrt{h_1^2 + h_2^2}\, \Omega(x, y, h_1, h_2)$$

with $\lim_{h_1, h_2 \to 0} \Omega(x, y, h_1, h_2) = 0$.

In that case, we see easily that

$$f_1 = \frac{\partial f}{\partial x} \quad \text{and} \quad f_2 = \frac{\partial f}{\partial y}$$

In general, f_1 and f_2 are complex-valued functions.

If $f = U + iV$ is differentiable with respect to x and y, then so are U and V. Indeed, one obtains easily, for the real part,

$$U(x + h_1, y + h_2) = \text{Re } f(x + h_1, y + h_2)$$
$$= \text{Re } f(x, y) + h_1 \text{ Re } f_1(x, y) + h_2 \text{ Re } f_2(x, y)$$
$$+ \sqrt{h_1^2 + h_2^2} \text{ Re } \Omega(x, y, h_1, h_2)$$
$$= U(x, y) + h_1 U_1(x, y) + h_2 U_2(x, y)$$
$$+ \sqrt{h_1^2 + h_2^2} \text{ Re } \Omega(x, y, h_1, h_2)$$

with $\lim_{h_1, h_2 \to 0} \text{Re } \Omega(x, y, h_1, h_2) = 0$.

The proof of the differentiability of V follows a similar line of reasoning. Clearly,

$$U_1(x, y) = \frac{\partial U}{\partial x} = \text{Re } \frac{\partial f}{\partial x} \quad \text{and} \quad U_2(x, y) = \frac{\partial U}{\partial y} = \text{Re } \frac{\partial f}{\partial y}$$

where

$$\frac{\partial f}{\partial x} = \frac{\partial U}{\partial x} + i\frac{\partial V}{\partial x} \quad \text{and} \quad \frac{\partial f}{\partial y} = \frac{\partial U}{\partial y} + i\frac{\partial V}{\partial y}$$

Conversely, if the real-valued functions U and V are both differentiable with respect to x and y, then so is $U + iV$. The proof is easy and is left to the reader. f is said to be differentiable with respect to x and y in the domain D if it is differentiable at each point of D.

THEOREM 1.6 *If the function $f = U + iV$ is complex differentiable in the domain D, then U and V are differentiable in D and the four partial derivatives U_x, U_y, V_x, V_y satisfy the Cauchy-Riemann equations*

$$U_x = V_y, \qquad U_y = -V_x$$

Conversely, if the real-valued functions U and V are differentiable in D and satisfy the Cauchy-Riemann equations, then $f = U + iV$ is complex differentiable.

PROOF

(i) Suppose f is complex differentiable in D. Then it is easy to see that f is differentiable with respect to x and y:

Let $z \in D$ and $h = h_1 + ih_2 \neq 0$. Since f is complex differentiable, we have

$$f(z + h) = f(z) + hf'(z) + h\Omega(z, h)$$

with $\lim_{h \to 0} \Omega(z, h) = 0$, according to relation R.

If, for convenience, we also use $f(x, y)$ to represent $f(z)$, that is, $f(z) = f(x, y)$, we have $f(z + h) = f(x + h_1, y + h_2)$, so that $f(z + h)$ may be written as

$$\begin{aligned} f(x + h_1, y + h_2) &= f(x, y) + (h_1 + ih_2)f'(z) + h\Omega(z, h) \\ &= f(x, y) + h_1 f'(z) + ih_2 f'(z) \\ &\quad + \sqrt{h_1^2 + h_2^2}\, \Omega^*(z, y, h_1, h_2) \end{aligned}$$

where $\Omega^* = (h/|h|)\Omega$. This shows that f is differentiable with respect to x and y, and that

$$f'(z) = \frac{\partial f}{\partial x}, \qquad if'(z) = \frac{\partial f}{\partial y} = i\frac{\partial f}{\partial x}$$

Now, the differentiability of f implies differentiability of $U = \operatorname{Re} f$ and of $V = \text{Im } f$, so that the four partial derivatives U_x, U_y, V_x, V_y exist. On the other hand, we have

$$\frac{\partial f}{\partial x} = U_x + iV_x, \qquad \frac{\partial f}{\partial y} = U_y + iV_y$$

so that the relation $\dfrac{\partial f}{\partial y} = i\dfrac{\partial f}{\partial x}$ implies $(U_y + iV_y) = i(U_x + iV_x) = iU_x - V_x$, and, consequently, $U_x = V_y$, $U_y = -V_x$.

(ii) Suppose now that U and V are differentiable in D and satisfy the Cauchy-Riemann equations. Then $f = U + iV$ is also differentiable in x and y, so that for sufficiently small h_1 and h_2 we have

$$f(x + h_1, y + h_2) = f(x, y) + h_1 f_1(x, y) + h_2 f_2(x, y)$$
$$+ \sqrt{h_1^2 + h_2^2}\, \Omega(x, y, h_1, h_2)$$

with $\lim_{h_1, h_2 \to 0} \Omega(x, y, h_1, h_2) = 0$, and therefore $\dfrac{\partial f}{\partial x} = f_1$, $\dfrac{\partial f}{\partial y} = f_2$. Now, since U and V satisfy the Cauchy-Riemann equations $U_x = V_y$ and $U_y = -V_x$, we have $U_y + iV_y = i(U_x + iV_x)$; that is, $\dfrac{\partial f}{\partial y} = i\dfrac{\partial f}{\partial x}$, and therefore $f_2(x, y) = if_1(x, y)$. Consequently, the previous relation can be rewritten as

$$f(x + h_1, y + h_2) = f(x, y) + (h_1 + ih_2)f_1(x, y)$$
$$+ \sqrt{h_1^2 + h_2^2}\, \Omega(x, y, h_1, h_2)$$

If we now set $h_1 + ih_2 = h$, $x + iy = z$, $f(x, y) = f(z)$ and $\dfrac{|h|}{h}\Omega(x, y, h_1, h_2) = \Omega(z, h)$, we obtain

$$f(z + h) = f(z) + hf_1(z) + h\Omega(z, h)$$

with $\lim_{h \to 0} \Omega(z, h) = 0$, which shows that f is complex differentiable with

$$f' = f_1 = \frac{\partial f}{\partial x} = -i\frac{\partial f}{\partial y} = U_x + iV_x = V_y - iU_y$$

DISCUSSION AND REMARKS

Let $f = U + iV$ be complex differentiable. Then U_x, U_y, V_x, V_y exist. Therefore, we may introduce the gradients of U and V by the definition

$$\text{grad } U = \left(\frac{\partial}{\partial x} + i\frac{\partial}{\partial y}\right)U = U_x + iU_y$$

$$\text{grad } V = \left(\frac{\partial}{\partial x} + i\frac{\partial}{\partial y}\right)V = V_x + iV_y$$

On the basis of the Cauchy-Riemann equations, we can then conclude that the scalar product

$$\text{grad } U \cdot \text{grad } V = U_x V_x + U_y V_y = 0$$

that is, that grad U and grad V are orthogonal to each other. By the same method, we deduce further that

$$|\text{grad } U| = \sqrt{U_x^2 + U_y^2} = \sqrt{V_y^2 + V_x^2} = |\text{grad } V|$$

Using $x = \dfrac{z + \bar{z}}{2}$ and $y = \dfrac{z - \bar{z}}{2i}$ and treating $f(x, y)$ as a function of z and $\bar{z}$,

$$f(x, y) = f(x(z, \bar{z}), y(z, \bar{z}))$$

we can compute $\dfrac{\partial f}{\partial z}$ and $\dfrac{\partial f}{\partial \bar{z}}$ as if z and $\bar{z}$ were independent variables. Thus, we find the expressions

$$\frac{\partial f}{\partial z} = \frac{1}{2}\left(\frac{\partial f}{\partial x} - i\frac{\partial f}{\partial y}\right) \quad \text{and} \quad \frac{\partial f}{\partial \bar{z}} = \frac{1}{2}\left(\frac{\partial f}{\partial x} + i\frac{\partial f}{\partial y}\right)$$

or, replacing $\dfrac{\partial f}{\partial x}$ by $U_x + iV_x$ and $\dfrac{\partial f}{\partial y}$ by $U_y + iV_y$,

$$\frac{\partial f}{\partial z} = \frac{1}{2}[(U_x + iV_x) - i(U_y + iV_y)] = \frac{1}{2}[(U_x + V_y) + i(V_x - U_y)]$$

$$\frac{\partial f}{\partial \bar{z}} = \frac{1}{2}[(U_x + iV_x) + i(U_y + iV_y)] = \frac{1}{2}[(U_x - V_y) + i(U_y + V_x)]$$

Therefore, if f is complex differentiable, the Cauchy-Riemann equations $U_x = V_y$, $U_y = -V_x$ imply that

$$\frac{\partial f}{\partial \bar{z}} = 0$$

which is a more condensed form of these equations.

4. Primitives

DEFINITION If $F \in H(D)$ and $F' = f$, then we say that F is a primitive of f.

If F is a primitive of f, then so is $F + a$, where a is a constant, since the derivative of a constant is 0. Conversely, let the primitive of $f \equiv 0$ be $F = F_1 + iF_2$, where F_1 and F_2 are real-valued functions of x

and y. Then we have both $F' = \dfrac{\partial F}{\partial x} = (F_1)_x + i(F_2)_x = 0$ and $F' = \dfrac{1}{i}\dfrac{\partial F}{\partial y} = \dfrac{1}{i}[(F_1)_y + i(F_2)_y] = 0$, which implies $(F_1)_x = (F_1)_y = 0$ and $(F_2)_x = (F_2)_y = 0$, so that F_1 and F_2 are two real constants and $F = F_1 + iF_2$ is a complex constant. It follows that two primitives of the same function differ by a constant.

EXAMPLES

1 The primitive of $f(z) = az + b$ is $\dfrac{a}{2}z^2 + bz + c$, where c is an arbitrary constant.

2 The primitive of $f(z) = z^n$, where n is an integer $\neq -1$ is $F(z) = \dfrac{1}{n+1}z^{n+1} + c$, with c an arbitrary constant.

3 The primitive of $f(z) = 1/z$ has not yet been defined in the complex domain, but later we will show that it is a multivalued function, conveniently identified as $\log z$ because it coincides with the natural logarithm of z in the case of $z \in \mathbf{R}^+$.

5. Geometric interpretation of the derivative

For a better understanding of what follows, let us first look at the geometric interpretation of the mapping $f(z) = az + b$:

(i) If $b = 0$ and a is real > 0, then $|az| = a|z|$, and $\arg az = \arg a + \arg z = \arg z$, since $\arg a = 0$. So az is a dilatation in the ratio a.

(ii) If $b = 0$ and $|a| = 1$, or $a = \operatorname{cis} \phi$, then $|az| = |a||z| = |z|$ and $\arg(az) = \arg a + \arg z = \phi + \arg z$, so az is a rotation through angle ϕ.

(iii) If $b = 0$ and a is an arbitrary complex number $a = r \operatorname{cis} \phi$, then az is a rotation through angle $\phi = \arg a$, combined with a dilatation in the ratio $r = |a|$.

(iv) If a and b are arbitrary complex numbers, then $az + b$ is a rotation through angle $\phi = \arg a$, combined with a dilatation in the ratio $r = |a|$, followed by a translation by vector b. In $a(z - c) + b$ we have a rotation about c through angle $\arg a$ and a dilatation in the ratio $|a|$, followed by a translation in the direction, and by the amount, of vector b.

Now let f be complex differentiable in the domain D and $c \in D$; then $f'(c)$ exists. Assume $f'(c) \neq 0$. If $\gamma = \gamma(t)$ is an arc passing through c, which possesses a tangent at $c = \gamma(t_o)$, then $\gamma'(t_o)$ exists and the direction of the tangent is given by arg $\gamma'(t_o)$. Denote by $\Gamma = f(\gamma)$ the image of γ by f; then $\Gamma(t) = f(\gamma(t))$ and $\Gamma'(t_o) = f'(c) \cdot \gamma'(t_o)$ exists, so Γ possesses a tangent at $f(c)$ whose direction is given by arg $\Gamma'(t_o)$. It follows from arg $\Gamma'(t_o) = $ arg $f'(c) + $ arg $\gamma'(t_o)$ that the tangent to γ is rotated through the angle arg $f'(c)$ by the mapping f. Since arg $f'(c)$ does not depend on γ, but only on c, the angle of rotation of the tangent to any γ through c will be the same. Thus we have conformality of the mapping f at points where $f' \neq 0$.

We obtain the same result by reasoning that, for $f'(c) \neq 0$, we have

$$\arg f'(c) = \arg \lim_{z \to c} \frac{f(z) - f(c)}{z - c}$$

$$= \lim_{z \to c} \arg \frac{f(z) - f(c)}{z - c}$$

$$= \lim_{z \to c} [\arg (f(z) - f(c)) - \arg (z - c)]$$

But $\lim_{\substack{z \to c \\ z \in \gamma}} \arg(z - c) = \arg \gamma'(t_o)$, defining the direction of the tangent to γ at c, while $\lim_{\substack{z \to c \\ z \in \gamma}} \arg(f(z) - f(c)) = \arg \Gamma'(t_o)$ determines the direction of the tangent to Γ at $f(c)$; hence arg $\Gamma'(t_o) = $ arg $f'(c) + $ arg $\gamma'(t_o)$.

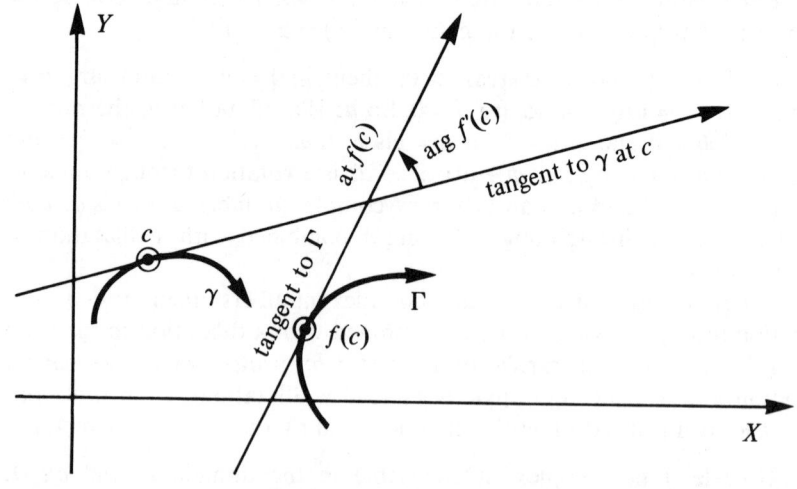

Figure 1.8

Now let us consider $|f'(c)|$. Then, by the definition of $f'(c)$ and the continuity of $|\cdot|$, we have

$$|f'(c)| = \lim_{z \to c} \left|\frac{f(z) - f(c)}{z - c}\right| = \lim_{z \to c} \frac{|f(z) - f(c)|}{|z - c|}$$

which is the dilatation coefficient of the mapping f at c. The dilatation coefficient does not depend on γ, but only on c. Intuitively, for $z \to c$, the mapping f reduces to a rotation through angle $\arg f'(c)$ and a dilatation in the ratio $|f'(c)|$. This is what we expect if we remember that

$$f(c + h) = f(c) + hf'(c) + h\Omega(c, h)$$

with $\lim_{h \to 0} \Omega(c, h) = 0$, or

$$f(z) = f(c) + (z - c)f'(c) + (z - c)\Omega(z, c)$$

with $\lim_{h \to 0} \Omega(z, c) = 0$, which shows that, for small values of $(z - c)$, the mapping f behaves very much like $f'(c)(z - c) + f(c)$, which defines a rotation by angle $\arg f'(c)$ and a dilatation in ratio $|f'(c)|$, followed by a translation by $f(c)$.

6. Isomorphisms

If $f \in H(D)$ and defines a homeomorphism between D and $f(D)$, then f^{-1} is also complex differentiable in $f(D)$. Indeed, since f is a homeomorphism, $z \neq c \Leftrightarrow f(z) \neq f(c)$ and $z \to c \Leftrightarrow f(z) \to f(c)$. It follows that

$$\lim_{f(z) \to f(c)} \frac{f^{-1}(f(z)) - f^{-1}(f(c))}{f(z) - f(c)} = \lim_{z \to c} \frac{z - c}{f(z) - f(c)}$$

$$= \lim_{z \to c} \frac{1}{\frac{f(z) - f(c)}{z - c}} = \frac{1}{f'(c)}$$

so that

$$(f^{-1})' = 1/f'.$$

Hence f^{-1} is complex differentiable. We call a complex differentiable homeomorphism an *isomorphism*.

If f is an isomorphism $D \to f(D)$, then we shall say that D and $f(D)$ are *isomorphic*. It is easy to see that *isomorphism* is an equivalence relation on the family of domains in $\mathbf{C}$. If $f(D) = D$, then we say that the isomorphism f is an *automorphism* of D.

Problems

1.31.: Show that the function $f(x, y) = \dfrac{x + iy}{x - iy}$ is not continuous at the origin.

1.32.: For what values of z are each of the following functions continuous?

(a) $f(z) = \dfrac{1}{z^3 + 1}$

(b) $f(z) = \dfrac{z^n - 1}{z^n + 1}$

(c) $f(z) = \dfrac{1}{z} + \dfrac{1}{z^2 + 1}$

Determine the singular points in each case.

1.33.: Show that the function $x^2 + iy^2$ is uniformly continuous in the open disc $d(0, 1)$, but not uniformly continuous in **C**.

1.34.: Show that the following functions are not complex differentiable.

(a) $|z|$ (b) Im z (c) $\bar{z}$

1.35.: Show that the function $x + 2iy$ is continuous in **C**. Determine for given c and $\varepsilon > 0$ the corresponding $\delta = \delta(c, \varepsilon) \ni f(d(c, \delta)) \subset d(f(c), \varepsilon)$.

1.36.: Using the Cauchy-Riemann equations, determine which of the following functions are complex differentiable.

(a) $ax + iby$ (b) $x^2 + y^2$ (c) $\dfrac{1}{x - iy}$

(d) $f(z) = e^y(\cos x + i \sin x)$

At which points do the Cauchy-Riemann conditions hold in each case?

1.37.: Show that the origin is the only point where $f(z) = z \operatorname{Re} z$ is complex differentiable.

1.38.: Let $f \in H(D)$. Show that Re f or Im f = constant $\Rightarrow f$ is constant.

1.39.: Show that

$$\frac{\partial}{\partial x} = \frac{\partial}{\partial z} + \frac{\partial}{\partial \bar{z}}, \qquad \frac{\partial}{\partial y} = i\left(\frac{\partial}{\partial z} - \frac{\partial}{\partial \bar{z}}\right)$$

and

$$\nabla^2 = \frac{\partial^2}{\partial x^2} + \frac{\partial^2}{\partial y^2} = 4 \frac{\partial^2}{\partial z \partial \bar{z}}$$

1.40.: Show that if $f \in H(D)$, then $f' = \dfrac{\partial f}{\partial x} = -i \dfrac{\partial f}{\partial y}$, and that the Jacobian of the functions $U = U(x, y) = \operatorname{Re} f$, $V = V(x, y) = \operatorname{Im} f$ at the point $z = x + iy$ is given by $|f'(z)|^2$.

1.41.: Consider the mapping $f(z) = z^2$. What are the transforms of lines parallel to the x-axis and lines parallel to the y-axis? Show by some examples that arg $f'(z)$ determines the angle of rotation of the tangent.

1.42.: Show that the Cauchy-Riemann equations in polar coordinates are given by

$$r\frac{\partial u}{\partial r} = \frac{\partial v}{\partial \phi}, \quad \frac{\partial u}{\partial \phi} = -r\frac{\partial v}{\partial r}$$

where $f(z) = f(r, \phi) = u(r, \phi) + iv(r, \phi)$.

chapter 2

Holomorphic Functions

A continuous function f is said to be holomorphic in the domain D if the integral $\int_\gamma f(z)\,dz = 0$ for any closed curve γ which can be shrunk to a point in D by continuous deformation in D. This property was first studied by the French mathematician Cauchy.

The most important result of this chapter is the equivalence of the two notions, complex differentiability and holomorphicity.

§1. Homotopy and line integrals

1. Homotopy of closed curves with respect to a domain D

First we shall give the general definition of homotopy:

DEFINITION 1 The closed curves γ_0 and γ_1 are called homotopic with respect to the domain D if $\exists$ a family of curves $\gamma(t, u)$, $0 \le t \le 1$, $0 \le u \le 1$, $\ni$:

(i) $\gamma(t, u)$ is a continuous mapping $[0, 1] \times [0, 1] \to D$;

(ii) $\forall u, 0 \le u \le 1$, $\gamma(t, u)$ is a closed curve, that is, $\gamma(0, u) = \gamma(1, u)$;

(iii) $\gamma(t, 0) = \gamma_0$ and $\gamma(t, 1) = \gamma_1$.

In 1.3.5, we defined a piecewise smooth, closed curve γ as the image of the interval $0 \le t \le 1$ in the complex plane by the continuous mapping $\gamma \colon [0, 1] \to \mathbf{C}$,

$$\gamma(t) = x(t) + iy(t)$$

with $\gamma(0) = \gamma(1)$, $\gamma'(t) = x'(t) + iy'(t) \ne 0$, and $\gamma'(t)$ continuous in $0 \le t \le 1$

except at a finite number of points where right and left limits of $\gamma'(t)$ exist, but are different.

We now consider families of piecewise smooth, closed curves $\gamma(t, u)$. We suppose that for each value of u, $0 \leq u \leq 1$, $\gamma(t, u)$ is a piecewise smooth, closed curve; $\gamma(0, u) = \gamma(1, u) \ \forall \ u, 0 \leq u \leq 1$; and, moreover, that for each value of t, $\gamma(t, u)$ is a piecewise smooth arc.

In order to simplify some later proofs, we shall introduce a definition which is less general than definition 1.

DEFINITION 2 We say that the closed curves $\gamma_0 = \gamma(t, 0)$ and $\gamma_1 = \gamma(t, 1)$ belonging to the family of piecewise smooth, closed curves $\gamma(t, u)$ are *homotopic with respect to the domain D*, or that γ_0 is homotopic to γ_1 with respect to D, if:

$\gamma(t, u)$ is a continuous mapping possessing continuous partial derivatives $\dfrac{\partial \gamma}{\partial t} \neq 0$, $\dfrac{\partial \gamma}{\partial u} \neq 0$, except at a finite number of points, and maps the square

$$S = \{(t, u) | 0 \leq t \leq 1, 0 \leq u \leq 1\}$$

into the domain D.

REMARK

Obviously, if γ_0 and γ_1 are homotopic with respect to D, then they are homotopic with respect to any domain $D' \supset D$.

DEFINITION If γ_0 is homotopic to γ_1 with respect to D, and γ_1 is a point, then we say that γ_0 *is homotopic to a point with respect to D*.

In that case, for $u = 1$, we allow $\dfrac{\partial \gamma}{\partial t} = \dfrac{\partial \gamma}{\partial u} = 0$.

If two piecewise smooth, closed curves

$$\gamma_0 = \gamma_0(t), \qquad \gamma_1 = \gamma_1(t), \qquad 0 \leq t \leq 1$$

and a domain D are given, and if we wish to show that γ_0 and γ_1 are homotopic with respect to D, then, using definition 2, we have to find a family of piecewise smooth, closed curves $\gamma(t, u)$ ∋:

(i) γ_0 and γ_1 are members of this family with $\gamma_0(t) = \gamma(t, 0)$ and $\gamma_1(t) = \gamma(t, 1)$;

(ii) $\gamma(t, u)$ is a continuous mapping $S \to \mathbf{C}$ with $\gamma(S) \subset D \ni \gamma(t, u)$ is a piecewise smooth closed curve $\forall u$, and $\gamma(t, u)$ is a piecewise smooth arc $\forall t$.

Intuitively, this means that we can transform γ_0 into γ_1 by continuous deformations given by $\gamma(t, u)$, such that during the process of deformation $\gamma(t, u)$, $0 \le u \le 1$, does not leave D.

This intuitive concept of homotopy enables us sometimes to visualize whether two given closed curves are homotopic even before finding the family $\gamma(t, u)$.

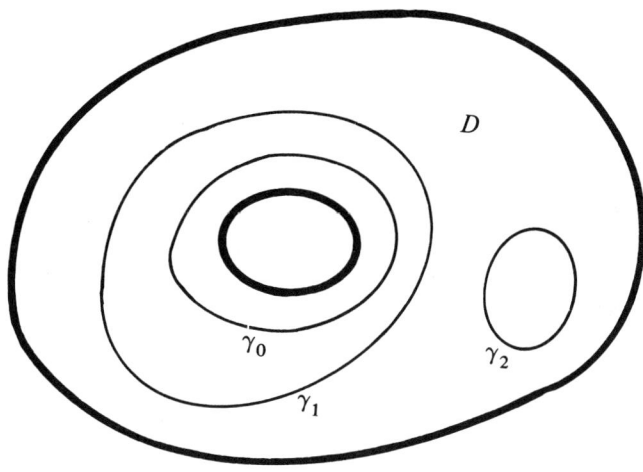

Figure 2.1

In Figure 2.1, γ_0 and γ_1 are homotopic with respect to D, while γ_0 and γ_2 are not; γ_2 is homotopic to a point with respect to D.

EXAMPLE

The circles
$$\gamma_0(t) = \tfrac{1}{2} \operatorname{cis} 2\pi t, \quad 0 \le t \le 1$$
and
$$\gamma_1(t) = \operatorname{cis} 2\pi t, \quad 0 \le t \le 1$$
are homotopic with respect to any domain D containing the closed unit disc $\bar{d}(0, 1)$, and both are homotopic to a point with respect to D.

Indeed, take $\gamma(t, u) = \dfrac{1 + u}{2} \operatorname{cis} 2\pi t$; then

(i) $\gamma(t, 0) = \tfrac{1}{2} \operatorname{cis} 2\pi t = \gamma_0(t)$

$\gamma(t, 1) = \operatorname{cis} 2\pi t = \gamma_1(t)$

(ii) $\gamma(t, u) \subset \overline{d}(0, 1) \subset D$
since
$$|\gamma(t, u)| = \frac{1 + u}{2} |\text{cis } 2\pi t| = \frac{1 + u}{2} \leq 1$$
for $0 \leq u \leq 1$, and

(iii) $\gamma(0, u) = \dfrac{1 + u}{2} = \gamma(1, u)$

since $\text{cis } 0 = \text{cis } 2\pi = 1$.

In order to see that $\gamma_0(t)$ is homotopic to a point with respect to D, take $\gamma(t, u) = \dfrac{1 - u}{2} \text{cis } 2\pi t$; then

(i) $\gamma(t, 0) = \tfrac{1}{2} \text{cis } 2\pi t = \gamma_0(t)$
$\gamma(t, 1) = 0 = \gamma_1(t)$

(ii) $\gamma(t, u) \subset \overline{d}(0, 1) \subset D$
since
$$|\gamma(t, u)| = \frac{1 - u}{2} |\text{cis } 2\pi t| = \frac{1 - u}{2} \leq 1$$
for $0 \leq u \leq 1$, and

(iii) $\gamma(0, u) = \dfrac{1 - u}{2} = \gamma(1, u)$

for $0 \leq u \leq 1$.

DEFINITION A domain D is called *simply connected* if every closed curve in D is homotopic to a point with respect to D.

THEOREM 2.1 *A starlike domain is simply connected.*

PROOF

Let D be a star domain with respect to c and let $\gamma_0 = \gamma_0(t)$, $0 \leq t \leq 1$, be a closed p.w.s. (piecewise smooth) curve in D. Then, consider the family
$$\gamma(t, u) = (1 - u)\gamma_0(t) + uc, \qquad 0 \leq u \leq 1$$
We have:

(i) $\gamma(t, 0) = \gamma_0, \qquad \gamma(t, 1) = c$

(ii) For any t, $0 \leq t \leq 1$, $\gamma(t, u)$ lies on the segment joining $\gamma_0(t)$ to c, and belongs therefore to D; obviously, $\gamma(t, u)$ is continuous.

(iii) $\gamma(0, u) = (1 - u)\gamma_0(0) + uc$
$= (1 - u)\gamma_0(1) + uc$
$= \gamma(1, u)$

for all u, $0 \leq u \leq 1$, since $\gamma_0(0) = \gamma_0(1)$.

Problem

2.1.: Show that homotopy with respect to a domain D is a transitive property.

---

2. Line integrals

Let $\gamma = \gamma(t)$, $0 \leq t \leq 1$, be a p.w.s. arc with initial point $\gamma(0) = a$ and endpoint $\gamma(1) = b$. If $0 = t_0 < t_1 < \cdots < t_n = 1$, then we call the $n + 1$ points $z_0 = \gamma(t_0)$, $z_1 = \gamma(t_1)$, ..., $z_n = \gamma(t_n)$ a *subdivision* D_n of γ. The number $v(D_n) = \max_i |z_i - z_{i-1}|$ is called the *norm* of the subdivision. Since γ is continuous on the compact set $[0, 1]$, it is uniformly continuous on $[0, 1]$, and $\max_i |t_i - t_{i-1}| < \delta(\varepsilon)$ implies $v(D_n) < \varepsilon$.

We know from real analysis that a p.w.s. arc is rectifiable. Therefore, if $L(\gamma)$ is the length of γ, we have

$$\sum_{i=1}^{n} |z_i - z_{i-1}| \leq L(\gamma)$$

for any subdivision D_n.

We define the vector $\zeta^{(n)} = (\zeta_1, \zeta_2, \ldots, \zeta_n)$ with complex coordinates ζ_i, where ζ_i is any point on the subarc (z_{i-1}, z_i), by $\zeta_i = \gamma(\tau_i)$, where $t_{i-1} \leq \tau_i \leq t_i$.

Now let f be continuous $\gamma \to \mathbf{C}$ ($f \in \mathbf{C}[\gamma]$). Then we define the Riemannian sum $S(D_n, \zeta^{(n)}, f)$ by:

$$S(D_n, \zeta^{(n)}, f) = \sum_{i=1}^{n} f(\zeta_i)(z_i - z_{i-1})$$

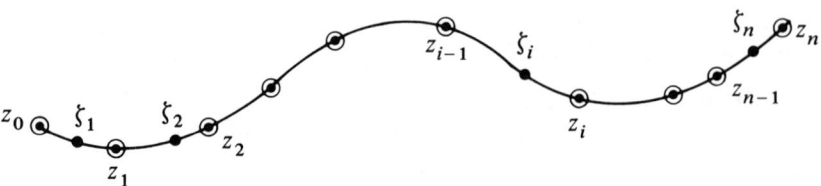

Figure 2.2

We shall say that the subdivision D_n is a *refinement* of the subdivision D_m if $n \geq m$ and if D_n contains all points of D_m. We get a refinement of D_m by subdividing each subarc (z_{i-1}, z_i) as follows:

$$z_{i-1} = z_{i,0}, z_{i,1}, \ldots, z_{i,k_i} = z_i \qquad (k_i \geq 1)$$

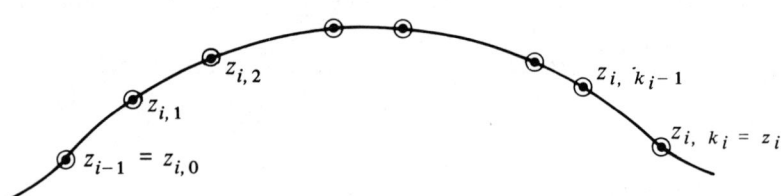

Figure 2.3

Now let $\{D_n\}$ be a sequence of subdivisions of $\gamma \ni D_{n+1}$ is a refinement of D_n for all n, and $\ni v(D_n) \to 0$. Then the corresponding sequence of complex numbers

$$\{S(D_n, \zeta^{(n)}, f)\}$$

is a Cauchy sequence.

Indeed, f is uniformly continuous on γ, since γ is a compact; hence, for $v(D_n)$ sufficiently small, we have $|f(z') - f(z'')| < \varepsilon$ for any pair z', z'' belonging to some subarc (z_{i-1}, z_i).

Now we group the terms of the difference

$$\Delta = S(D_{n+p}, \zeta^{(n+p)}, f) - S(D_n, \zeta^{(n)}, f)$$

according to the subarcs of D_n. This gives

$$|\Delta| = \left| \sum_{i=1}^{n} \left(\sum_{r=1}^{k_i} f(\zeta_{i,r})(z_{i,r} - z_{i,r-1}) - f(\zeta_i)(z_i - z_{i-1}) \right) \right|$$
$$= \left| \sum_{i=1}^{n} \sum_{r=1}^{k_i} (f(\zeta_{i,r}) - f(\zeta_i))(z_{i,r} - z_{i,r-1}) \right|$$

since $\sum_{r=1}^{k_i} (z_{i,r} - z_{i,r-1}) = z_i - z_{i-1}$. Therefore we have

$$|\Delta| < \varepsilon \sum_{i=1}^{n} \sum_{r=1}^{k_i} |z_{i,r} - z_{i,r-1}| \leq \varepsilon L(\gamma).$$

Denote the limit of the sequence $\{S(D_n, \zeta^{(n)}, f)\}$ by $\int_\gamma f(z)\,dz$. We see that $\int_\gamma f(z)\,dz$ does not depend on the choice of $\zeta^{(n)}$. It is easy to show that if $\{D_n'\}$ is another sequence of subdivisions with $v(D_n') \to 0$, even without requiring that D_{n+1}' is a refinement of D_n', then $\{S(D_n', \zeta^{(n)}, f)\}$ is again a Cauchy sequence with the same limit.

Indeed, if D_n and D_n' are two subdivisions and $D_{n,n'}$ a common refinement, then in the same way as above we can show:

$$|S(D_n, \zeta^{(n)}, f) - S(D_{n,n'}, \zeta^{(n,n')}, f)| \leq \varepsilon L(\gamma)$$

as well as

$$|S(D_{n'}, \zeta^{(n')}, f) - S(D_{n,n'}, \zeta^{(n,n')}, f)| \leq \varepsilon L(\gamma)$$

which gives

$$|S(D_n, \zeta^{(n)}, f) - S(D_{n'}, \zeta^{(n')}, f)| \leq 2\varepsilon L(\gamma)$$

so that

$$\lim_{v(D_n) \to 0} S(D_n, \zeta^{(n)}, f) = \lim_{v(D_{n'}) \to 0} S(D_{n'}, \zeta^{(n')}, f) = \int_\gamma f(z)\,dz$$

It can be easily shown that the choice of $\gamma = \gamma(t)$ in the set of representations of γ has no influence on the integral $\int_\gamma f(z)\,dz$, since the only important property of $\gamma(t)$ we have used in the proof is the uniform continuity of $\gamma(t)$ in $[0, 1]$, which holds for all representations.

Clearly

$$S(D_n, \zeta^{(n)}, f) = \sum_{i=1}^{n} f(z(\tau_i))(z(t_i) - z(t_{i-1}))$$

which gives

$$\lim_{v(D_n) \to 0} S(D_n, \zeta^{(n)}, f) = \int_0^1 f(z(t))z'(t)\,dt = \int_\gamma f(z)\,dz$$

This can also be shown by separating $S(D_n, \zeta^{(n)}, f)$ into real and imaginary parts and using the mean value theorem of real analysis.

For the Riemannian sum $S(D_n, \zeta^{(n)}, f)$, we have the inequality

$$|S(D_n, \zeta^{(n)}, f)| = \left|\sum_{i=1}^n f(\zeta_i)(z_i - z_{i-1})\right| \le \sum_{i=1}^n |f(\zeta_i)||z_i - z_{i-1}|$$

$$\le M \sum_{i=1}^n |z_i - z_{i-1}| \le ML(\gamma)$$

where $M = \max\{|f(z)| \,|\, z \in \gamma\}$. The expressions $\left|\sum_{i=1}^n f(\zeta_i)(z_i - z_{i-1})\right|$ and $\sum_{i=1}^n |f(\zeta_i)||z_i - z_{i-1}|$ are equal iff $\arg f(\zeta_i)(z_i - z_{i-1})$ remains constant for $i = 1, 2, \ldots, n$. (See problem 1.9(a).)

From the above inequality for Riemannian sums, we derive a very important inequality for integrals:

$$\left|\int_\gamma f(z)\, dz\right| \le \int_\gamma |f(z)||dz| = \int_0^1 |f(z(t))||z'(t)|\, dt \le ML(\gamma)$$

Here $\int_\gamma |f(z)||dz|$ is defined by

$$\int_\gamma |f(z)||dz| = \lim_{v(D_n)\to 0} \sum_{i=1}^n |f(\zeta_i)||z_i - z_{i-1}|$$

The two values $\left|\int_\gamma f(z)\, dz\right|$ and $\int_\gamma |f(z)||dz|$ are equal iff $\arg f(z(t))z'(t)$ remains a constant for $0 \le t \le 1$, a consequence of the relation

$$|a + b| = |a| + |b| \Leftrightarrow \arg a = \arg b$$

(see problem 1.9(a)).

If γ is a point, then $\int_\gamma f(z)\, dz = 0$ since $L = 0$.

EXAMPLES

1 Let γ be a piecewise smooth regular arc joining a and b and $f(z) \equiv 1$; then $S(D_n) = \sum_{i=1}^n f(\zeta_i)(z_i - z_{i-1}) = \sum_{i=1}^n (z_i - z_{i-1}) = b - a$. It follows that

$$\int_a^b f(z)\, dz = b - a$$

2 Let γ be, as in 1, a piecewise smooth arc from a to b, and let $f(z) = z$; then $S(D_n) = \sum_{i=1}^{n} \zeta_i(z_i - z_{i-1})$. Since $\lim_{v(D_n) \to 0} S(D_n)$ does not depend on the choice of ζ_i on the arc (z_{i-1}, z_i), we set $\zeta_i = z_{i-1}$, and then we form another sum $S(D_n)$ setting $\zeta_i = z_i$.

In the first case, we get

$$S_1 = \sum_{i=1}^{n} z_{i-1}(z_i - z_{i-1})$$

and in the second case,

$$S_2 = \sum_{i=1}^{n} z_i(z_i - z_{i-1})$$

Now we form the sum

$$S_1 + S_2 = \sum_{i=1}^{n} [z_{i-1}(z_i - z_{i-1}) + z_i(z_i - z_{i-1})]$$

$$= \sum_{i=1}^{n} (z_i^2 - z_{i-1}^2) = b^2 - a^2$$

Then we have, writing v for $v(D_n)$,

$$\lim_{v \to 0} S(D_n) = \lim_{v \to 0} \frac{S_1 + S_2}{2} = \frac{b^2 - a^2}{2} = \int_{\gamma} z \, dz$$

In both examples, we see that the integral $\int_{\gamma} f(z) \, dz$ does not depend on the particular choice of γ, but only on the endpoints a and b. If γ is a closed curve, then $a = b$, and both $\int_{\gamma} dz = 0$ and $\int_{\gamma} z \, dz = 0$.

EXAMPLE

3 Let γ be the unit circle $z = \text{cis } t, 0 \le t \le 2\pi$. Consider the line integral

$$\int_{\gamma} \frac{dz}{z}$$

Divide γ into n equal parts by taking $\alpha = \pi/n$ and $z_0 = 1$, $z_1 = \operatorname{cis} 2\alpha = \cos 2\alpha + i \sin 2\alpha$, ..., $z_k = \operatorname{cis} 2k\alpha$, ..., $z_n = \operatorname{cis} 2n\alpha = 1$. Then take $\zeta_k = \operatorname{cis}(2k-1)\alpha$, $k = 1, 2, \ldots, n$. It follows that

$$S(D_n) = \sum_{k=1}^{n} \frac{1}{\zeta_k} (z_k - z_{k-1})$$

$$= \sum_{k=1}^{n} \operatorname{cis}[-(2k-1)\alpha](\operatorname{cis} 2k\alpha - \operatorname{cis}(2k-2)\alpha)$$

$$= \sum_{k=1}^{n} (\operatorname{cis} \alpha - \operatorname{cis}(-\alpha)) = 2i \sum_{k=1}^{n} \sin \alpha = 2in \sin \frac{\pi}{n}$$

Thus, we have

$$\lim_{n \to \infty} S(D_n) = 2i \lim_{n \to \infty} n \sin \frac{\pi}{n} = 2i\pi = \int_{\gamma} \frac{dz}{z}$$

In contrast to examples 1 and 2, where the integrals taken over any arbitrary closed curve were zero, we have here an example of an integral taken over a closed curve whose value is different from zero.

Note that the functions $f(z) \equiv 1$ and $f(z) = z$ are both differentiable in the entire complex plane, whereas the function $f(z) = 1/z$ is singular at the origin, which lies inside γ.

3. Properties of line integrals

The line integral $\int_{\gamma} f(z)\, dz$ is a function of the integrand f and of the arc γ. Let us first consider the integral as a function of the integrand f. So we fix γ, and consider the set $C[\gamma]$ of all complex-valued functions which are continuous on γ. We have seen that $(C[\gamma], \|\cdot\|)$ with

$$\|f\| = \operatorname{l.u.b.}\{|f(z)| \,|\, z \in \gamma\}$$

is a Banach space, since γ is compact.

THEOREM 2.2 Denote $\int_{\gamma} f(z)\, dz$ by $I(f)$. Then I is a linear continuous mapping $C[\gamma] \to \mathbf{C}$; that is,
 (i) (linearity of I)

$$I(f + g) = I(f) + I(g)$$

and

$$I(af) = aI(f) \quad \forall a \in \mathbf{C}$$

(ii) *(continuity of I)* If $f_1, f_2, \ldots, f_n, \ldots$ is a sequence of functions belonging to $C[\gamma]$ which converge uniformly to f on γ, then $I(f_n)$ converges to $I(f)$.

PROOF

The proof of (i) is easy. Using the Riemannian sums:

$$S(D_n, \zeta^{(n)}, f + g) = \sum_{i=1}^{n} (f(\zeta_i) + g(\zeta_i))(z_i - z_{i-1})$$

$$\sum_{i=1}^{n} f(\zeta_i)(z_i - z_{i-1}) + \sum_{i=1}^{n} g(\zeta_i)(z_i - z_{i-1})$$

$$= S(D_n, \zeta^{(n)}, f) + S(D_n, \zeta^{(n)}, g)$$

It follows that

$$I(f + g) = \lim_{v \to 0} S(D_n, \zeta^{(n)}, f + g)$$

$$= \lim_{v \to 0} S(D_n, \zeta^{(n)}, f) + \lim_{v \to 0} S(D_n, \zeta^{(n)}, g)$$

$$= I(f) + I(g)$$

Similarly, we can show that

$$S(D_n, \zeta^{(n)}, af) = aS(D_n, \zeta^{(n)}, f)$$

which implies $I(af) = aI(f)$.

To prove (ii), choose $N(\varepsilon) \ni n > N(\varepsilon) \Rightarrow |f_n(z) - f(z)| < \varepsilon, \forall z \in \gamma$. Then

$$|I(f_n) - I(f)| = \left| \int_\gamma (f_n(z) - f(z)) \, dz \right| \leq \varepsilon L(\gamma)$$

An immediate consequence of Theorem 2.2 is:

THEOREM 2.3 Let $f(z, u)$, $z \in \gamma$, $u \in$ compact $K \subset \mathbf{C}$, be a continuous function $\gamma \times K \to \mathbf{C}$. Then

$$F(u) = \int_\gamma f(z, u) \, dz$$

is a continuous function of u.

PROOF

Since $\gamma \times K$ is compact, the continuity of f on $\gamma \times K$ implies its uniform continuity on $\gamma \times K$. It follows that if $u_1, u_2, \ldots, u_n, \ldots$ is a sequence of complex numbers $\in K$, converging to u_0, then the sequence

$$f(z, u_1) = f_1(z), \quad f(z, u_2) = f_2(z), \ldots, f(z, u_n) = f_n(z), \ldots$$

converges uniformly on γ to $f(z, u_0) = f(z)$.

Hence, by Theorem 2.2,

$$F(u_n) = \int_\gamma f(z, u_n)\, dz = \int_\gamma f_n(z)\, dz$$

converges to $F(u_0) = \int_\gamma f(z, u_0)\, dz$, which proves that $F(u)$ is continuous.

We now consider the line integral as a function of the arc γ.

DEFINITION If $\gamma_1 = \gamma_1(t)$ and $\gamma_2 = \gamma_2(t)$, $0 \le t \le 1$, are two p.w.s. arcs with $\gamma_1(1) = \gamma_2(0)$, then we define $\gamma = \gamma_1 + \gamma_2$ by

$$\gamma(t) = \begin{cases} \gamma_1(2t), & 0 \le t \le \tfrac{1}{2} \\ \gamma_2(2t - 1), & \tfrac{1}{2} \le t \le 1 \end{cases}$$

It is clear that $\gamma(t)$ defines a p.w.s. arc.

We want to show that

$$I(\gamma) = \int_\gamma f(z)\, dz$$

is a homogeneous, additive, and continuous function of γ.

THEOREM 2.4 Let $f \in C(D)$, and $\gamma, \gamma_1, \gamma_2 \subset D$. Then we have:

(i) $\quad \int_{-\gamma} f(z)\, dz = -\int_\gamma f(z)\, dz \qquad$ (I is homogeneous)

(ii) $\quad \int_{\gamma_1 + \gamma_2} f(z)\, dz = \int_{\gamma_1} f(z)\, dz + \int_{\gamma_2} f(z)\, dz \qquad$ (I is additive)

Also, if $\gamma(t, u) \subset D$, $0 \le t \le 1$, $0 \le u \le 1$, is a family of p.w.s. arcs in D, and if the two functions $\gamma(t, u)$ and $\dfrac{\partial}{\partial t}\gamma(t, u)$ are continuous on the square $S = \{(t, u) \mid 0 \le t \le 1, 0 \le u \le 1\}$, then

(iii) $\quad \int_{\gamma(t, u)} f(z)\, dz \quad$ is a continuous function of u.

Holomorphic Functions

PROOF

(i) Since

$$S(D_n, \zeta^{(n)}, -\gamma) = \sum_{i=0}^{n-1} f(\zeta_{n-1})(z_{n-i-1} - z_{n-i})$$

$$= -\sum_{i=1}^{n} f(\zeta_i)(z_i - z_{i-1}) = -S(D_n, \zeta^{(n)}, \gamma)$$

it follows in the limit that

$$\int_{-\gamma} f(z) \, dz = -\int_{\gamma} f(z) \, dz$$

(ii) In $S(D_n, \zeta^{(n)}, \gamma_1 + \gamma_2)$, we choose the subdivision $D_n \ni$ one of the points, say z_k, of the subdivision is $\gamma_1(1) = \gamma_2(0)$. It follows that

$$S(D_n, \zeta^{(n)}, \gamma_1 + \gamma_2) = S(D_k, \zeta^{(k)}, \gamma_1) + S(D_{n-k}, \zeta^{(n-k)}, \gamma_2)$$

which, in the limit, becomes

$$\int_{\gamma_1 + \gamma_2} f(z) \, dz = \int_{\gamma_1} f(z) \, dz + \int_{\gamma_2} f(z) \, dz$$

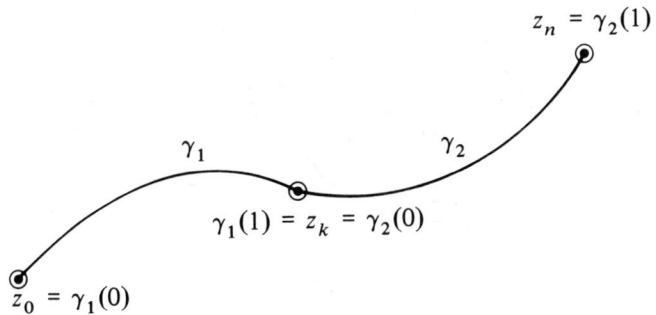

Figure 2.4

(iii) Since

$$\int_{\gamma(t, u)} f(z) \, dz = \int_0^1 f(\gamma(t, u)) \frac{\partial \gamma(t, u)}{\partial t} \, dt$$

and $f(\gamma(t, u)) \dfrac{\partial \gamma(t, u)}{\partial t}$ is continuous, it follows by Theorem 2.3 that

$$F(u) = \int_0^1 f(\gamma(t, u)) \frac{\partial \gamma(t, u)}{\partial(t)} \, dt$$

is a continuous function of u.

EXAMPLE

If Δ is a counterclockwise-oriented triangle, and we divide it into four counterclockwise-oriented triangles $\Delta_1^{(1)}$, $\Delta_1^{(2)}$, $\Delta_1^{(3)}$, $\Delta_1^{(4)}$, by joining the midpoints of the sides of Δ (see Figure 2.5), then

$$\int_\Delta f(z)\,dz = \sum_{i=1}^4 \int_{\Delta_1^{(i)}} f(z)\,dz$$

The integrals taken over interior sides vanish, being taken twice in opposite directions.

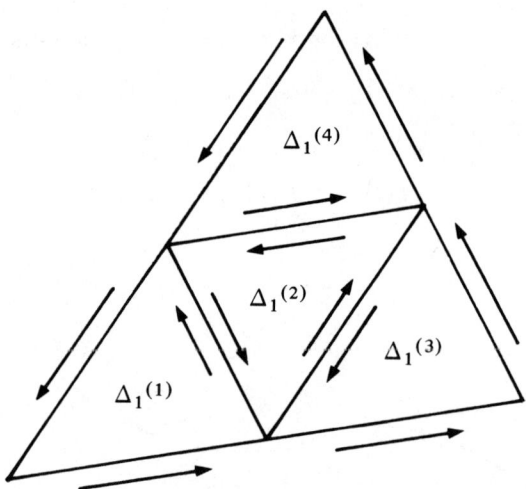

Figure 2.5

If D is a domain which contains Δ and its interior, and if $\gamma = \gamma(t, u)$ is a continuously differentiable mapping $D \to \mathbf{C}$ then we have

$$\int_{\gamma(\Delta)} f(z)\,dz = \sum_{i=1}^4 \int_{\gamma(\Delta_1^{(i)})} f(z)\,dz$$

provided that f is continuous on $\gamma(D)$. $\Big($The mapping $\gamma(t, u)$ is said to be continuously differentiable if $\dfrac{\partial \gamma}{\partial t}$ and $\dfrac{\partial \gamma}{\partial u}$ exist and are continuous.$\Big)$

Problems

2.2.: Show that homotopy is an equivalence relation. (*Hint*: See 0.1.2 for (*ii*) and use $\gamma(t, 1-u)$; for (*iii*), use $\gamma_1\left(t, \dfrac{u}{2}\right)$ and $\gamma_2\left(t, \dfrac{u+1}{2}\right)$).

2.3.: Show that the circle $\partial(0, r)$ and the ellipse centered at the origin with axes a and b parallel to the x-axis and y-axis, respectively, are homotopic with respect to the open square $\{(x, y)\,|\,-c < x < c,\ -c < y < c\}$, where $c > \max(r, a, b)$. (See Figure 2.6.) (*Hint*: Take $\gamma(t, u) = [au + r(1-u)]\cos t + i[bu + r(1-u)]\sin t$.)

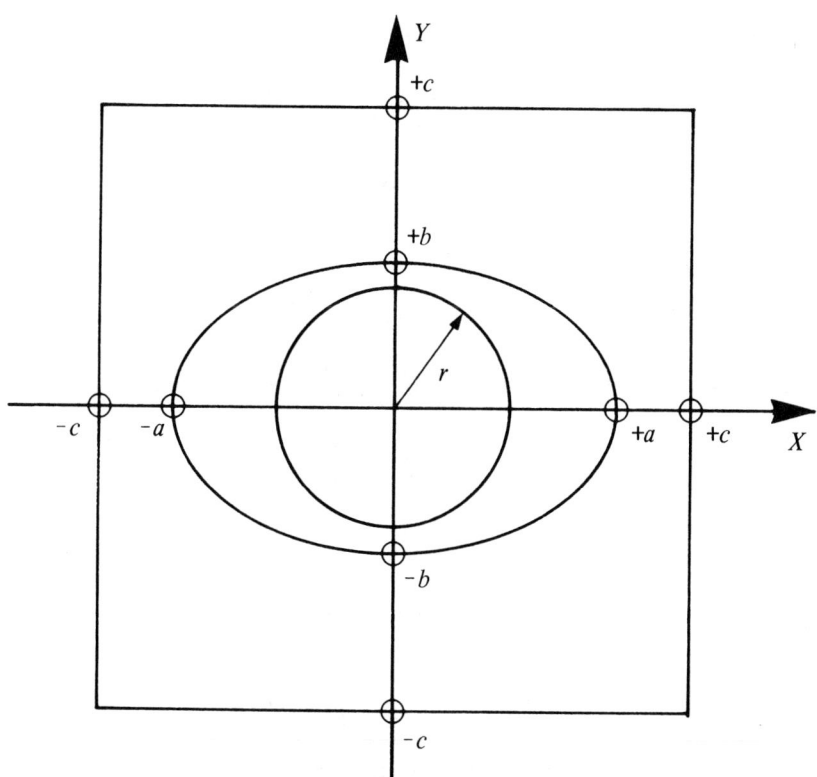

Figure 2.6

2.4.: Show that the circle $\gamma_0(\phi) = a + r \text{ cis } \phi$, $0 \leq \phi \leq 2\pi$, and the closed curve formed by

$$\gamma_1(\phi) = \begin{cases} R \text{ cis } \phi & \text{for } 0 \leq \phi \leq \pi \quad \text{(half-circle)} \\ R \cos \phi & \text{for } \pi \leq \phi \leq 2\pi \quad \text{(interval } [-R, R]) \end{cases}$$

are homotopic with respect to any disc which contains both γ_0 and γ_1.

2.5.: With respect to which domain are the circles $\gamma_0 = \text{cis } \phi$ and $\gamma_1 = \frac{1}{2} \text{cis } \phi$, $0 \leq \phi \leq 2\pi$, homotopic?

2.6.: Show that if two domains are homeomorphic, and one of them is simply connected, then so is the other.

2.7.: Prove

$$\left| \int_\gamma (x^2 - iy^2) \, dz \right| \leq 2.5$$

if:
a) γ is the interval $[-i, i]$ on the y-axis
b) γ is the semicircle $z = \text{cis } \phi$, $-\pi/2 \leq \phi \leq \pi/2$.

2.8.: Compute

$$\int_1^{2+i} (x^2 + iy) \, dz$$

if we choose as the path of integration
a) the line $y = x - 1$
b) the line $y = (x - 1)^2$.

2.9.: Compute

$$\int_0^{1+i} (z^2 + z) \, dz$$

Choose two different paths of integration and show that the results are the same.

2.10.: Prove that

$$\left| \int_\partial \frac{dz}{z^2 + 1} \right| \leq \frac{4\pi}{3}$$

if ∂ is the circle

$$\partial(\phi) = 2 \text{ cis } \phi, \quad 0 \leq \phi \leq 2\pi$$

2.11.: Prove that

$$\left| \int_2^i (z + 1)^2 \, dz \right| \leq 9\sqrt{5}$$

(*Hint:* Take as the path of integration the line

$$z = i + (2 - i)t, \quad 0 \leq t \leq 1$$

and determine $\max |(z + 1)^2|$ on this line.)

2.12.: Let $f_n(z) = \dfrac{1}{z^2 + n^2}$; determine

$$\lim_{n \to \infty} n^2 \int_\gamma \frac{dz}{z^2 + n^2}$$

where γ is the semicircle $\gamma(t) = \text{cis } t$, $0 \leq t \leq \pi$. $\left(\text{Hint: Use the uniform convergence of } \dfrac{1}{z^2 + n^2} \text{ on } \gamma.\right)$

2.13.: Let γ be a closed n-sided counterclockwise-oriented polygonal line whose inside is starlike with respect to the point P. Join P to all the vertices of γ and denote by $\Delta_1, \Delta_2, \ldots, \Delta_n$ the counterclockwise oriented triangles you get. Show that

$$\int_\gamma f(z)\, dz = \sum_{i=1}^{n} \int_{\Delta_i} f(z)\, dz$$

2.14.: Let $\{\gamma_n\}$ be the sequence of arcs,

$$\gamma_n(t) = \left(1 - \frac{1}{n+1}\right) + \frac{2t}{n+1} + it, \qquad 0 \leq t \leq 1, \quad n = 1, 2, \ldots$$

and $f(z) = z^2 + z + 1$. Determine

$$\lim_{n \to \infty} \int_{\gamma_n} f(z)\, dz$$

---

§2. Cauchy's theorem

DEFINITION We say that the continuous function is *holomorphic on D* if $\int_\gamma f(z)\, dz = 0$ for any closed p.w.s. curve $\gamma \subset D$ which is homotopic to a point with respect to D.

1. Cauchy's theorem for starlike domains

Cauchy's theorem is the fundamental theorem of the theory of complex-valued functions. It says that complex differentiability in D of a function implies holomorphicity in D.

Morera's theorem shows that the converse of Cauchy's theorem is also true—namely, that holomorphicity in D implies complex differentiability in D. A combination of these two theorems proves the equivalence of the two notions, "complex differentiable in D," and "holomorphic in D."

Since the proof of Cauchy's theorem is rather difficult, we shall first present some special cases.

THEOREM 2.5 *A continuous function f which possesses a primitive F in D is holomorphic in D.*

PROOF

Let $\gamma = \gamma(t)$, $0 \leq t \leq 1$, be a piecewise smooth arc with endpoints $\gamma(0) = a$ and $\gamma(1) = b$. Then

$$\int_\gamma f(z)\,dz = \int_0^1 f(\gamma(t))\gamma'(t)\,dt = \int_0^1 F'(\gamma(t))\gamma'(t)\,dt$$

$$= \int_0^1 \frac{d}{dt} F(\gamma(t))\,dt = F(\gamma(1)) - F(\gamma(0))$$

$$= F(b) - F(a)$$

This shows us that if f is continuous and possesses a primitive, then $\int_\gamma f(z)\,dz$ does not depend on γ, but only on its endpoints a and b.

It follows that if γ is a closed curve, that is, if $a = b$, then $\int_\gamma f(z)\,dz = F(a) - F(a) = 0$.

So we may conclude that all complex differentiable functions which possess a primitive in D are holomorphic. We shall see later that a continuous function which possesses a primitive in D is differentiable in D.

EXAMPLES

1   The polynomials $P(z) = a_n z^n + a_{n-1} z^{n-1} + \cdots + a_0$ are holomorphic functions in $\mathbf{C}$; in particular, a_0 and $a_1 z + a_0$ are holomorphic throughout $\mathbf{C}$, since each of them possesses a primitive in $\mathbf{C}$.

2   The functions $\dfrac{1}{(z-c)^n}$ with integer $n > 1$ are holomorphic in $\mathbf{C}\backslash\{c\}$. The same holds for any polynomial in $\dfrac{1}{z-c}$ where each term is of degree >1, since such a polynomial possesses a primitive in $\mathbf{C}\backslash\{c\}$.

Note that, for $n > 1$, $\int_\gamma \dfrac{1}{(z-c)^n}\,dz = 0$ for any closed curve γ on which $\dfrac{1}{(z-c)^n}$ is continuous—that is, for any closed γ which does not pass through c—even if γ is not homotopic to a point with respect to $\mathbf{C}\backslash\{c\}$ (for example, if γ is any circle with center at c).

Holomorphic Functions

In Theorem 2.5, we saw that Cauchy's theorem is true for special complex-differentiable functions, namely, for functions which possess a primitive, with γ being an arbitrary closed curve.

In the following theorem, we shall show that Cauchy's theorem is true for arbitrary complex-differentiable functions in D if we restrict γ to be a triangle $\Delta \subset D$.

THEOREM 2.6 [*Cauchy-Goursat*] *Let f be complex differentiable in D, and let Δ be a triangle which is homotopic to a point in D; in other words, a triangle which lies with its interior in D. Then $\int_\Delta f(z)\,dz = 0$.*

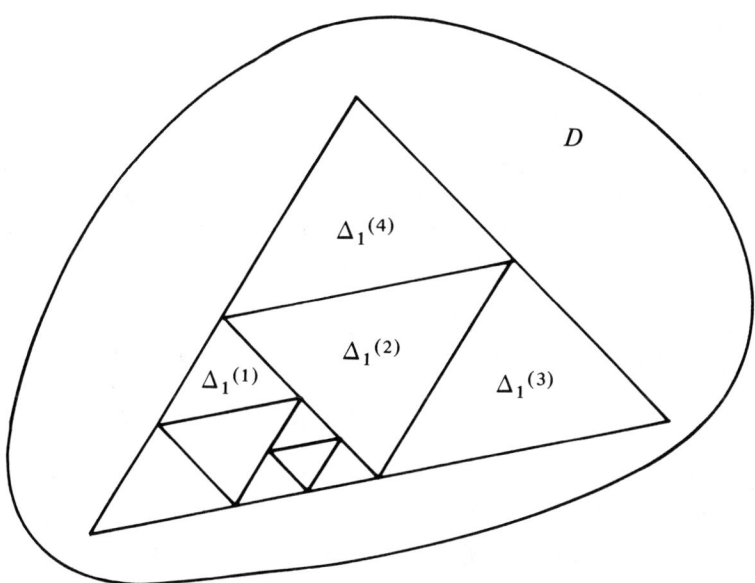

Figure 2.7

PROOF

Let $I_0 = \int_\Delta f(z)\,dz$ and let L be the perimeter of the triangle. Join the midpoints of the sides of Δ (Figure 2.7) to obtain four triangles $\Delta_1^{(1)}$, $\Delta_1^{(2)}$, $\Delta_1^{(3)}$, $\Delta_1^{(4)}$, each with perimeter $L/2$.

If we integrate over the four triangles $\Delta_1^{(1)}$, $\Delta_1^{(2)}$, $\Delta_1^{(3)}$, $\Delta_1^{(4)}$, always in the positive sense (that is, counterclockwise), then $I_0 = I_1^{(1)} + I_1^{(2)} + I_1^{(3)} + I_1^{(4)}$, where $I_1^{(i)} = \int_{\Delta_1^{(i)}} f(z)\,dz$. With no loss of generality, we assume that the four integrals on the right side are indexed according to decreasing magnitude of their absolute values, and that the indices of the triangles are chosen in the same order, so that

$$|I_1^{(1)}| \geq |I_1^{(2)}| \geq |I_1^{(3)}| \geq |I_1^{(4)}|$$

Then, setting $I_1 = I_1^{(1)}$, we have

$$|I_0| \leq 4|I_1|$$

Repeating the same process with Δ_1, namely, joining the midpoints of the sides of Δ_1 and indexing the resulting four triangles $\Delta_2^{(1)}$, $\Delta_2^{(2)}$, $\Delta_2^{(3)}$, $\Delta_2^{(4)}$ so that the corresponding integrals $I_2^{(1)}$, $I_2^{(2)}$, $I_2^{(3)}$, $I_2^{(4)}$ satisfy the relation

$$|I_2^{(1)}| \geq |I_2^{(2)}| \geq |I_2^{(3)}| \geq |I_2^{(4)}|$$

we obtain

$$|I_1| \leq 4|I_2|$$

Repeating this process, we obtain a sequence of triangles $\Delta_0 \supset \Delta_1 \supset \cdots \supset \Delta_n \supset \cdots$, where the perimeter of Δ_n is $L/2^n$, and a sequence of integrals $I_0, I_1, \ldots, I_n$, which satisfies the inequalities

$$|I_0| \leq 4|I_1| \leq 4^2|I_2| \leq \cdots \leq 4^n|I_n| \leq \cdots$$

Note that the diameter of a triangle (the length of the greater side) is no greater than its perimeter, so $\lim_{n \to \infty} \text{diam } \Delta_n = 0$. It follows by Theorem 1.1 that $\bigcap_{n=1}^{\infty} \Delta_n$ contains at least one point, say z_0, which $\in D$. (Here we use the fact that Δ is homotopic to a point with respect to D.) Since f is complex differentiable in D, we have

$$f(z) = f(z_0) + (z - z_0)f'(z_0) + (z - z_0)\Omega(z, z_0)$$

with

$$\lim_{z \to z_0} \Omega(z, z_0) = 0$$

For $|z - z_0| < H(\varepsilon)$ with $H(\varepsilon)$ sufficiently small and $d(z_0, H(\varepsilon)) \subset D$, we get $|\Omega(z, z_0)| < \varepsilon$, and for $n > N(H(\varepsilon))$, all the triangles Δ_n lie in $d(z_0, H(\varepsilon))$. Hence, for $n > N(H(\varepsilon))$ we get

$$I_n = \int_{\Delta_n} f(z)\, dz = \int_{\Delta_n} [f(z_0) + (z - z_0)f'(z_0)]\, dz$$
$$+ \int_{\Delta_n} (z - z_0)\Omega(z, z_0)\, dz$$

Now, $f(z_0) + (z - z_0)f'(z_0)$ is a polynomial of the first degree in z. Hence $\int_{\Delta_n} [f(z_0) + (z - z_0)f'(z_0)] \, dz = 0$. Therefore,

$$I_n = \int_{\Delta_n} (z - z_0)\Omega(z, z_0) \, dz$$

Now, $|z - z_0| \leq$ (diameter of Δ_n) $\leq$ (perimeter of Δ_n) $= L/2^n$, and $|\Omega(z, z_0)| < \varepsilon$. Since the perimeter of Δ_n is $L \cdot 2^{-n}$, we find

$$|I_n| \leq \varepsilon(2^{-n}L)(2^{-n}L) = \varepsilon L^2 4^{-n}$$

Consequently, we have

$$|I_0| \leq 4^n |I_n| \leq 4^n \varepsilon L^2 4^{-n} = \varepsilon L^2$$

for arbitrarily small $\varepsilon > 0$. Hence, $|I_0| = 0$; that is, $I_0 = 0$.

One can generalize this result by taking for γ a closed polygon. Cutting the closed polygon into triangles $\Delta_1, \Delta_2, \ldots, \Delta_k$, we get $\int_\gamma f(z) \, dz = \sum_{i=1}^{k} \int_{\Delta_i} f(z) \, dz = 0$. We could approximate an arbitrary closed piecewise smooth curve by such a polygon and show that the integral along this curve vanishes. But we shall use another method to establish that result. It is based on the Cauchy-Goursat Theorem 2.6 for triangles, and on Theorem 2.5.

THEOREM 2.7 [Cauchy's theorem for starlike domains] *If D is starlike with respect to c, and if f is a complex differentiable function in D ($f \in H(D)$), then f is holomorphic in D.*

In fact, we shall prove that if D is starlike with respect to c and f is continuous with $\int_\Delta f(z) \, dz = 0$ for all triangles lying with their interior in D and having one vertex at c, then f is holomorphic in D.

PROOF

We will show that f possesses a primitive in D.

Let $z \in D$, and let $F(z) = \int_c^z f(u) \, du$, the integral taken over the

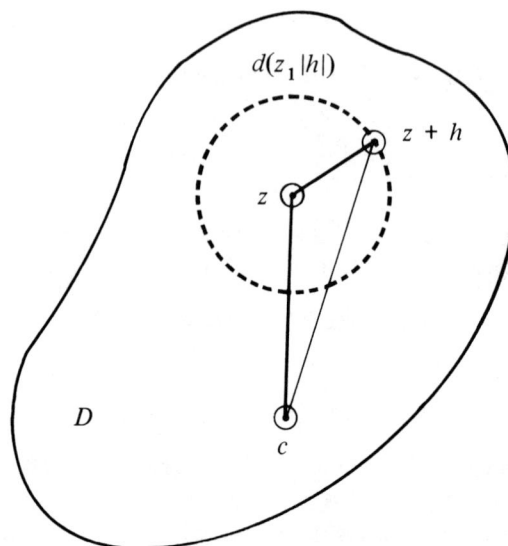

Figure 2.8

segment joining c to z. Then $F(z)$ is a primitive of $f(z)$ in D. Indeed, if h is sufficiently small, so that $d(z, h) \subset D$, then

$$\left| \frac{F(z+h) - F(z)}{h} - f(z) \right| = \left| \frac{1}{h} \left[\int_c^{z+h} f(u)\, du - \int_c^z f(u)\, du \right] - f(z) \right|$$

Since $f \in H(D)$, the integral over the triangle $(c, z, z+h)$ is equal to 0 (Cauchy-Goursat theorem); that is,

$$\int_c^{z+h} \cdots + \int_{z+h}^z \cdots + \int_z^c f(u)\, du = 0$$

or

$$\int_c^{z+h} f(u)\, du - \int_c^z f(u)\, du = \int_z^{z+h} f(u)\, du$$

Moreover,

$$\int_z^{z+h} f(z)\, du = f(z)(z+h-z) = hf(z)$$

It follows that

$$\left| \frac{F(z+h) - F(z)}{h} - f(z) \right| = \left| \frac{1}{h} \int_z^{z+h} (f(u) - f(z))\, du \right| \leq M$$

where $M = \max|f(u) - f(z)|$ for $\forall u$ on the segment $[z, z+h]$. Clearly, $h \to 0 \Rightarrow M \to 0$ because of the continuity of f. Hence

$$\lim_{h \to 0} \frac{F(z+h) - F(z)}{h} = f(z)$$

which shows that F is a primitive of f in D.

Now, using Theorem 2.5, it follows that f is holomorphic in D.

DEFINITION We say that f is *locally holomorphic in D* if with every point $z \in D$ we can associate a disc $d(z, r) \ni f$ is holomorphic in $d(z, r)$.

Obviously, if f is complex differentiable in domain D, then f is locally holomorphic in D. Indeed, since D is open, for any $z \in D$ $\exists$ an open disc $d(z, r) \subset D$. Now $d(z, r)$ is convex, and thereby starlike. So f possesses a primitive in $d(z, r)$ and is holomorphic in $d(z, r)$.

2. Cauchy's theorem for general domains; residues

THEOREM 2.8 *A function f which is complex differentiable in a domain D is holomorphic in D.*

In fact, we shall show that a locally holomorphic function is holomorphic in D.

The proof is based on the following lemma:

LEMMA 2.8 *If f is locally holomorphic in D, and if the piecewise smooth closed curves γ_0 and γ_e are homotopic with respect to D, then*

$$\int_{\gamma_0} f \, dz = \int_{\gamma_e} f \, dz$$

PROOF

Let $\gamma(t, u)$ be the family of p.w.s. closed curves which define the homotopy between γ_0 and γ_e.

Divide the interval $0 \le u \le 1$ into n equal intervals $[u_{i-1}, u_i]$ with $u_i - u_{i-1} = 1/n$, $u_0 = 0$, and $u_n = 1$. Define γ_i by $\gamma_i = \gamma(t, u_i)$. We shall show that, for sufficiently large n,

$$\int_{\gamma_{i-1}} f(z) \, dz = \int_{\gamma_i} f(z) \, dz, \quad i = 1, 2, \ldots, n$$

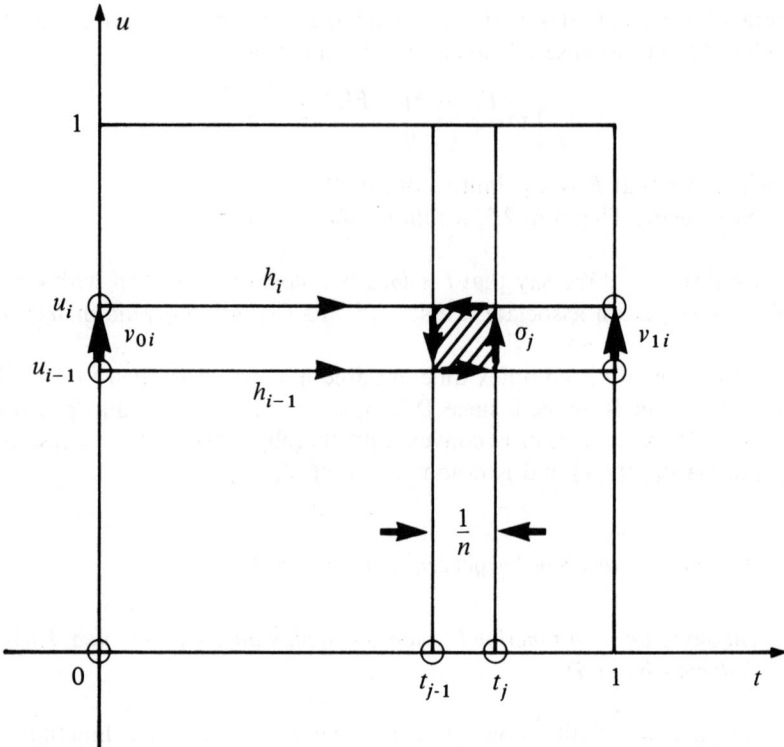

Figure 2.9

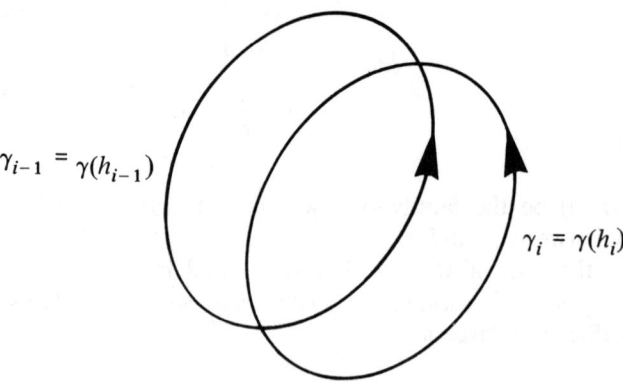

Figure 2.10

Denote by h_i the horizontal segment $u = u_i$, $0 \leq t \leq 1$, $i = 0, 1, \ldots, n$; by $V_{0,i}$ the vertical segment $t = 0$, $u_{i-1} \leq u \leq u_i$; and by $V_{1,i}$ the vertical segment $t = 1$, $u_{i-1} \leq u \leq u_i$.

Then divide the interval $0 \leq t \leq 1$ into n equal intervals $[t_{i-1}, t_i]$ with $t_i - t_{i-1} = 1/n$. We obtain n small squares between the horizontal lines h_{i-1} and h_i, with sides of length $1/n$, defined by $u_{i-1} \leq u \leq u_i$, i fixed, and $t_{j-1} \leq t \leq t_j$, $j = 1, 2, \ldots, n$.

Denote the boundaries of these n small squares, described in counterclockwise sense, by $\sigma_1, \sigma_2, \ldots, \sigma_n$. Then we have:

$$h_{i-1} = h_i + \sigma_1 + \sigma_2 + \cdots + \sigma_n - v_{1,i} + v_{0,i}$$

If we denote by $\gamma = \gamma(t, u)$ the mapping $S \to \mathbf{C}$, where $S = \{(t, u) | 0 \leq t \leq 1, 0 \leq u \leq 1\}$, then we find $\gamma(h_{i-1}) = \gamma(h_i) + \gamma(\sigma_1) + \gamma(\sigma_2) + \cdots + \gamma(\sigma_n) - \gamma(v_{1,i}) + \gamma(v_{0,i})$. Since $\gamma(h_{i-1}) = \gamma(t, u_{i-1}) = \gamma_{i-1}$, $\gamma(h_i) = \gamma(t, u_i) = \gamma_i$, and $\gamma(0, u) = \gamma(1, u) \Rightarrow \gamma(v_{0,i}) = \gamma(v_{1,i})$, the above relation becomes

$$\gamma_{i-1} = \gamma_i + \sum_{j=1}^{n} \gamma(\sigma_j)$$

Thus we find

$$\int_{\gamma_{i-1}} f(z)\, dz = \int_{\gamma_i} f(z)\, dz + \sum_{j=1}^{n} \int_{\gamma(\sigma_j)} f(z)\, dz$$

In order to prove the lemma, it is therefore sufficient to show that $\int_{\gamma(\sigma_j)} f(z)\, dz = 0$, for $j = 1, \ldots, n$. This can be done by using the local holomorphicity of f in D.

In fact, since f is locally holomorphic on D, each point $z \in D$ is the center of a disc $d(z, r_z)$ in which f is holomorphic. The discs $d(z, r_z/2)$, $z \in \gamma(S)$, cover $\gamma(S)$. Since $\gamma(S)$ is compact as image of the compact S by the continuous mapping γ (see Theorem 0.7), a finite number of these discs, say $d(z_1, r_1/2)$, $d(z_2, r_2/2)$, $\ldots$, $d(z_k, r_k/2)$, cover $\gamma(S)$. Now let $r = \min(r_1/2, r_2/2, \ldots, r_k/2)$. Since the continuity of γ on the compact S implies its uniform continuity on S, we can cover S by a net of squares $S_1, S_2, \ldots, S_{n^2}$, with sides of length $1/n$ sufficiently small so that the distance between the images of any two points in one of the squares is smaller than r. Obviously, the image of any one of the squares lies entirely in one of the discs $d(z_1, r_1)$, $d(z_2, r_2)$, $\ldots$, $d(z_k, r_k)$ in which f is holomorphic. Thus, if σ_j is the boundary of S_j, we have

$$\int_{\gamma(\sigma_j)} f(z)\, dz = 0, \quad j = 1, 2, \ldots, n^2$$

Consequently we arrive at

$$\int_{\gamma_{i-1}} f(z)\,dz = \int_{\gamma_i} f(z)\,dz \quad \text{for} \quad i = 1, 2, \ldots, n$$

or

$$\int_{\gamma_o} f(z)\,dz = \int_{\gamma_e} f(z)\,dz$$

PROOF OF THEOREM 2.8

Now if γ_o is homotopic to a point $\gamma_e = p$ with respect to D, then $\int_{\gamma_o} f(z)\,dz = \int_{\gamma_e} f(z)\,dz = 0$. This proves the theorem.

COROLLARY *If γ is a p.w.s. closed curve lying in the simply connected domain D, and if $f \in H(D)$, then $\int_\gamma f(z)\,dz = 0$.*

This follows immediately from the theorem, since γ is homotopic to a point with respect to D.

It can also be shown that $\int_\gamma f(z)\,dz = 0$ if the p.w.s. closed curve γ is the boundary of a simply connected domain D, and if $f \in H(D)$ and $f \in C(\overline{D})$. Indeed, if $\{\gamma_n\}$ is a sequence of p.w.s. closed curves in D with $\lim_{n \to \infty} \gamma_n = \gamma$, then $\int_{\gamma_n} f(z)\,dz = 0$. Using theorem 2.4, we find $\int_\gamma f(z)\,dz = 0$.

EXAMPLE

We have seen (example 3, 2.1.2) that

$$\int_{\partial(0,\,1)} \frac{dz}{z} = 2\pi i$$

Now any circle $\partial(0, r)$ centered at the origin is homotopic to $\partial(0, 1)$ with respect to $\mathbf{C}\setminus\{0\}$. It follows that

$$\int_{\partial(0,\,r)} \frac{dz}{z} = 2\pi i$$

for arbitrary r.

In fact, the circle $\partial(0, 1)$ can be replaced by any other circle, a square, a closed polygon or an ellipse, provided that the replacement contains the origin in its interior. It is easy to establish the homotopy of these closed curves with the circle $\partial(0, 1)$, so if γ is such a closed curve, we have

$$\frac{1}{2\pi i}\int_\gamma \frac{dz}{z} = 1$$

We recall that a point c at which f is not complex differentiable is called a singular point or a singularity of f. A singular point c is called an *isolated singular point* if $\exists$ a punctured disc $\dot{d}(c, r)$ in which f is complex differentiable.

EXAMPLE

The point c is an isolated singularity of $\dfrac{1}{(z-c)^n}$, where the integer n is ≥ 1.

Let $f \in H(D)$ and let c be an isolated singularity of f; then $\exists$ an $r \ni \dot{d}(c, r) \subset D$ with $\dot{d}(c, r) = d(c, r)\backslash\{c\}$. All circles $\partial(c, \rho)$ described in counterclockwise sense with $\rho < r$ are homotopic with respect to D. It follows that the integral

$$\frac{1}{2\pi i}\int_{\partial(c, \rho)} f(z)\, dz$$

does not depend on ρ. We call the value of that integral the *residue* of f at c and denote it by $\text{Res}(f, c)$. It is clear that in the integral we can replace the circle $\partial(c, \rho)$ by any other closed curve which is homotopic to $\partial(c, \rho)$ with respect to D.

EXAMPLE

$$\text{Res}\left(\frac{1}{(z-c)^n}, c\right) = \begin{cases} 0 & \text{for } n \text{ an integer} > 1 \\ 1 & \text{for } n = 1 \end{cases}$$

In many applications, the determination of the residue at isolated singularities is of great importance. For rational functions whose singularities are the zeros of the denominator for which the numerator is nonzero, the determination of the residue is particularly simple since it is sufficient to decompose the rational functions into partial fractions.

EXAMPLE

The function

$$f(z) = \frac{z-2}{z^2(z-1)} = \frac{1}{z} + \frac{2}{z^2} - \frac{1}{z-1}$$

has two isolated singularities, namely, 0 and 1. $\mathrm{Res}(f, 0) = \mathrm{Res}(1/z, 0) = 1$ since the contribution of $2/(z^2)$ and of $-1/(z-1)$ to the residue at the origin is zero. Similarly, we find $\mathrm{Res}(f, 1) = \mathrm{Res}(-1/(z-1), 1) = -1$.

THEOREM 2.9 *A function f which is holomorphic in a simply connected domain D possesses a primitive in D.*

(Compare the proof of Theorem 2.7.)

PROOF

Let a be fixed and $z \in D$, and join a to z by some piecewise smooth arc γ. Consider the integral

$$\int_a^z f(u)\, du = \int_\gamma f(u)\, du = F(z)$$

Since f is holomorphic and D is simply connected, $\int_\gamma f(u)\, du$ does not depend on γ, but only on z. Indeed, if γ^* is any other piecewise smooth arc joining a to z, then $\gamma + (-\gamma^*)$ is a piecewise smooth closed curve in D, which is homotopic to a point. It follows that

$$\int_\gamma f(u)\, du + \int_{-\gamma^*} f(u)\, du = 0$$

or

$$\int_\gamma f(u)\, du = \int_{\gamma^*} f(u)\, du$$

We want to show that F is a primitive of f. We choose h sufficiently small so that $d(z, h) \subset D$. Then denote by $h\Omega(z, h)$ the expression:

$$h\Omega(z, h) = F(z + h) - F(z) - hf(z)$$

$$= \int_a^{z+h} f(u)\, du - \int_a^z f(u)\, du - \int_z^{z+h} f(z)\, du$$

$$= \int_z^{z+h} (f(u) - f(z))\, du$$

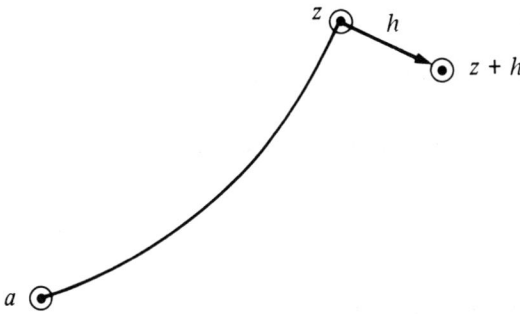

Figure 2.11

The function F is complex differentiable with derivative f provided we can show that

$$\lim_{h \to 0} \Omega(z, h) = 0$$

Since

$$|h\Omega(z, h)| = \left| \int_z^{z+h} (f(u) - f(z))\, du \right| \leq M|h|$$

where

$$M = \max_u \{|f(u) - f(z)| \, | \, u \in \text{segment}[z, z+h]\}$$

and clearly $\lim_{h \to 0} M = 0$, we have

$$\lim_{h \to 0} \Omega(z, h) = 0$$

Problems

2.15.: Let D be the domain defined by:
$$D = \{z \,|\, a < \max(|\text{Re } z|, |\text{Im } z|) < b\}$$
and suppose $f \in H(D)$ and $f \in C(\overline{D})$. State the Cauchy-Goursat theorem for D.

2.16.: Prove the Cauchy-Goursat theorem for a starlike polygon.

2.17.: Prove the Cauchy-theorem with the aid of Green's theorem under the assumption that f' is continuous on D.

§3. Cauchy's formula

If γ is a p.w.s. closed curve in the domain D, and if $f \in H(D)$, then it is possible to express the value of f at a point z interior to γ as a function of its values on γ (γ is traced only once in counterclockwise sense about z) by the formula

$$f(z) = \frac{1}{2\pi i} \int_\gamma \frac{f(u)}{u - z} \, du$$

This expression is called *Cauchy's formula*. It is the most important formula of complex analysis. For its proof, we need the notions of log z and of the winding number $I(\gamma, c)$. These will be discussed in the following sections.

1. The multivalued function log z

In real analysis, we define log x for positive values of x as a primitive of $1/x$:

$$\log x = \int_1^x \frac{du}{u}$$

Analogously, we shall define log z for complex values of z as a primitive of $1/z$. Suppose $z \neq 0$, and consider the circle $\partial(0, |z|)$ passing through z and intersecting the x-axis at $x = |z|$. Let γ be an arc on this circle which joins $|z|$ with z. We define

$$\log z = \int_1^{|z|} \frac{du}{u} + \int_\gamma \frac{du}{u}$$

The first integral taken along the real axis gives the real integral

$$\int_1^{|z|} \frac{du}{u} = \log |z|$$

To determine the value of the second integral, set

$$u = |z| \operatorname{cis} t = |z|(\cos t + i \sin t)$$

which gives

$$du = |z|(-\sin t + i \cos t) \, dt = i|z|(\cos t + i \sin t) \, dt$$
$$= i|z| \operatorname{cis} t \, dt$$

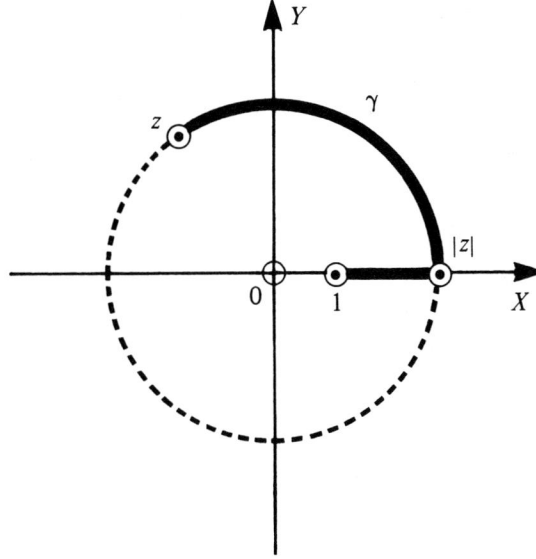

Figure 2.12

and

$$\int_\gamma \frac{du}{u} = \int_0^{\arg z} i\, dt = i \arg z$$

So we have, finally,

$$\log z = \log|z| + i \arg z$$

Since arg z is multivalued, so is log z. In fact, to every value of z there corresponds an infinite set of values of log z which depend on the choice of γ and differ from each other by multiples of $2\pi i$.

EXAMPLES

$\log 1 = 2k\pi i$; $\log 2i = \log|2i| + i \arg 2i = \log 2 + i\frac{\pi}{2} + 2k\pi i$; $\log(-1) = \log 1 + i \arg(-1) = (2k+1)\pi i$, where k is an arbitrary integer.

We define a principal value for log z, denoted by Log z, in the same way as for arg z:

$$\text{Log } z = \log|z| + i \text{ Arg } z$$

The principal value of log z is real for $z \in \mathbf{R}^+$. Log z is complex differentiable in the domain $\mathbf{C} \backslash \mathbf{R}_0^-$ (recall $\mathbf{R}_0^- = \mathbf{R}^- \cup \{0\}$) with the derivative $1/z$. Log z is not even continuous on $\mathbf{R}^-$ because its imaginary part Im(Log z) = Arg z, jumps from $-\pi$ to π on $\mathbf{R}^-$.

$\log_k z = \text{Log } z + 2k\pi i$ is called the *determination* k of log z. Any determination of log z is complex differentiable in the domain $\mathbf{C} \backslash \mathbf{R}_0^-$ with derivative $1/z$.

If D is a simply-connected domain which does not contain the origin, then we can define in D a primitive F of $1/z$ (see Theorem 2.9) in the following way: Choose $a \in D$, and a p.w.s. arc $\gamma \subset D$ joining a to $z \in D$. Then

$$F(z) = \int_\gamma \frac{du}{u} + \log a \qquad (F(z) = \log z)$$

F is called a *branch* of log on D. Clearly, if F is a branch of log on D, then, because of the term log a, so is $F + 2k\pi i$.

EXAMPLE

If $D = \mathbf{C} \backslash \mathbf{R}_0^+$ (recall $\mathbf{R}_0^+ = \mathbf{R}^+ \cup \{0\}$), then a branch of log in D is

$$\log z = \log|z| + i \arg z$$

with $0 \leq \arg z < 2\pi$.

If F is a branch of log on D, then F is a homeomorphism between D and $F(D)$: It is clear that F is single valued once a determination of log a is chosen, and if $F(z_1) = F(z_2)$, that is, log z_1 = log z_2, then Re log $z_1 = \log|z_1| = \log|z_2|$ = Re log z_2, which implies $|z_1| = |z_2|$, and Im log z_1 = arg z_1 = arg z_2 = Im log z_2, which implies arg z_1 = arg z_2; so $z_1 = z_2$. Hence a branch of log on D defines a one-to-one correspondence between D and $F(D)$. Thus $F(D)$ cannot contain two points which differ by a multiple of $2\pi i$, because $F(z_2) = F(z_1) + 2k\pi i$ would imply $z_1 = z_2$, and F would not be single valued. By 1.4.6, the inverse function, F^{-1}, is also complex differentiable. So F is an isomorphism between D and $F(D)$.

Intuitively, we see that if the domain D winds several times around the origin, then F takes on the values of several determinations of log z. In Figure 2.13, $F(z_1)$ and $F(z_2)$ belong to different determinations of log z. Indeed if we start at z_1 and move to z_2 on an arc $\gamma \subset D$, then the argument increases by 2π.

If we want to uniformize log z, that is, to establish a one-to-one

Holomorphic Functions 109

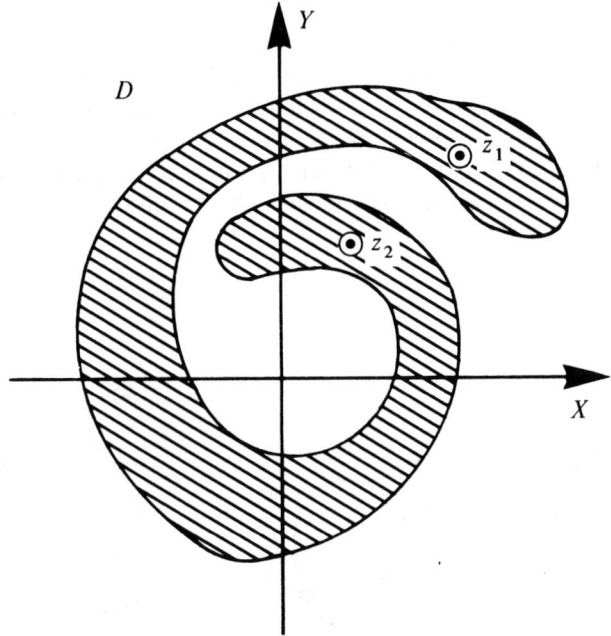

Figure 2.13

correspondence between the domain of definition $\mathbf{C}\backslash\{0\}$ and the range, we modify the domain of definition by introducing the *Riemann surface* of log z.

Intuitively the Riemann surface of log z is defined as follows: For every determination $\log_k z$, we provide one plane which we denote by $\mathbf{C}_k$. The plane $\mathbf{C}_0\backslash\{0\}$ is the domain of definition of the principal value Log $z = \log_0 z$. We cut each of these planes at the negative x-axis, which gives us two edges on each plane at the negative x-axis, namely, the upper edge and the lower edge. We attach the upper edge of $\mathbf{C}_k$ to the lower edge of $\mathbf{C}_{k+1}$. Winding around the origin on $\mathbf{C}_k$ in the counterclockwise sense, we leave $\mathbf{C}_k$ at the negative x-axis and arrive on $\mathbf{C}_{k+1}$, where k is any integer. Thus, if D is a simply connected domain which does not contain the origin, and which winds around the origin several times, then D is lying on several successive planes $\mathbf{C}_r, \mathbf{C}_{r+1}, \ldots, \mathbf{C}_{r+i}$. In Figure 2.13, D is lying on two planes, $\mathbf{C}_r$ and $\mathbf{C}_{r+1}$.

The set of planes $\mathbf{C}_k$, $k = 0, \pm 1, \pm 2, \ldots$, cut at the negative x-axis and attached to each other as described above, is called the Riemann surface of log z.

2. Index of closed curve

An intuitive definition of *index* is as follows: Let γ be a p.w.s. closed curve and $c \notin \gamma$. The *index* of γ with respect to c, denoted by $I(\gamma, c)$, indicates how many times γ winds around c, and in which sense. It follows that $I(\gamma, c)$ is a positive integer if γ winds counterclockwise around c, and a negative integer if γ winds clockwise around c. $I(\gamma, c) = 0$ if c is not inside γ.

EXAMPLE

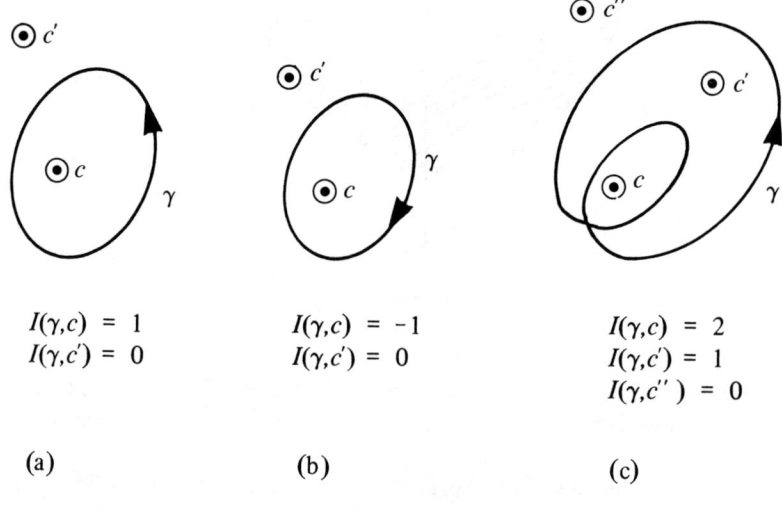

$I(\gamma,c) = 1$ $\qquad\qquad$ $I(\gamma,c) = -1$ $\qquad\qquad$ $I(\gamma,c) = 2$
$I(\gamma,c') = 0$ $\qquad\qquad$ $I(\gamma,c') = 0$ $\qquad\qquad$ $I(\gamma,c') = 1$
$\qquad\qquad\qquad\qquad\qquad\qquad\qquad\qquad\qquad\qquad$ $I(\gamma,c'') = 0$

(a) $\qquad\qquad\qquad\qquad$ (b) $\qquad\qquad\qquad\qquad$ (c)

Figure 2.14

In order to develop an analytic definition of the index, let $\gamma = \gamma(t)$, $0 \leq t \leq 1$, $\gamma(0) = \gamma(1)$, be a p.w.s. closed curve, and let $c \notin \gamma$. Consider the mapping $\arg(\gamma(t) - c)$: $[0, 1] \to \mathbf{R}$. The image of $[0, 1]$ covers a certain segment on the real axis. Since $\gamma(0) = \gamma(1)$, we have $\arg(\gamma(1) - c) - \arg(\gamma(0) - c) = 2k\pi$, where $k \in \mathbf{Z}$. k is called the *index of γ with respect to c*. Instead of $\arg(\gamma(t) - c)$, we can use $\log(\gamma(t) - c) = \log|\gamma(t) - c| + i\arg(\gamma(t) - c)$. Since $\gamma(0) = \gamma(1)$, we have

$$\log(\gamma(1) - c) - \log(\gamma(0) - c)$$
$$= i[\arg(\gamma(1) - c) - \arg(\gamma(0) - c)]$$
$$= 2k\pi i$$

It follows that

$$k = \frac{1}{2\pi i}(\log(\gamma(1) - c) - \log(\gamma(0) - c)) = \frac{1}{2\pi i}\int_\gamma \frac{dz}{z - c}$$

Therefore, the analytic definition of the index is

$$I(\gamma, c) = \frac{1}{2\pi i}\int_\gamma \frac{dz}{z - c}$$

PROPERTIES OF THE INDEX

1 $I(-\gamma, c) = -I(\gamma, c)$
2 If γ_0 and γ_1 are homotopic with respect to $\mathbf{C}\backslash\{c\}$, then $I(\gamma_0, c) = I(\gamma_1, c)$.
3 *The index is a continuous function of c on any compact set $K \subset \mathbf{C}$ which does not intersect γ.*

Property **1** is obvious (see Theorem 2.4); property **2** is a consequence of Lemma 2.8. Property **3** follows from Theorem 2.3, since $z \notin K \Rightarrow 1/(z - c) \in c[K]$ (compare with the remark in 2.1.1).

COROLLARY   *If γ lies in a simply connected domain D which does not contain c, then $I(\gamma, c) = 0$.*

Indeed, γ is homotopic to a point with respect to D, and since $D \subset \mathbf{C}\backslash\{c\}$, γ is homotopic to a point with respect to $\mathbf{C}\backslash\{c\}$, by the remark in 2.1.1.

DEFINITION   The *interior* of γ, denoted by int γ, is the set of all points $z \notin \gamma$ with $I(\gamma, z) \neq 0$.

THEOREM 2.10   int γ *is an open set.*

PROOF

Indeed, if $c \notin \gamma$, then $\exists$ an open disc $d(c, r) \ni d(c, r) \cap \gamma = \emptyset$. It follows that, since $I(\gamma, z)$ is an integer and a continuous function of z, $I(\gamma, z) = I(\gamma, c)\ \forall z \in d(c, r)$.

DEFINITION   We call γ *simple* if int γ is connected, and if $|I(\gamma, c)| = 1$ $\forall c \in$ int γ.

EXAMPLE

The circle $\partial = \partial(t) = a + r \operatorname{cis} t$, $0 \le t \le 2\pi$, is simple, since $\operatorname{int} \partial = \partial(a, r)$ is connected and $I(\partial, c) = 1 \forall c \in d(a, r)$. Clearly, $-\partial = -\partial(t) = a + r \operatorname{cis}(-t)$, $0 \le t \le 2\pi$, is also simple with $I(-\partial, c) = -1$. The circle $\partial = \partial(t) = a + r \operatorname{cis} t$, $0 \le t \le 4\pi$, is not simple, since $I(\partial, c) = 2$ for all points $c \in \operatorname{int} \partial$.

Problem

2.18.: If γ is the boundary of a semidisc, described counterclockwise, show that $I(\gamma, c) = 1$ for all points inside γ. Show also that the same holds true for a rectangle.

---

3. Linear integral transformations

Let D be a domain, $C(D)$ the set of all functions which are continuous on D, γ a p.w.s. arc, $C[\gamma]$ the set of all functions continuous on γ, and $k(z, u)$ a continuous mapping from the Cartesian product $D \times \gamma$ into $\mathbb{C}$. Then the integral

$$F(z) = \int_\gamma k(z, u) f(u) \, du$$

defines a function F in D.

The relationship between f and F will be denoted by

$$F = kf$$

The function $k(z, u)$ will be called the *kernel of the integral transformation k*.

THEOREM 2.11 k *is a linear mapping* $C[\gamma] \to C(D)$.

PROOF

(i) The linearity of k is obvious:

$$k(f + g) = kf + kg \quad \text{and} \quad kaf = akf$$

for $f, g \in C[\gamma]$ and $a \in \mathbb{C}$.

(ii) Let K be a compact subset of D; then $K \times \gamma$ is compact and, since $g(z, u) = k(z, u)f(u)$ is a continuous mapping $D \times \gamma \to \mathbf{C}$, it follows from Theorem 2.3 that $F \in C[K]$. For any point $z \in D$, $\exists$ compact $K \ni z \in K \subset D$. Hence $F \in C(D)$.

Clearly, $(C[\gamma], \|\cdot\|)$ and $(C[K], \|\cdot\|)$ are Banach spaces if $\|\cdot\|$ is defined by $\|f\| = \text{l.u.b.}\{|f(z)| \,|\, z \in \gamma\}$ or $\text{l.u.b.}\{|f(z)| \,|\, z \in K\}$, respectively.

DEFINITION Let B_1 and B_2 be two Banach spaces and let k be a linear mapping $B_1 \to B_2$. k is called *bounded* if $\exists M \ni \|kf\| \le M\|f\| \; \forall f \in B_1$.

THEOREM 2.12 *The integral transformation k,*

$$k = \int_\gamma k(z, u) \cdot du$$

is bounded if $k(z, u)$ is continuous in $K \times \gamma$.

PROOF

$$|F(z)| = \left| \int_\gamma k(z, u) f(u) \, du \right| \le m \|f\| L(\gamma)$$

where $m = \max\{|k(z, u)| \,|\, z \in K, u \in \gamma\}$ and $L(\gamma)$ is the length of γ. Set $M = mL(\gamma)$; then it follows that

$$\|kf\| = \text{l.u.b.}\{|F(z)| \,|\, z \in K\} \le M\|f\|$$

So k is bounded.

COROLLARY *The linearity and boundedness of k imply its continuity.*

Indeed, let $\{f_n\}$ be a sequence of functions belonging to $C[\gamma]$ which converge uniformly to f. Set $F_n = kf_n$, $n = 1, 2, \ldots$, and $F = kf$. Then

$$\|F_n - F\| = \|k(f_n - f)\| \le M\|f_n - f\| \le M\varepsilon$$

for $n > N(\varepsilon)$. Hence the sequence $\{F_n\}$ converges uniformly on K to $F = kf$. Hence k is continuous.

DEFINITION Let B_1 and B_2 be two Banach spaces and k a continuous mapping $B_1 \to B_2$. We say that k is *completely continuous* if it transforms bounded sets into precompact sets (see 1.3.5).

THEOREM 2.13 *The mapping $k: C[\gamma] \to C[K]$ is completely continuous.*

PROOF

The proof leans on Arzela's theorem (Theorem 0.21) and consists of two steps.

Let $B[\gamma]$ be a bounded subset of $C[\gamma]$; that is, $f \in B[\gamma] \Rightarrow \|f\| \leq M'$ where M' does not depend on f.

(i) *The family of functions $kB[\gamma] \subset C[K]$ is equibounded*: Indeed, $f \in B[\gamma] \Rightarrow \|kf\| \leq M\|f\| \leq MM' \forall f \in B[\gamma]$,

(ii) $kB[\gamma]$ *is equicontinuous*: Since $k(z, u)$ is uniformly continuous on $K \times \gamma$, we have $|k(z', u) - k(z, u)| < \varepsilon$ for $|z' - z| < \delta(\varepsilon)$, and

$$|F(z') - F(z)| = \left| \int_\gamma (k(z', u) - k(z, u)) f(u)\, du \right|$$

$$\leq \varepsilon \|f\| L(\gamma) \leq \varepsilon M' L(\gamma)$$

$$\forall f \in B[\gamma].$$

Now using Arzela's theorem, it follows that $kB[\gamma]$ is precompact, and so k is completely continous.

THEOREM 2.14 *If the kernel $k(z, u)$ is complex differentiable with respect to z, and if the derivative $k_z(z, u)$ is continuous throughout $D \times \gamma$, then the function $F(z)$ is complex differentiable and its derivative is given by*

$$F'(z) = \int_\gamma k_z(z, u) f(u)\, du$$

that is, you may differentiate under the integral sign.

PROOF

Let $z \in D$ and choose r sufficiently small $\ni d(z, r) \subset D$, and let $|h| < r$. Denote by $h\Omega(z, h)$ the expression

$$h\Omega(z, h) = F(z + h) - F(z) - h \int_\gamma k_z(z, u) f(u)\, du$$

$$= \int_\gamma (k(z + h, u) - k(z, u)) f(u)\, du - h \int_\gamma k_z(z, u) f(u)\, du$$

Let σ be the segment joining z with $z + h$. Then

$$h\Omega(z, h) = \int_\gamma \int_\sigma k_z(\zeta, u) f(u)\, d\zeta\, du - \int_\gamma \int_\sigma k_z(z, u) f(u)\, d\zeta\, du$$

$$= \int_\gamma \int_\sigma (k_z(\zeta, u) - k_z(z, u)) f(u)\, d\zeta\, du$$

It follows that

$$|h||\Omega(z, h)| \le \int_\gamma |\varepsilon h M||du| \le \varepsilon |h| ML(\gamma)$$

where $M = \max\{|f(u)|\,|\,u \in \gamma\}$ and $\varepsilon = \max\{|k_z(\zeta, u) - k_z(z, u)|\,|\,\zeta \in \sigma\}$; that is, $|\Omega(z, h)| < \varepsilon ML(\gamma)$ for $u \in \gamma$.

Since $k_z(z, u)$ is uniformly continuous on the Cartesian product $\overline{d}(z, r) \times \gamma$, it follows that $\lim_{h \to 0} \varepsilon = 0$, and, consequently, that $\lim_{h \to 0} \Omega(z, h) = 0$. This proves that F is complex differentiable, and also shows that its derivative is found by differentiating under the integral sign.

DEFINITION The kernel $\dfrac{1}{u - z}$ is called the *Cauchy kernel*.

The Cauchy kernel and its derivatives of every order are complex differentiable with respect to u and z, provided that $u \ne z$. It follows that if $f \in C[\gamma]$, then $F(z) = \displaystyle\int_\gamma \dfrac{f(u)}{u - z}\, du$ possesses derivatives of every order, provided $z \in \mathbf{C}\setminus\gamma$, and we find:

$$F^{(n)}(z) = n! \int_\gamma \frac{f(u)}{(u - z)^{n+1}}\, du \qquad (m = 1, 2, \ldots)$$

THEOREM 2.15 *Let g be holomorphic in $D\setminus\{c\}$ and continuous in D. Then g is holomorphic in D.*

PROOF

We shall prove that g is holomorphic in a disc $d(c, r) \subset D$. Using the proof of Theorem 2.7, it is sufficient to show that $\int_\Delta f(z)\, dz = 0$ for any triangle with vertices (c, z_1, z_2) where $z_1, z_2 \in d(c, r)$.

Determine z_1' and $z_2' \ni z_1' - c = \dfrac{1}{n}(z_1 - c)$ and $z_2' - c = \dfrac{1}{n}(z_2 - c)$. Then if Δ' is the triangle (c, z_1', z_2'), we have $\int_{\Delta'} f(z)\, dz = \int_\Delta f(z)\, dz$, since obviously $\int_T f(z)\, dz = 0$, where T is the trapezoid with vertices z_1', z_1, z_2, z_2'. Clearly, $\lim_{n \to \infty} \int_{\Delta'} f(z)\, dz = 0$, since f is continuous and the

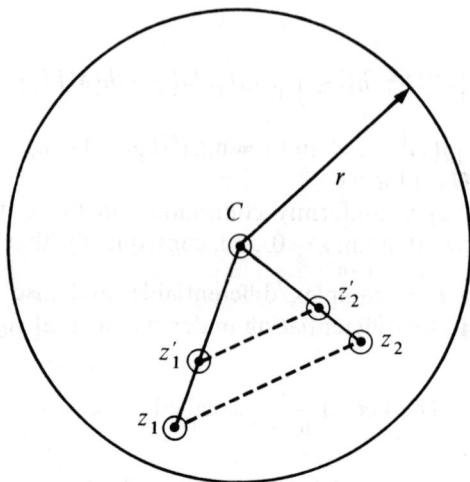

Figure 2.15

perimeter of Δ' converges to 0. It follows that $\int_\Delta f(z)\,dz = 0$. The argument used in the proof of Theorem 3.6 shows that f possesses a primitive in $d(c, r)$. Hence f is holomorphic in $d(c, r)$ and in D.

4. Cauchy's formula

THEOREM 2.16 *Let f be holomorphic in D, $z \in D$, and γ a piecewise smooth closed curve with $I(\gamma, z) = 1$. Then*

$$f(z) = \frac{1}{2\pi i} \int_\gamma \frac{f(u)}{u - z}\,du \qquad \text{(Cauchy's formula)}$$

PROOF

We consider the auxiliary function g:

$$g(u) = \frac{f(u) - f(z)}{u - z} \qquad (z \text{ is fixed, } u \text{ variable})$$

Since $f(u) - f(z)$ and $u - z$ are both complex differentiable in D, the quotient $g(u) = \dfrac{f(u) - f(z)}{u - z}$ is complex differentiable except eventually at $u = z$, where $g(u)$ is not defined. We define $g(z)$ by continuity:

$$g(z) = \lim_{u \to z} \frac{f(u) - f(z)}{u - z} = f'(z)$$

Then g is continuous on D and, by virtue of Theorem 2.15, g is holomorphic on D. It follows that $\int_\gamma g(u)\,du = 0$ or

$$\int_\gamma \frac{f(u)}{u-z}\,du = \int_\gamma \frac{f(z)}{u-z}\,du = f(z)\int_\gamma \frac{du}{u-z} = f(z)I(\gamma, z) \cdot 2\pi i$$

Hence

$$I(\gamma, z)f(z) = \frac{1}{2\pi i}\int_\gamma \frac{f(u)}{u-z}\,du$$

which proves Cauchy's formula for an arbitrary index $I(\gamma, z)$. For $I(\gamma, z) = 1$, we obtain

$$f(z) = \frac{1}{2\pi i}\int_\gamma \frac{f(u)}{u-z}\,du$$

We may replace γ in Cauchy's formula by any other piecewise smooth closed curve homotopic to γ with respect to $D\backslash\{z\}$.

There are two most important corollaries to Cauchy's formula. Keeping in mind Theorem 2.14, we find:

COROLLARY 1 *If f is holomorphic in D, then f possesses derivatives of every order. These derivatives may be found by differentiating under the integral:*

$$f'(z) = \frac{1}{2\pi i}\int_\gamma \frac{f(u)}{(u-z)^2}\,du$$

$$f''(z) = \frac{2}{2\pi i}\int_\gamma \frac{f(u)}{(u-z)^3}\,du$$

$$\vdots$$

$$f^{(n)}(z) = \frac{n!}{2\pi i}\int_\gamma \frac{f(u)}{(u-z)^{n+1}}\,du$$

$$\vdots$$

(Note: $I(\gamma, z) = 1$.)

COROLLARY 2 [*Morera's theorem*] *A holomorphic function is complex differentiable.*

Indeed, if f is holomorphic in D, then f possesses a local primitive F in any open disc $d(z, r) \in D$. Now if $F \in H(d(z, r))$, then F possesses complex derivatives of every order. But $F' = f$. Hence f is complex differentiable throughout D.

Thus, f is holomorphic $\Rightarrow f$ is complex differentiable by Morera's theorem, and f is complex differentiable $\Rightarrow f$ is holomorphic by Cauchy's theorem.

Thus the two notions "holomorphic" and "complex differentiable" are equivalent.

There is still another interesting result of Cauchy's formula: If f is complex differentiable in the domain D, then the values of f in the interior of a disc $d(c, r) \subset D$ are determined by the values of f on the circle $\partial(c, r)$; more generally, if γ is a piecewise smooth closed curve which lies with its interior in D, then the values of f inside γ are determined by the values of f on γ.

THEOREM 2.17 [*Cauchy's formula for the annulus*] Let $f \in H(A)$ where A is the annulus

$$A = \{z \mid r < |z - c| < R\}$$

Let $r < r' < R' < R$, and let Γ' be the circle $\Gamma'(t) = c + R'$ cis $2\pi t$, and γ' the circle $\gamma'(t) = c + r'$ cis $2\pi t$, $0 \leq t \leq 1$. Let A' be the subannulus $A' = \{z \mid r' < |z - c| < R'\}$. Then

$$f(z) = \frac{1}{2\pi i} \int_{\Gamma'} \frac{f(u)}{u - z} \, du - \frac{1}{2\pi i} \int_{\gamma'} \frac{f(u)}{u - z} \, du, \qquad \forall z \in A'$$

PROOF

Clearly, Γ' and γ' are homotopic with respect to A. It follows that if $z \in A'$, and $g(u) = \dfrac{f(u) - f(z)}{u - z}$ with $g(z) = f'(z)$, then $g(u)$ is holomorphic in A, and

$$\int_{\Gamma'} g(u) \, du = \int_{\gamma'} g(u) \, du$$

or

$$\int_{\Gamma'} \frac{f(u)}{u - z} \, du - \int_{\gamma'} \frac{f(u)}{u - z} \, du = f(z) \left[\int_{\Gamma'} \frac{du}{u - z} - \int_{\gamma'} \frac{du}{u - z} \right]$$

Since

$$\int_{\Gamma'} \frac{du}{u - z} = 2\pi i \quad \text{and} \quad \int_{\gamma'} \frac{du}{u - z} = 0$$

we find
$$f(z) = \frac{1}{2\pi i} \int_\Gamma \frac{f(u)}{u-z} du - \frac{1}{2\pi i} \int_{\gamma'} \frac{f(u)}{u-z} du$$
which is Cauchy's formula for the annulus.

REMARK

If f is not complex differentiable on D, but if $\frac{\partial f}{\partial \bar{z}}$ exists and is continuous, then we have a generalized Cauchy formula which we shall give here without proof:
$$f(z) = \frac{1}{2\pi i} \int_\gamma \frac{f(u)}{u-z} du - \frac{1}{\pi} \iint_{\text{int } \gamma} \frac{\partial f}{\partial \bar{z}} \cdot \frac{1}{u-z} dx\, dy$$
where $u = x + iy$, $z \in \text{int } \gamma$, and $I(\gamma, z) = 1$.

In case f is complex differentiable, then $\frac{\partial f}{\partial \bar{z}} = 0$, and the preceding expression reduces to Cauchy's formula.

Problems

2.19.: Compute the value of $f(z) = 2z + 3$ for $z = 0$ by the Cauchy formula, using for γ the circle
$$\gamma(t) = \text{cis } t, \qquad 0 \le t \le 2\pi$$

2.20.: (a) What are the possible values of $\int_\gamma \frac{dz}{z^2 + 1}$ if γ is a p.w.s. closed curve?
$$\left(\text{Hint: } \frac{1}{z^2 + 1} = \frac{A}{z - i} + \frac{B}{z + i}\right)$$
In particular, what are the values if γ is given by:
  (i) $|z - i| = 1$
  (ii) $|z + i| = 1$
  (iii) $|z| = \frac{1}{2}$
  (iv) $|z| = 4$?
(b) What are the possible values of
$$\int_\gamma \frac{(z+1)\, dz}{z^2(z-2)}$$
if γ is a circle $\gamma(t) = a + r \text{ cis } t$, $0 \le t \le 2\pi$?

2.21.: Determine the value of
$$\int_\gamma \left(z + \frac{1}{z}\right) dz$$
if γ is the unit circle $\gamma(t) = \operatorname{cis} t$, $0 \le t \le 2\pi$.

2.22.: Using Cauchy's formula for $f^{(n)}(z)$, determine the value of
$$\int_\partial \frac{P(z)}{(z-c)^k} dz$$
where ∂ is the circle $\partial(t) = c + R \operatorname{cis} \phi$ and where $P(z)$ is a polynomial and k a positive integer.

2.23.: Compute
$$\int_\gamma \frac{z^2 + 3z + 5}{z+1} dz$$
where γ is given by $\gamma(t) = a + R \operatorname{cis} t$, $0 \le t \le 2\pi$, where $R > 0$ and $a \in \mathbf{C}$.

2.24: Compute:
$$\int_\gamma \frac{z}{z^4 - 1} dz$$
where $\gamma(t) = R \operatorname{cis} t$, $0 \le t \le 2\pi$ ($R > 0$).

2.25.: Determine the residues of the functions

(a) $f(z) = \dfrac{2z + i}{z^2 - 3z + 2}$

(b) $f(z) = \dfrac{z^2 + 1}{z^4 + z^2 + 1}$

at their singularities.

2.26.: Compute the integral
$$\int_\gamma \frac{2z - i}{z^2 - 2z + 5} dz$$
in the positive sense around the contour γ of a simply connected domain. Discuss all possibilities that can arise in the case.

---

§4. Compactness of a bounded and closed subset of $H(D)$

The set $H(D)$ of all complex differentiable functions on D is an algebra on the complex field $\mathbf{C}$ (see 1.4.2) and a subalgebra of the set $C(D)$ of all continuous functions on D.

In order to introduce a metric on $C(D)$, we need the concept of a *compact exhaustion* of a domain D.

DEFINITION   We say that the sequence of compact sets $\{K_n\}$ is a *compact exhaustion* of D if

(i) $K_n \subset D$, $n = 1, 2, \ldots$

(ii) $K_n \subset K_{n+1}$, $n = 1, 2, \ldots$

(iii) For any compact subset $K \subset D$, $\exists$ a K_r in the sequence $\ni K \subset K_r$.

THEOREM 2.18   *Every domain $D \subset \mathbf{C}$ possesses a compact exhaustion.*

PROOF

Cover $\mathbf{C}$ by a net S_n of squares with sides of length 2^{-n}, parallel to the x and y axes, and vertices $\dfrac{k + il}{2^n}$, where $k, l \in \mathbf{Z}$. Denote by K_n the union of all closed squares of S_n with $k \leq n2^n$, $l \leq n2^n$, which lie in D. Then $\{K_n\}$ is a compact exhaustion of D:

Clearly, $K_n \subset D$ is a compact set, and $K_n \subset K_{n+1}$. Now let K be any compact subset of D. If $z \in K$, then $\exists$ an open square $\sigma(z)$ centered at z with length of side 2^{-n_z}, $n_z \in \mathbf{P}$, $\ni \sigma(z) \subset D$. For each $z \in K$, $\exists$ open square $\sigma'(z)$ centered at z with the length of side $2^{-n_z - 2}$. These squares cover K, and since K is compact, a finite number of them, say $\sigma'(z_1), \sigma'(z_2), \ldots, \sigma'(z_n)$ cover K. Choose m so that $m \geq |z| \forall z \in K$, and 2^{-m} is less than the length of side of the smallest $\sigma'(z_j)$. Then $K_m \supset K$. Indeed, if $c \in K$, then c is contained in some $\sigma'(z_j)$, $1 \leq j \leq l$, and $\exists$ a closed square of length of side 2^{-m} of the net S_m which covers c; but that square lies in $\sigma(z_j) \subset D$. Hence $c \in K_m$.

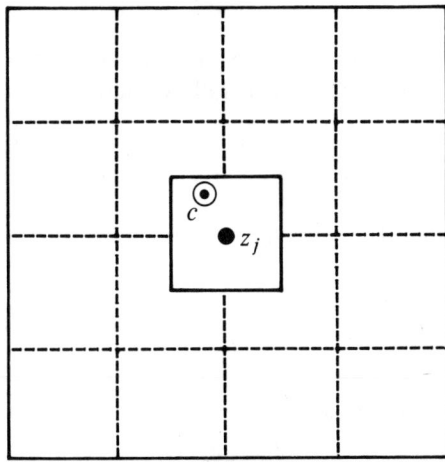

Figure 2.16

Let $\{K_n\}$ be a given compact exhaustion of D. The set $C[K_n]$ of continuous functions on K_n is a Banach-space with $\|f\|_n =$ l.u.b.$\{|f(z)| \,|\, z \in K_n\}$.

Now we define a metric on $C(D)$ by setting

$$d(f, g) = \sum_{n=1}^{\infty} \frac{1}{2^n} \cdot \frac{\|f-g\|_n}{1 + \|f-g\|_n}$$

for any pair $f, g \in C(D)$.

The series for $d(f, g)$ converges, since

$$\frac{\|f-g\|_n}{1 + \|f-g\|_n} < 1$$

and $\sum_{n=i}^{\infty} \frac{1}{2^n}$ converges.

It remains to show that $d(f, g)$ satisfies all conditions of a distance (see. problem 0.17):

(i) Clearly, $d(f, g) \geq 0$ for any pair $f, g \in C(D)$. Furthermore, $f = g$ on $D \Rightarrow \|f-g\|_n = 0 \Rightarrow d(f,g) = 0$, and, conversely, $d(f,g) = 0 \Rightarrow \|f-g\|_n = 0$ for all $n \in \mathbf{P} \Rightarrow f = g$ on all $K_n \Rightarrow f = g$ on D.

(ii) $d(f, g) = d(g, f)$ since $\|f-g\|_n = \|g-f\|_n$ for $n \in \mathbf{P}$.

(iii) $d(f, h) \leq d(f, g) + d(g, h)$ by virtue of $\|f-h\|_n \leq \|f-g\|_n + \|g-h\|_n$ and the inequality

$$\frac{c}{1+c} \leq \frac{a}{1+a} + \frac{b}{1+b}$$

for any three nonnegative numbers, a, b, c with $c \leq a + b$. The student may easily check this inequality. It follows that

$$\frac{\|f-h\|_n}{1 + \|f-h\|_n} \leq \frac{\|f-g\|_n}{1 + \|f-g\|_n} + \frac{\|g-h\|_n}{1 + \|g-h\|_n}$$

and so $d(f, h) \leq d(f, g) + d(g, h)$.

Note that

$$\|f-g\|_{n+p} \geq \|f-g\|_n \quad \text{for} \quad n, p \in \mathbf{P}$$

since $K_{n+p} \supset K_n$.

If $d(f, g) \leq \varepsilon$, then obviously

$$\frac{\|f-g\|_n}{1 + \|f-g\|_n} \leq 2^n \varepsilon$$

It follows that $\|f-g\|_n(1-2^n\varepsilon) \leq 2^n\varepsilon$, and for $\varepsilon < 2^{-n-1}$ we find $\|f-g\|_n \leq 2^{n+1}\varepsilon$.

It follows that if $\{f_r\}$ is a Cauchy sequence in $(C(D), d)$, then $\{f_r\}$ is a Cauchy sequence on K_n. Now $(C[K_n], \|\cdot\|_n)$ is a complete Banach space (see 0.4.5); therefore $\exists$ a function $f \in C[K_n] \ni \|f_r - f\|_n \to 0$, or f_r converges uniformly on K_n to f. Clearly, if $\{f_r\}$ converges uniformly on K_n, then it converges uniformly to f on any subset of K_n. Hence the sequence $\{f_r\}$ converges uniformly to f on each compact $K \subset D$. Indeed, always $\exists$ an $m \ni K_m \supset K$. Furthermore, if the sequence f_r, $r \in \mathbf{P}$, converges uniformly to f on each compact $K \subset D$, then $d(f_r, f) \to 0$. Indeed, if we choose

$$n_0 = n_0(\varepsilon) \ni \sum_{i=n_0+1}^{\infty} \frac{1}{2^i} < \frac{\varepsilon}{2}$$

then

$$\sum_{i=n_0+1}^{\infty} \frac{1}{2^i} \cdot \frac{\|f_r - f\|_i}{1 + \|f_r - f\|_i} < \frac{\varepsilon}{2}$$

Taking $r_0 = r_0(\varepsilon) \ni$ for $i = 1, 2, \ldots, n_0$ we have $\|f_r - f\|_i < \varepsilon/2$, we have, for $r \geq r_0$,

$$d(f_r, f) = \sum_{i=1}^{n_0} \frac{1}{2^i} \frac{\|f_r - f\|_i}{1 + \|f_r - f\|_i} + \sum_{i=n_0+1}^{\infty} \frac{1}{2^i} \frac{\|f_r - f\|_i}{1 + \|f_r - f\|_i}$$

$$\leq \frac{\varepsilon}{2} \sum_{i=1}^{n_0} \frac{1}{2^i} + \frac{\varepsilon}{2} < \varepsilon$$

This means that $(C(D), d)$ is a complete metric space. Now let us study the algebra $H(D) \subset C(D)$.

THEOREM 2.19 *If the sequence $\{f_r\}$, $f_r \in H(D)$, converges to f in $(C(D), d)$, that is, if the sequence $\{f_r\}$, $f_r \in H(D)$, converges uniformly on each compact $K \subset D$ to the function f, then $f \in H(D)$. In other words, $H(D)$ is closed in $C(D)$.*

PROOF

Since $C(D)$ is complete, f is continuous. Let γ be any closed piecewise smooth curve homotopic to a point in D. Then, since f_r is holomorphic, we have

$$\int_\gamma f_r(z)\, dz = 0$$

Since γ is compact, the sequence $\{f_r\}$ converges uniformly on γ to f. It follows that

$$\int_\gamma f(z)\, dz = \int_\gamma \lim_{r\to\infty} f_r(z)\, dz$$

$$= \lim_{r\to\infty} \int_\gamma f_r(z)\, dz = 0$$

Hence $f \in H(D)$ and $H(D)$ is closed.

THEOREM 2.20 *If the sequence $\{f_r\}$, $f_r \in H(D)$, converges uniformly to f on every compact $K \subset D$, then the sequence $\{f_r'\}$ of derivatives also converges uniformly to the derivative f' of f on each compact $K \subset D$. That is, the mapping $H(D) \to H(D)$ which associates with every function $f \in H(D)$ its derivative $f' \in H(D)$ is continuous:*

$$\left(\lim_{n\to\infty} f_n\right)' = \lim_{n\to\infty} f_n'$$

PROOF

Let K be a compact subset of D and $c \in K$. Then, since D is open, $\exists$ an open disc $d(c, r_c) \ni \overline{d}(c, r_c) \subset D$. Denote its boundary by ∂: $\partial(t) = c + r_c \operatorname{cis} t$, $0 \le t \le 2\pi$. Then for any point $z \in d(c, r_c)$, we have

$$f'(z) = \frac{1}{2\pi i} \int_\partial \frac{f(u)}{(u-z)^2}\, du$$

denoted by $f' = k'f$.

The kernel $1/(u-z)^2$ is continuous on $d(c, r) \times \partial$; it follows that k' is a continuous mapping $C[\partial] \to C[\overline{d}(c, r_c/2)]$. Since the sequence $\{f_r\}$ converges uniformly to f on each compact $K \subset D$, it converges uniformly on ∂ to f. Hence, by Theorem 2.12 and its corollary, $f_r' = k'f_r$ converges uniformly on $\overline{d}(c, r_c/2)$ to f'. The open discs $d(c, r_c/2)$ with $c \in K$ cover K, and since K is compact a finite number of them cover K, say $d_1, d_2, \ldots, d_s$. Clearly the corresponding closed discs $\overline{d}_1, \overline{d}_2, \ldots, \overline{d}_s$ cover K. Since the functions $f_r' = k'f_r$, $r \in \mathbf{P}$, converge uniformly on each of these discs, they converge uniformly on the union $\bigcup_{i=1}^{s} \overline{d}_i$, and hence on K.

DEFINITION We say that a family $B(D) \subset C(D)$ is *bounded* if, for each compact $K \subset D$, $\exists$ a $M(K) \ni \|f\|_K \le M(K)$ for all $f \in B(D)$. We denote by $B(K)$ the set of the restrictions of f to K, for $f \in B(D)$.

THEOREM 2.21 *If the family $B(D) \subset H(D)$ is bounded, then it is precompact in $(C(D), d)$. If, moreover, $B(D)$ is closed, then it is compact.*

PROOF

Let K be a compact subset of D and $c \in K$. Then $\exists$ an open disc $d(c, r_c) \subset D$ with boundary ∂, and for any $z \in d(c, r_c)$, we have:

$$f(z) = \frac{1}{2\pi i} \int_\partial \frac{f(u)}{u - z} du$$

The kernel $1/(u - z)$ is continuous on $d(c, r_c) \times \partial$. It follows by Theorem 2.13 that the family of functions $B(D)$ is equicontinuous on $\bar{d}(c, r_c/2)$. Since K is compact, a finite number of the discs $\bar{d}(c, r_c/2)$, $c \in K$, cover K, say $\bar{d}_1, \bar{d}_2, \ldots, \bar{d}_s$. Obviously, equicontinuity on a finite number of compact sets implies equicontinuity on their union. It follows that $B(K)$ is equicontinuous and equibounded on K and, by Arzela's theorem (Theorem 0.21), $B(K)$ is precompact. If, moreover, $B(K)$ is closed, then $B(K)$ is compact. Note that if $B(D)$ is closed, so is $B(K)$.

Now consider the compact exhaustion $\{K_n\}$ of D and let $B(D)$ be closed and $S \subset B(D)$ be a countably infinite subset of $B(D)$. We wish to show that it contains a subsequence which is uniformly convergent on each of the K_n, $n \in \mathbf{P}$. Since $B(K_1)$ is compact, S contains a subsequence $S_1 = \{f_{1,m}\}$, $m \in \mathbf{P}$, which is uniformly convergent on K_1. Since $B(K_2)$ is compact, a subsequence $S_2 = \{f_{2,m}\}$ of S_1 converges uniformly on K_2, and, since $B(K_r)$ is compact, a subsequence $S_r = \{f_{r,m}\}$ of S_{r-1} converges uniformly on K_r.

The sequence $\{f_{r,r}\}$, $r \in \mathbf{P}$, is a subsequence of S and converges uniformly on all K_n. Indeed, for $r > n$ the elements $f_{r,r} \in K_n$. It follows that $B(D)$ is compact.

Problem

2.27.: Let f be a continuous function on $[0, 1]$ and define $F(z)$ by

$$F(z) = \int_0^1 \frac{f(u)}{u - z} du$$

Prove that F is holomorphic in $\mathbf{C}\backslash[0, 1]$ and that $\lim_{z \to \infty} zF(z)$ is finite.

§5. Harmonic functions

DEFINITION   A real or complex valued function g is said to be *harmonic* on the domain D if it possesses continuous second partial derivatives $\dfrac{\partial^2 g}{\partial x^2}$ and $\dfrac{\partial^2 g}{\partial y^2}$ on D which satisfy

$$\nabla^2 g = \frac{\partial^2 g}{\partial x^2} + \frac{\partial^2 g}{\partial y^2} = 0 \qquad (\nabla^2 = \Delta)$$

∇^2 is called the *Laplacian differential operator*. Obviously, if $g(x, y) = U(x, y) + iV(x, y)$ is harmonic, then so are U and V, and conversely (U and V are real-valued). More generally, since ∇^2 is a linear operator, the set of functions which are harmonic on D form a vector space over **C**.

The student may easily check that the Laplacian operator may be expressed by:

$$\nabla^2 = 4 \frac{\partial}{\partial z}\left(\frac{\partial}{\partial \bar{z}}\right)$$

(see problem 1.39).

THEOREM 2.22 f is holomorphic on $D \Rightarrow f$ is harmonic on D.

PROOF

We have shown (see 1.4.3) that if $f \in H(D)$, then $\dfrac{\partial f}{\partial \bar{z}} = 0$. It follows that $\nabla^2 f = 0$.

COROLLARY   If $f = U + iV$ is holomorphic on D, then $\operatorname{Re} f = U$ and $\operatorname{Im} f = V$ are harmonic on D.

Note that the converse is not true.

EXAMPLE

x and $-y$ are harmonic functions on **C**, but $x + i(-y) = \bar{z}$ is not holomorphic, since

$$\frac{\partial \bar{z}}{\partial \bar{z}} = 1 \neq 0$$

Because of the great importance of the corollary to Theorem 2.22 for applications, we shall give a second proof of it:

Let f be holomorphic. Then its derivatives of any order, $f, f', f'', \ldots$, exist. (See the first corollary to Theorem 2.16.)

$$f' = U_x + iV_x = \frac{1}{i}(U_y + iV_y)$$

and

$$f'' = U_{xx} + iV_{xx} = \left(\frac{1}{i}\right)^2(U_{yy} + iV_{yy}) = -(U_{yy} + iV_{yy})$$

Therefore,

$$U_{xx} + U_{yy} = \nabla^2 U = 0 \quad \text{and} \quad V_{xx} + V_{yy} = \nabla^2 V = 0$$

The second derivatives are continuous since all higher derivatives of f exist.

If two harmonic functions U and V satisfy the Cauchy-Riemann equations

$$U_x = V_y \quad \text{and} \quad U_y = -V_x$$

then we say that V is a *harmonic conjugate* of U. Obviously, if V is a harmonic conjugate of U, then so is $V + c$, where c is a real constant.
It follows that U is a harmonic conjugate of $-V$.

EXAMPLE

$\log z = \log|z| + i \arg z$ is holomorphic in $\mathbf{C} \backslash \mathbf{R}_0^-$. It follows that $\log|z|$ and $\arg z$ are harmonic in $\mathbf{C} \backslash \mathbf{R}_0^-$. $\log|z|$ is harmonic even in $\mathbf{C} \backslash \{0\}$, and $\arg z$ is a harmonic conjugate of $\log|z|$.

THEOREM 2.23 *With every function U which is harmonic in the disc $d(c, r)$, we can associate a conjugate harmonic function $V \ni f = U + iV$ is holomorphic in $d(c, r)$.*

PROOF

Indeed, for $z = x + iy \in d(c, r)$, let σ be the segment joining c to z and define V by:

$$V(x, y) = \int_\sigma \left(-\frac{\partial U}{\partial y} d\xi + \frac{\partial U}{\partial x} d\eta\right) = \int_c^z (-U_y \, d\xi + U_x \, d\eta)$$

We want to show that $V_y = U_x$. Set $c = a + ib$. Since U is harmonic we have, by Green's theorem,

$$\int_c^{x+bi} \cdots + \int_{x+ib}^{x+iy} \cdots + \int_{x+iy}^{c} (-U_y \, d\xi + U_x \, d\eta) = \iint_\Delta (U_{xx} + U_{yy}) \, d\xi \, d\eta = 0$$

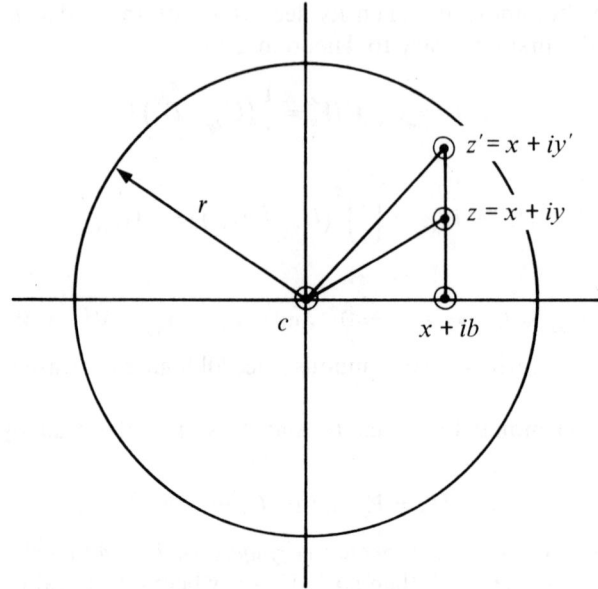

Figure 2.17

where Δ denotes the triangle with vertices c, $x + bi$, $x + iy$ over whose sides the left-hand integrals are taken. Hence, it follows that

$$\int_c^{x+iy} (-U_y\, d\xi + U_x\, d\eta) = \int_c^{x+ib} \cdots + \int_{x+ib}^{x+iy} (-U_y\, d\xi + U_x\, d\eta)$$

Now consider:

$$\left| \frac{1}{y' - y}(V(x, y') - V(x, y)) - U_x(x, y) \right|$$

$$= \left| \frac{1}{y' - y}\left(\int_c^{x+iy'} \cdots - \int_c^{x+iy} (-U_y\, d\xi + U_x\, d\eta) \right) - U_x(x, y) \right|$$

$$= \left| \frac{1}{y' - y}\left(\int_c^{x+ib} \cdots + \int_{x+ib}^{x+iy'} \cdots - \int_c^{x+ib} \cdots \right. \right.$$

$$\left. \left. - \int_{x+ib}^{x+iy} (-U_y\, d\xi + U_x\, d\eta) \right) - U_x(x, y) \right|$$

$$= \left| \frac{1}{y' - y}\int_{x+iy}^{x+iy'} (-U_y\, d\xi + U_x\, d\eta) - U_x(x, y) \right|$$

and this, because $\int_{x+iy}^{x+iy'} -U_y\, d\xi = 0$, is equal to

$$\left| \frac{1}{y'-y} \left(\int_{x+iy}^{x+iy'} U_x(x, \eta)\, d\eta - \int_{x+iy}^{x+iy'} U_x(x, y)\, d\eta \right) \right|$$

$$\leq \left| \frac{1}{y'-y} \right| \max_\eta |U_x(x, \eta) - U_x(x, y)|\, |y'-y|$$

$$= \max_\eta |U_x(x, \eta) - U_x(x, y)|$$

Since $\lim_{y' \to y} \max_\eta |U_x(x, \eta) - U_x(x, y)| = 0$ by continuity of U_x, it follows that $\lim_{y' \to y} 1/(y'-y)(V(x, y') - V(x, y)) = U_x(x, y)$, or $V_y = U_x$.

In the same way, by using the triangle with vertices c, $x + iy$, $a + iy$, we can show that $V_x = -U_y$. The relation $V_{xx} + V_{yy} = -U_{yx} + U_{xy} = 0$ indicates that V is a conjugate harmonic function of U.

Note that two conjugate harmonic functions of a harmonic function U differ by a constant, since $F_x = 0$ and $F_y = 0$ imply F is a constant.

Theorem 2.23 can easily be generalized for simply connected domains.

COROLLARY *Every real-valued harmonic function U is the real part of a holomorphic function f.*

§6. The mean-value property and maximum modulus principle

DEFINITION Let $f(z) = f(x, y)$ be a real or complex valued function on the domain D. We say that f has the mean-value property on D if, for any closed disc $\bar{d}(c, r) \subset D$, the value of f at c is the mean of the values of f taken over the boundary $\partial(c, r)$ of $\bar{d}(c, r)$:

$$f(c) = \frac{1}{2\pi} \int_0^{2\pi} f(c + r \operatorname{cis} \phi)\, d\phi$$

It is easy to show that the set of functions which have the mean-value property on D is a complex vector space. Moreover, if the complex-valued function $f = U + iV$ has the mean-value property, then the real-valued functions U and V have it too, and conversely.

THEOREM 2.24 *If $f \in H(D)$, then f possesses the mean-value property on D.*

PROOF

Let $\overline{d}(c, r) \subset D$. Then, using Cauchy's formula,

$$f(c) = \frac{1}{2\pi i} \int_\partial \frac{f(u)}{u - c} \, du$$

where $\partial = \partial(c, r)$ is the boundary of $\overline{d}(c, r)$. Take $u = c + r \operatorname{cis} t$; then

$$f(c) = \frac{1}{2\pi i} \int_0^{2\pi} \frac{f(c + r \operatorname{cis} t)}{r \operatorname{cis} t} \, ir \operatorname{cis} t \, dt = \frac{1}{2\pi} \int_0^{2\pi} f(c + r \operatorname{cis} t) \, dt$$

It follows that every real-valued harmonic function has the mean-value property, since any real-valued harmonic function is the real part of some holomorphic function (see Theorem 2.23). The converse—namely, that a function which possesses the mean-value property on D is harmonic on D—is true, but we shall not prove it here.

LEMMA 2.25 *Let f be a continuous real or complex valued function which has the mean-value property on D, and suppose $|f|$ has a relative maximum at $c \in D$—that is, $\exists\, r > 0 \ni |f(z)| \leq |f(c)|$ for all $z \in \overline{d}(c, r) \subset D$. Then f is a constant on $\overline{d}(c, r)$.*

PROOF

We shall distinguish three cases to prove the lemma.
(a) $f(c) = 0$. Then obviously $f(z) = 0$ for all $z \in \overline{d}(c, r) \subset D$.
(b) Let $f(c)$ be positive. Then take an auxiliary real-valued function h defined by $h(z) = f(c) - \operatorname{Re} f(z)$. Now the constant $f(c)$ has the mean-value property, and so does $\operatorname{Re} f$. Hence h has the mean value property on D. It follows that

$$h(c) = \frac{1}{2\pi} \int_0^{2\pi} h(c + \rho \operatorname{cis} \phi) \, d\phi \qquad \text{for any } \rho \leq r$$

On the other hand, it is easy to see that $h(z) \geq 0$, $z \in \overline{d}(c, r)$. If $h(z) > 0$ at a point on the circle $\partial(c, \rho)$, it would follow that $h(c) > 0$ (since h is continuous on D), in contradiction with $h(c) = f(c) - \operatorname{Re} f(c) = 0$. Hence $h(c + \rho \operatorname{cis} \phi) = 0 \ \forall$ values of $\rho < r$ and $0 \leq \phi < 2\pi$. Thus, $\operatorname{Re} f(z) = f(z)$ $\forall z \in d(c, r) \Rightarrow f(z)$ is constant on $\overline{d}(c, r)$.

(c) $f(c)$ is an arbitrary complex number $f(c) = R \operatorname{cis} \phi$. Then we introduce the auxiliary function g defined by $g(z) = \operatorname{cis}(-\phi)f(z)$, which permits us to reduce case (c) to case (b).

Indeed, g has the mean-value property on D and has a relative maximum at c. Since $g(c) = R > 0$, the function g is a constant in $\bar{d}(c, r)$. It follows that f is a constant in $\bar{d}(c, r)$.

THEOREM 2.25 [*The maximum principle*] *Let f be a real or complex valued function which is continuous on the compact $\bar{D}$, and let f have the mean-value property on the domain D.*

Let $M = \max\{|f(z)| \,|\, z \in \bar{D}\}$. If at some point $c \in D$ $|f(c)| = M$, then $f(z)$ is a constant on $\bar{D}$.

PROOF

Denote by S the set $S = \{z \in D \,|\, f(z) = f(c)\}$. S is not empty, since $c \in S$. S is open in the relative topology of D, since f has a relative maximum at z if $z \in S$, and, by virtue of Lemma 2.25, is therefore constant in some disc $d(z, \rho) \subset D$, which then belongs to S.

On the other hand, S is closed in the relative topology of D. Indeed, if $z_1, z_2, \ldots, z_n, \ldots$ is a sequence of points in S, converging to $z_0 \in D$, then $f(z_0) = \lim_{n \to \infty} f(z_n) = f(c)$ because of the continuity of f. It follows by the connectedness argument that $S = D$. But f is continuous on $\bar{D}$, so that $f(z) = f(c)\ \forall\ z \in \bar{D}$.

COROLLARY *If f satisfies the conditions of Theorem 2.25 and is not a constant on $\bar{D}$, then the maximum of the modulus of f is attained on the boundary of D.*

Problems

2.28.: Let U and V be conjugate harmonic functions and

$$U^2(x, y) + V^2(x, y) = f(x)$$

Show that $f(x) = e^{\alpha x + \beta}$ with $\alpha, \beta \in \mathbf{R}$, and determine $U(x, y)$ and $V(x, y)$. (*Hint*: Let $\phi(z) = U + iV$; then use $U^2 + V^2 = |\phi|^2$ and the fact that $\log |\phi|$ and $\arg \phi$ are conjugate harmonic functions.)

2.29.: Let U and V be conjugate harmonic functions and

$$\frac{V(x, y)}{U(x, y)} = f(x)$$

Show that $f(x) = \tan(\alpha x + \beta)$ with $\alpha, \beta \in \mathbf{R}$, and determine $U(x, y)$ and $V(x, y)$.

2.30.: Show that if f and g are two harmonic conjugate functions, then so are—
  (a) $F = f^2 - g^2$ and $G = 2fg$
  (b) $\phi = e^f \cos g$ and $\psi = e^f \sin g$.

2.31.: Find the conjugate harmonic function of—

  (a) $U(x, y) = \dfrac{x}{x^2 + y^2}$

  (b) $U(x, y) = \sinh x \sin y$

2.32.: Let γ_0 and γ_1 be two p.w.s. closed curves whose equations in polar coordinates (r, ϕ) are given by

$$\gamma_0 : r_0 = r_0(\phi)$$

and

$$\gamma_1 : r_1 = r_1(\phi)$$

where r_0 and r_1 are two single-valued functions $[0, 2\pi] \to \mathbf{R}^+$, with $r_0(0) = r_0(2\pi)$ and $r_1(0) = r_1(2\pi)$. Show that γ_0 and γ_1 are homotopic with respect to any domain which is starlike with respect to the origin and which contains both γ_0 and γ_1. (*Hint*: $\gamma_0(\phi) = r_0(\phi)\text{cis } \phi$, $\gamma_1(\phi) = r_1(\phi)\text{cis } \phi$, $0 \le \phi \le 2\pi$; set $\gamma(t, u) = \gamma_0(t)u + \gamma_1(t)(1 - u)$.) Compare with problem 2.3.

chapter 3

Analytic Functions

We say that a function f is *analytic* in the domain D if we can associate with each point $c \in D$ a power series $\sum_{n=0}^{\infty} a_n(z-c)^n$ which converges in an open disc $d(c, r)$ with $r > 0$, and which represents there the function f, that is,

$$f(z) = \sum_{n=0}^{\infty} a_n(z-c)^n$$

$\forall\, z \in d(c, r)$.

The German mathematician Weierstrass was the first to study functions from this point of view and define analytic functions locally by power series.

We shall show in this chapter that the notions "holomorphicity in D" and "analyticity in D" are equivalent.

§1. Power series

1. Numerical series

We assume that the reader is familiar with the basic elements of the theory of infinite series of real numbers.

The series $\sum_{n=0}^{\infty} c_n$, where the c_n are arbitrary complex numbers, is said to be *convergent* if $\lim_{k \to \infty} \sum_{n=0}^{k} c_n$ exists. The limit of the partial sums $\sum_{n=0}^{k} c_n$ for $k \to \infty$, if it exists, is called the sum of the series and is denoted by $\sum_{n=0}^{\infty} c_n$.

EXAMPLE

Let us consider the series

$$\sum_{n=0}^{\infty} \frac{i}{(n+1)(n+2)} = \frac{i}{1.2} + \frac{i}{2.3} + \frac{i}{3.4} + \cdots.$$

The partial sums

$$\sum_{n=0}^{k} \frac{i}{(n+1)(n+2)}$$

$$= \frac{i}{1.2} + \frac{i}{2.3} + \frac{i}{3.4} + \cdots + \frac{i}{(k+1)(k+2)}$$

$$= i\left(\frac{1}{1} - \frac{1}{2}\right) + i\left(\frac{1}{2} - \frac{1}{3}\right) + \cdots + i\left(\frac{1}{k+1} - \frac{1}{k+2}\right) = i\left(\frac{1}{1} - \frac{1}{k+2}\right)$$

has for $k \to \infty$ the limit i. Thus

$$\sum_{n=0}^{\infty} \frac{i}{(n+1)(n+2)}$$

is a convergent series with the sum equal to i.

We say that the series $\sum_{n=0}^{\infty} c_n$ *converges absolutely* if $\sum_{n=0}^{\infty} |c_n|$ converges.

EXAMPLES

1 Consider the series $1 - \frac{i}{2} + \frac{i^2}{2^2} - \frac{i^3}{2^3} + \cdots$ with $c_n = (-i)^n \frac{1}{2^n}$. This series converges and its sum is equal to $\frac{2}{5}(2-i)$. Indeed,

$$1 - \frac{i}{2} + \cdots + \left(-\frac{i}{2}\right)^k = \frac{1 - \left(-\frac{i}{2}\right)^{k+1}}{1 + \frac{i}{2}}$$

and

$$\lim_{k \to \infty} \frac{1 - \left(-\frac{i}{2}\right)^{k+1}}{1 + \frac{i}{2}} = \frac{1}{1 + \frac{i}{2}} = \frac{2}{2+i} = \frac{2}{5}(2-i)$$

The series converges absolutely because

$$1 + \frac{1}{2} + \frac{1}{2^2} + \cdots = \frac{1}{1 - \frac{1}{2}} = 2$$

converges also.

2 The series $1 - \frac{1}{2} + \frac{1}{3} \cdots$ converges as an alternating series with decreasing terms, but it does not converge absolutely because $1 + \frac{1}{2} + \frac{1}{3} + \cdots$ does not converge.

CAUCHY'S CRITERION *The series $\sum_{n=0}^{\infty} c_n$ converges iff $\forall \, \varepsilon > 0 \; \exists \, N(\varepsilon) \ni$*

$$\left| \sum_{i=1}^{p} c_{n+i} \right| < \varepsilon \text{ for } n > N(\varepsilon) \text{ and } \forall \text{ integer } p > 0.$$

For $p = 1$, we find $|c_n| < \varepsilon$ for $n > N(\varepsilon)$, which implies that a necessary condition for convergence of a series is that the general term c_n converges to zero: $\lim_{n \to \infty} c_n = 0$. This condition is not sufficient, since the series $\sum_{n=1}^{\infty} (1/n)$ is not convergent, although $\lim_{n \to \infty} c_n = \lim_{n \to \infty} (1/n) = 0$.

The proof of Cauchy's criterion is the same as in the real case. Also as in the real case, we can show that absolute convergence implies ordinary convergence with the aid of the inequality

$$\left| \sum_{i=1}^{p} c_{n+i} \right| \leq \sum_{i=1}^{p} |c_{n+i}|$$

If a series does not converge, we say it is *divergent* or it *diverges*.

EXAMPLE

The series $\sum_{n} (1/n)$ is divergent.

DEFINITION Let $\{a_n\}$ and $\{b_n\}$ be two sequences. Then we call the sequence $\{c_n\}$, with

$$c_n = \sum_{i=0}^{n} a_i b_{n-i}$$

the *convolution* of the two sequences $\{a_n\}$ and $\{b_n\}$ and denote it by $\{c_n\} = \{a_n\} * \{b_n\}$.

THEOREM 3.1   If the two series $\sum_{n=0}^{\infty} a_n$ and $\sum_{n=0}^{\infty} b_n$ are absolutely convergent, then (i) the series $\{c_n\} = \{a_n\} * \{b_n\}$ is absolutely convergent, and (ii) $\sum_{n=0}^{\infty} c_n = \sum_{n=0}^{\infty} a_n \cdot \sum_{n=0}^{\infty} b_n$.

PROOF

(i) Let $A_{|n} = \sum_{i=0}^{n} |a_i|$, $B_{|n} = \sum_{i=0}^{n} |b_i|$, and $C_{|n} = \sum_{i=0}^{n} |c_i|$. Then it is easy to check that

$$C_{|n} \leq A_{|n} \cdot B_{|n}$$

since the product $A_{|n} \cdot B_{|n}$ contains all the terms of $C_{|n}$.

Obviously, $A_{|n} \leq \sum_{i=0}^{\infty} |a_i|$ and $B_{|n} \leq \sum_{i=0}^{\infty} |b_i|$; hence $C_{|n}$ is monotonically increasing and bounded, and therefore convergent.

(ii) Let $A_n = \sum_{i=0}^{n} a_i$, $B_n = \sum_{i=0}^{n} b_i$, and $C_n = \sum_{i=0}^{n} c_i$. Then $\{C_n\}$ is convergent, since the absolute convergence of a series implies its convergence. On the other hand, we can check that

$$|A_{2n} B_{2n} - C_{2n}| \leq |A_{|2n} \cdot B_{|2n} - A_{|n} \cdot B_{|n}|$$

and, since for sufficiently great n the right-hand term is $\leq \varepsilon$ (Cauchy criterion), it follows that

$$\lim_{n \to \infty} A_{2n} B_{2n} = \lim_{n \to \infty} C_{2n}$$

or

$$\sum_{i=0}^{\infty} a_i \cdot \sum_{i=0}^{\infty} b_i = \sum_{i=0}^{\infty} c_i$$

2. Series of functions and uniform convergence

DEFINITION   The series of functions $\sum_{n=0}^{\infty} f_n$ converges uniformly on the compact set K to the function $f(z)$ if $\forall \varepsilon > 0 \; \exists \; N(\varepsilon) \ni$

$$\left| \sum_{v=0}^{n} f_v(z) - f(z) \right| < \varepsilon$$

for all $n > N(\varepsilon)$ and $z \in K$.

EXAMPLE

The series $1 + z + z^2 + \cdots$ converges uniformly to the function $\dfrac{1}{1-z}$ on every closed disc $\bar{d}(0, r)$ with $0 < r < 1$:

$$\left| \sum_{v=0}^{n} f_v(z) - f(z) \right| = \left| \frac{1 - z^{n+1}}{1 - z} - \frac{1}{1 - z} \right|$$

$$= \left| \frac{z^{n+1}}{1 - z} \right| \leq \frac{r^{n+1}}{1 - r} \quad \text{for } |z| \leq r$$

The expression $r^{n+1}/(1 - r)$ is independent of z and converges to zero with $n \to \infty$, so that

$$\frac{r^{n+1}}{1 - r} < \varepsilon$$

for a sufficiently large n.

WEIERSTRASS' CRITERION OF THE NUMERICAL MAJORANT *If $\forall z \in K$ we have $|f_n(z)| \leq a_n$, $n = 0, 1, 2, \ldots$, then the convergence of the series $\sum_{n=0}^{\infty} a_n$ implies the uniform convergence of $\sum_n f_n$ on K.*

The proof depends on using Cauchy's criterion:

$$\left| \sum_{i=1}^{p} f_{n+i}(z) \right| \leq \sum_{i=1}^{p} |f_{n+i}(z)| \leq \sum_{i=1}^{p} a_{n+i}$$

and $\sum_{i=1}^{p} a_{n+i} < \varepsilon$ for n sufficiently large $(n > N(\varepsilon))$.

In the above example $1 + z + z^2 + \cdots$, we have in fact used Weierstrass' criterion by comparing the series $1 + z + z^2 + \cdots$ with the numerical majorant $1 + r + r^2 + \cdots$. Obviously, Weierstrass' criterion remains true if we have $|f_n(z)| \leq a_n \ \forall n$ with exception of a finite number of terms.

The set of points where $\sum f_n(z)$ converges is called its *domain of convergence*, even if it is neither open nor connected.

3. Power series

A *power series with center c* has the form

$$a_0 + a_1(z - c) + a_2(z - c)^2 + \cdots,$$

where c and the coefficients $a_0, a_1, a_2, \ldots$ are given complex numbers.

EXAMPLE

In the power series $1 + 2z + 3z^2 + \cdots$, we have $c = 0$ and $a_n = n + 1$.

For the domain of convergence of the power series $a_0 + a_1(z - c) + a_2(z - c)^2 + \cdots$ there are three possibilities:

1. The series converges only for $z = c$.
2. The series converges in the entire complex plane.
3. $\exists$ a disc $d(c, R) \ni$ the series converges inside the disc and diverges outside the disc. R is then called the *radius of convergence*.

1 and 2 are special cases of 3 with $R = 0$ and $R = \infty$, respectively.

THEOREM 3.2 [*Fundamental Theorem Concerning Power Series*] *With each power series $\sum_n a_n(z - c)^n$ we can associate a real number R, $0 \leq R \leq \infty$, called the radius of convergence, which has the following properties:*

1. *The series converges absolutely in the open disc $d(c, R)$, called the disc of convergence. Moreover, the series converges uniformly in every closed disc $\overline{d}(c, r)$ with $r < R$; that is, the series converges uniformly on every compact $K \subset d(c, R)$.*

2. *The series is divergent outside the closed disc $\overline{d}(c, R)$.*

3. *The sum of the series is a function f which is holomorphic in the open disc $d(c, R)$.*

4. *The series $\sum_{n=1}^{\infty} na_n(z - c)^{n-1}$, which we get by differentiating the terms of the original series, has the same disc of convergence $d(c, R)$, and its sum is $f'(z)$.*

Let us call the circle $|z - c| = R$ the circle of convergence. Then inside this circle the series converges, outside it diverges. At points of the circle of convergence the series may be convergent or divergent.

For the proof of Theorem 3.2, the concept of $\overline{\lim}$ is needed.

DEFINITION Let $\{r_n\}$ be a sequence of real numbers. If the sequence is not bounded above, we say that $\overline{\lim} \, r_n = \infty$.

If the sequence is bounded above, let S be the set of real numbers which are surpassed by at most a finite number of elements of the sequence. Then we define

$$\overline{\lim} \, r_n = \text{g.l.b. } S$$

EXAMPLE

$$r_n = (-1)^n + \frac{1}{n}, \qquad n = 1, 2, \ldots$$

We see immediately that $\overline{\lim}\, r_n = 1$.

PROOF OF THEOREM 3.2

1: Let R be $= \dfrac{1}{\overline{\lim}\, \sqrt[n]{|a_n|}}$. We want to show that R is the radius of convergence. If $R = 0$—that is, if $\overline{\lim}\, \sqrt[n]{|a_n|} = \infty$—then, no matter how small we take $r = |z - c|$, the expression $\sqrt[n]{|a_n|}\, r$ will be greater than an arbitrary number g for an infinite number of terms. That is, $|a_n|r^n > g^n$. Choosing $g > 1$, we see immediately that the sequence $|a_n||z - c|^n$ does not converge to zero for $n \to \infty$, and neither does the sequence $a_n(z - c)^n$. Hence, the series $\sum_{n=0}^{\infty} a_n(z - c)^n$ diverges for $|z - c| \neq 0$ and converges to a_0 only for $z = c$.

If $R > 0$—that is, if $\overline{\lim}\, \sqrt[n]{|a_n|} = 1/R$—then for any r, $0 < r < R$, we can choose $r_o \ni 0 < r < r_o < R$, and we have for sufficiently large n, say $n > N$,

$$\sqrt[n]{|a_n|} < \frac{1}{r_o} \qquad \left(\frac{1}{r_o} > \frac{1}{R}\right)$$

or $|a_n|r_o^n < 1$, so that

$$|a_n|r^n = |a_n|r_o^n \left(\frac{r}{r_o}\right)^n < \left(\frac{r}{r_o}\right)^n$$

Thus, for $|z - c| \leq r$, the terms of the series $\sum_{n=0}^{\infty} a_n(z - c)^n$, with the exception of a finite number of them, are smaller in modulus than the corresponding terms of the convergent series $\sum_{n=0}^{\infty} (r/r_o)^n$. It follows that the series $\sum_{n=0}^{\infty} a_n(z - c)^n$ converges absolutely inside the disc $d(c, R)$. Using the Weierstrass criterion, we find that the series $\sum_{n=0}^{\infty} a_n(z - c)^n$ converges uniformly on every closed disc $\overline{d}(c, r)$, $r < R$. The same series

$\sum_{n=0}^{\infty} a_n(z-c)^n$ converges uniformly on any compact $K \subset d(c, R)$, since K lies in a closed disc $\bar{d}(c, r)$ with $r < R$ conveniently chosen.

If $R = \infty$—that is, $\overline{\lim} \sqrt[n]{|a_n|} = 0$—then, however large we take $r = |z - c|$, we always have $\lim_{n \to \infty} \sqrt[n]{|a_n|} r = 0$, or $|a_n| r^n < \varepsilon^n$ for n sufficiently large and any $0 < \varepsilon < 1$. Therefore, $\sum_{n=0}^{\infty} \varepsilon^n$ is a convergent majorant for the series $\sum_{n=0}^{\infty} a_n(z-c)^n$. Consequently, $\sum_{n=0}^{\infty} a_n(z-c)^n$ converges absolutely $\forall$ values of z, and uniformly on any compact subset K of the complex plane $\mathbf{C}$, and defines an entire function (see 1.4.2), as we shall see in 3.

2: Suppose $|z - c| = r > R$. Then for an infinite number of terms of the series we have

$$\sqrt[n]{|a_n|} > \frac{1}{r} \quad \text{or} \quad |a_n||z - c|^n = |a_n| r^n > 1$$

Hence $\sum_{n=0}^{\infty} a_n(z-c)^n$ diverges because in a convergent series the general term must tend to zero; that is, $\lim_{n \to \infty} a_n(z-c)^n$ must be zero, by Cauchy's criterion.

3: The partial sums of the series are polynomials in z, and polynomials are entire functions; that is, holomorphic in the entire complex plane. We saw in the first part of this proof that these polynomials converge uniformly on each closed disc $\bar{d}(c, r)$ with radius $r < R$, so that $f(z) = \sum_{n=0}^{\infty} a_n(z-c)^n$ is also holomorphic on the same disc (see Theorem 2.19); that is, f is holomorphic in $d(c, R)$.

This also implies that a function which is analytic in D is locally holomorphic in D (see 2.2.1) and hence, by Lemma 2.8, holomorphic in D. In particular, if $R = \infty$, the series represents an entire function.

4: We may differentiate each term of the series $\sum_{n=0}^{\infty} a_n(z-c)^n = f(z)$ in order to get the derivative f' of the function f by virtue of Theorem 2.20. The radius of convergence R' of the differentiated series $\sum_{n=1}^{\infty} n a_n(z-c)^{n-1}$ is certainly not greater than R because $n|a_n| \geq |a_n|$, $n > 1$. Consequently, $\overline{\lim} \sqrt[n]{|n a_n|} \geq \overline{\lim} \sqrt[n]{|a_n|}$. This means $R' \leq R$. On the other hand, the radius $R' \geq R$ by Theorem 2.20. Hence $R' = R$.

EXAMPLES

1 For $\sum_{n=1}^{\infty} n^n z^n$, we have

$$R = \frac{1}{\overline{\lim} \sqrt[n]{n^n}} = \frac{1}{\overline{\lim} n} = 0$$

The series converges only for $z = 0$. No function is defined in this case.

2 We define e^z by the series

$$e^z = 1 + \frac{z}{1!} + \frac{z^2}{2!} + \cdots = \sum_{n=0}^{\infty} \frac{z^n}{n!}$$

We can easily show that the radius of convergence of this series is $R = \infty$. Indeed:

If a is an arbitrary positive integer and $n > a$, then $n! \geq a! \, a^{n-a}$; hence

$$\sqrt[n]{n!} \geq \sqrt[n]{\frac{a!}{a^a} a^n} = a \sqrt[n]{\frac{a!}{a^a}}$$

Now $\lim_{n} \sqrt[n]{c} = 1$ for any constant $c > 0$; it follows that for sufficiently large n we have

$$\sqrt[n]{\frac{a!}{a^a}} > \frac{1}{2}$$

and

$$\sqrt[n]{n!} > \frac{a}{2} \quad \text{or} \quad \lim_{n \to \infty} \sqrt[n]{n!} = \infty$$

since a is arbitrarily large. Thus $R = \infty$; that is, e^z is an entire function.

It can easily be shown that, for any two complex numbers z_1 and z_2, we have $e^{z_1 + z_2} = e^{z_1} \cdot e^{z_2}$.

3 For $\sum_{n=0}^{\infty} z^n = 1 + z + z^2 + \cdots$

$$R = \frac{1}{\overline{\lim} \sqrt[n]{|a_n|}} = 1$$

This series plays an important role in the expansion of an arbitrary holomorphic function into a Taylor series. The series $\sum_{n=1}^{\infty} nz^{n-1}$ has the same radius of convergence.

4. A function holomorphic on D is analytic on D

THEOREM 3.3 *Let f be holomorphic on D. Then we can expand f in a power series $f(z) = \sum_{n=0}^{\infty} a_n(z-c)^n$ with a positive radius of convergence about each point $c \in D$.*

PROOF

Let $c \in D$ and let $d(c, R)$ be the greatest open disc belonging to D. Since D is open, R is positive. Now pick r and $r_o \ni 0 < r < r_o < R$ and denote by ∂ the boundary of $d(c, r_o)$: $\partial(t) = c + r_o \operatorname{cis} t$, $0 \leq t \leq 2\pi$.

By Cauchy's formula, we have for each $z \in \bar{d}(c, r)$:

$$f(z) = \frac{1}{2\pi i} \int_{\partial} \frac{f(u)}{u - z} \, du$$

Transform the Cauchy kernel $1/(u - z)$ as follows:

$$\frac{1}{u-z} = \frac{1}{(u-c)-(z-c)} = \frac{1}{u-c} \cdot \frac{1}{1 - \dfrac{z-c}{u-c}}$$

with

$$|u - c| = r_o, \qquad |z - c| \leq r, \quad \text{and} \quad \left|\frac{z-c}{u-c}\right| \leq \frac{r}{r_o} < 1$$

Then

$$\frac{1}{1 - \dfrac{z-c}{u-c}} = 1 + \frac{z-c}{u-c} + \left(\frac{z-c}{u-c}\right)^2 + \cdots$$

and

$$\frac{1}{u-z} = \frac{1}{u-c}\left[1 + \frac{z-c}{u-c} + \left(\frac{z-c}{u-c}\right)^2 + \cdots\right]$$

Since $1 + \dfrac{r}{r_o} + \cdots + \left(\dfrac{r}{r_o}\right)^n + \cdots$ is a numerical majorant, the geometric series in brackets converges uniformly with respect to u on ∂ and $z \in \bar{d}(c, r)$. Now $f(u)$ is bounded on ∂ since it is a continuous function on the compact ∂. Therefore, if we multiply $1/(u - z)$ by $f(u)$, the expression

$$\frac{f(u)}{u - z} = \frac{f(u)}{u - c} + (z - c)\frac{f(u)}{(u - c)^2} + (z - c)^2 \frac{f(u)}{(u - c)^3} + \cdots$$

converges also uniformly, and we can integrate the right-hand side $f(u)/(u - z)$ term by term to get

$$f(z) = \frac{1}{2\pi i} \int_\partial \frac{f(u)}{u - z} du = \frac{1}{2\pi i} \int_\partial \frac{f(u)}{u - c} du$$

$$+ (z - c)\frac{1}{2\pi i} \int_\partial \frac{f(u)}{(u - c)^2} du + (z - c)^2 \frac{1}{2\pi i} \int_\partial \frac{f(u)}{(u - c)^3} du + \cdots$$

If we set

$$a_n = \frac{1}{2\pi i} \int_\partial \frac{f(u)}{(u - c)^{n+1}} du, \qquad n = 0, 1, 2, ,\ldots$$

we find $f(z) = a_0 + a_1(z - c) + a_2(z - c)^2 + \ldots$.

This shows that a function which is holomorphic on D is analytic on D.

Noting that

$$a_n = \frac{1}{2\pi i} \int_\partial \frac{f(u)}{(u - c)^{n+1}} du = \frac{1}{n!} f^{(n)}(c)$$

we find

$$f(z) = \sum_{n=0}^\infty \frac{1}{n!} f^{(n)}(c)(z - c)^n$$

The right-hand side of this equation is called the Taylor series representation of f about c. This series converges uniformly with respect to z in the closed disc $\bar{d}(c, r)$, for any $r < R$.

About each point c of the domain D of holomorphicity, the function f possesses a Taylor series expansion. The coefficients $a_n = \dfrac{f^{(n)}(c)}{n!}$ depend on f and on c. Let us recall that $d(c, R)$ is the largest open disc belonging to D, the domain of holomorphicity. Thus, the radius of convergence about c is at least equal to R.

If in the complex plane we know all the singular points of f, and if c is a point where f is holomorphic, then the radius of convergence of its Taylor series about c is the distance R from c to the nearest singularity.

This is because the radius of convergence about c cannot be greater than R, for if it were we would have singular points inside the convergence disc, which is impossible, by part 3 of Theorem 3.2.

EXAMPLE

Let $f(z)$ be $1/z$. We know already that the domain of holomorphicity of $1/z$ is $\mathbf{C}\backslash\{0\}$. Therefore, if we develop $1/z$ about the point c in a Taylor series, the radius of convergence is equal to $|c|$. For $c = 1$, we find

$$\frac{1}{z} = \frac{1}{1 + (z-1)} = 1 - (z-1) + (z-1)^2 + \cdots$$

a series which converges for $|z - 1| < 1$. Hence the radius of convergence equals 1.

THEOREM 3.4 [*Uniqueness Theorem*] *A given holomorphic function f possesses a unique power series expansion about a given point c.*

PROOF

Indeed, let us suppose $f(z) = \sum_n a_n(z-c)^n = \sum_n b_n(z-c)^n$. If both series converge uniformly on the closed disc $\bar{d}(c, r)$, and if γ is a circle $|z - c| = r$, then we find, after dividing by $(z-c)^{n+1}$ and integrating over γ:

$$a_n = \frac{1}{2\pi i} \int_\gamma \frac{f(u)}{(u-c)^{n+1}} \, du = b_n, \qquad n = 0, 1, 2, \ldots$$

Thus the Taylor series about a point c of a given function is unique.

REMARK

It is easy to check that if f and g are holomorphic in a neighborhood of c, and if $f(z) = \sum_{n=0}^{\infty} a_n(z-c)^n$ with radius of convergence R_f and $g(z) = \sum_{n=0}^{\infty} b_n(z-c)^n$ with radius of convergence R_g, then:

(i) $f(z) + g(z) = \sum_{n=0}^{\infty} (a_n + b_n)(z-c)^n$ with radius of convergence $R_{f+g} \geq \min(R_f, R_g)$.

(ii) $f(z)g(z) = \sum_{n=0}^{\infty} c_n(z-c)^n$ with radius of convergence $R_{fg} \geq \min(R_f, R_g)$ and $\{c_n\} = \{a_n\} * \{b_n\}$ (see 3.1.1 and problem 3.4).

5. Cauchy's inequalities

Cauchy's inequalities furnish us with upper bounds for the coefficients of the Taylor series. Let f be holomorphic in $d(c, r)$ and continuous on $\bar{d}(c, r)$. Let $M(r)$ be the least upper bound of f on the circle $\partial(c, r)$. We have found previously

$$a_n = \frac{1}{2\pi i} \int_\partial \frac{f(u)}{(u - c)^{n+1}} \, du$$

Consequently, we have

$$|a_n| \leq \frac{1}{2\pi} \frac{M(r)}{r^{n+1}} \cdot 2r\pi = \frac{M(r)}{r^n}$$

The above inequalities, that is,

$$|a_n| \leq \frac{M(r)}{r^n}, \qquad n = 0, 1, 2, \ldots$$

are called *Cauchy's inequalities*.

THEOREM 3.5 [*Liouville's Theorem*] *An entire function bounded in the complex plane is a constant.*

PROOF

Let $|f(z)| < M \; \forall \, z \in \mathbf{C}$, then for an arbitrary r we certainly have $M(r) \leq M$, so that $\lim_{r \to \infty} (M(r)/r^n) = 0$ for $n = 1, 2, \ldots$. By Cauchy's inequalities, we have

$$|a_n| \leq \frac{M(r)}{r^n}$$

for arbitrary r, which means that $|a_n| \leq \varepsilon$ for $n = 1, 2, \ldots$, where ε is arbitrarily small. It follows that $a_n = 0$, $n = 1, 2, \ldots$, and $f(z) = a_0$.

As an application of Liouville's theorem, we present the theorem of Gauss-d'Alembert.

THEOREM 3.6 *A polynomial with complex coefficients, $P(z) = a_0 + a_1 z + \cdots + a_n z^n$, with $a_n \neq 0$ and $n \geq 1$, possesses at least one complex root.*

PROOF

Consider the function $1/P(z)$; we see immediately that $\lim_{z \to \infty} (1/P(z)) = 0$. Indeed,

$$\lim_{z \to \infty} \left|\frac{1}{P(z)}\right| = \lim_{z \to \infty} \left|\frac{1}{z^n}\right| \frac{1}{\left|\frac{a_0}{z^n} + \frac{a_1}{z^{n-1}} + \cdots + a_n\right|} = \frac{1}{|a_n|} \lim_{z \to \infty} \left|\frac{1}{z^n}\right| = 0$$

Therefore, $|1/P(z)| < \varepsilon$ for $|z| > R(\varepsilon)$.

Suppose that $P(z)$ has no zeros, then $1/P(z)$ is holomorphic in the entire plane. Now for $|z| \leq R(\varepsilon)$, $1/P(z)$ is bounded, being a continuous function on the compact $\overline{d}(0, R(\varepsilon))$. Thus $1/P(z)$ is an entire function and bounded in the entire plane. Hence, by Liouville's theorem, $1/P(z)$ is a constant, in contradiction with the hypothesis that $P(z)$ is a nonconstant polynomial.

6. Schwarz's lemma

Another application of the Taylor series is Schwarz's Lemma:

THEOREM 3.7 Let f be analytic in the open disc $d(0, 1)$, with $f(0) = 0$ and $|f(z)| < 1$ for $z \in d(0, 1)$. Then $|f(z)| \leq |z|$ for $|z| < 1$. If for one $z_0 \in d(0, 1)$ we have $|f(z_0)| = |z_0|$, then $f(z) = z \operatorname{cis} t$ for a constant t, and $|f(z)| = |z| \forall z$.

PROOF

Let $f(z) = a_1 z + a_2 z^2 + \ldots (a_0 = f(0) = 0)$. Then the function $\phi(z) = f(z)/z = a_1 + a_2 z + \ldots$ is also holomorphic in $d(0, 1)$. Using the maximum principle (Theorem 2.25) for the disc $d(0, r)$ with $r < 1$, we find $|\phi(z)| < 1/r$ if $z \in d(0, r)$. Since this is true for any $r < 1$, but arbitrarily close to 1, we find $|\phi(z)| \leq 1$, or $|f(z)| \leq |z|$ for $|z| < 1$. If we have, for one z_0, $|\phi(z_0)| = 1$, then $|\phi(z)|$ attains a relative maximum at z_0. Applying the maximum principle again, we find that $\phi(z)$ is a constant with module 1, or $\phi(z) = \operatorname{cis} t$. Hence $f(z) = z \operatorname{cis} t$. Generally, we have $|f'(0)| = |a_1| = |\phi(0)| \leq 1$, but if $|f'(0)| = |\phi(0)| = 1$, then $f(z) = z \operatorname{cis} t$.

7. Zeros of an analytic function

DEFINITION

Let f be analytic in a neighborhood of c and let $\sum_n a_n(z - c)^n$ be its

Taylor series about the point c. We say that c is a *zero* of f if $f(c) = 0$. If $f(c) = f'(c) = \cdots = f^{(p-1)}(c) = 0$ and $f^{(p)}(c) \neq 0$, then we have $a_0 = a_1 = \cdots = a_{p-1} = 0$, and we say that c is a *zero of order p* of f. A zero of order one is called a *simple zero*.

If c is a zero of order p, then the Taylor series of f may be written:

$$f(z) = \sum_{n=p}^{\infty} a_n(z-c)^n = (z-c)^p \sum_{n=p}^{\infty} a_n(z-c)^{n-p}$$

$$= (z-c)^p \sum_{n=0}^{\infty} a_{n+p}(z-c)^n$$

Let us denote by $g(z)$ the expansion

$$g(z) = \sum_{n=0}^{\infty} a_{n+p}(z-c)^n$$

Then $g(c) = a_p = \dfrac{f^{(p)}(c)}{p!} \neq 0$. Now g is analytic and therefore continuous. Since $g(c) \neq 0$, $g(z)$ will be different from zero in a certain neighborhood of c. The factor $(z-c)^p = 0$ only for $z = c$, so that $f(z) \neq 0$ in a certain neighborhood of c except at c. We may conclude that the zeros of finite order of an analytic function are isolated.

Now we can prove the following theorem:

THEOREM 3.8 *If f is holomorphic in D, and c is a zero of infinite order that is, if*

$$f(c) = f'(c) = f''(c) = \cdots = 0$$

then $f(z) \equiv 0$ on D.

PROOF

Let $E \subset D$ be the set of all points z with $f(z) = f'(z) = \cdots = f^{(n)}(z) = \cdots = 0$. Then E is open in the relative topology. For if we develop f in a Taylor series about a point z of E, then all coefficients of the series in question are zero, and $f \equiv 0$ in the disc of convergence about the point z. On the other hand, by virtue of the continuity of the functions f, f', f'', $\ldots$, $f^{(n)}\ldots$, E is also closed. Thus E is open and closed in the relative topology and nonempty. Therefore $E = D$, which shows that $f(z) \equiv 0$ on D.

COROLLARY *If f is holomorphic on D and $f(z) \equiv 0$ on a subdomain D_1 of D, then $f(z) \equiv 0$ on D.*

PROOF

Each point of D_1 is a zero of order infinity for f.

COROLLARY *Let f be holomorphic on D and $c \in D$. If a subsequence of the zeros of f converges to c, then $f(z) \equiv 0$ on D.*

PROOF

If f is holomorphic and $\neq 0$ on D, then the zeros of finite order of f are isolated points. Thus, c must be a zero of infinite order, and $f(z) \equiv 0$.

EXAMPLE

Let us define the functions $\sin z$ and $\cos z$ in terms of power series:

$$\sin z = z - \frac{z^3}{3!} + \frac{z^5}{5!} + \cdots$$

and

$$\cos z = 1 - \frac{z^2}{2!} + \frac{z^4}{4!} + \cdots$$

We see immediately that the function $\sin z$ has a simple zero at the origin.

$$\sin z = z\left(1 - \frac{z^2}{3!} + \frac{z^4}{5!} \cdots\right)$$

and

$$\sin 0 = 0$$
$$\cos 0 = (\sin z)'_{z=0} \neq 0$$

The function $1 - \cos z$ has a zero of order two at the origin. Indeed, $1 - \cos 0 = 0$, $\sin 0 = (1 - \cos z)'_{z=0} = 0$ and $(1 - \cos z)''_{z=0} = \cos 0 = 1$. We can also see this by writing

$$1 - \cos z = z^2\left(\frac{1}{2!} - \frac{z^2}{4!} + \frac{z^4}{6!} \cdots\right)$$

Both functions $\sin z$ and $\cos z$ are entire functions because the radius of convergence of both series is infinite. (See 4.1.3, example 2.)

For real values of z, the functions $\sin z$ and $\cos z$ coincide, of course, with the trigonometric functions as defined in real analysis. On the basis of the infinite series it is now easy to verify the relation

$$\cos z + i \sin z = e^{iz}$$

for all values of z, and in particular for real $z = \phi$, in which it reduces to Euler's formula:

$$\text{cis } \phi = \cos \phi + i \sin \phi = e^{i\phi}$$

Let us consider now the function $\sin(1/z)$, which is holomorphic in the entire plane except at the origin. Its zeros are $c_n = 1/(n\pi)$, $n = 1, 2, \ldots$, because $f(c_n) = \sin n\pi = 0$. The sequence $c_1, c_2, \ldots, c_n, \ldots$ belongs to the domain of holomorphicity of $\sin(1/z)$. The c_n's are converging to zero, but this point is a boundary point of the domain of holomorphicity of f and does not belong to it. But even on the boundary of the domain of holomorphicity of f, f may not have too many zeros. Indeed, we have the following theorem, which we give without proof:

THEOREM *If f is holomorphic on D whose boundary is a Jordan curve, and if $\lim f(z) = 0$ as z approaches anyone of the points of an arc γ of the boundary, then $f(z) \equiv 0$ in D.*

COROLLARY 3 *If g and h are analytic on D, and if on the sequence of points $c_1, c_2, \ldots, c_n$, converging to $c \in D$, we have $g(c_n) = h(c_n)$, then $g(z) \equiv h(z)$ on D.*

To see this it is sufficient to consider $f(z) = g(z) - h(z)$.

Corollary 3 reveals the interesting fact that a function which is holomorphic on D is uniquely determined by its values on a sequence of points having a limit point on D.

8. Analytic continuation

DEFINITION Let f be holomorphic on D, and g holomorphic on D_1 with $D_1 \supset D$. If $f(z) \equiv g(z)$ on D, we shall say that g is an *analytic continuation* of f. (Fig. 3.1).

EXAMPLE

$f(z) = \sum_{n=0}^{\infty} z^n$. This series converges for $|z| < 1$. Thus f is defined and holomorphic on $\mathscr{d}(0, 1)$. Consider, on the other hand, the function $g(z) = \dfrac{1}{1-z}$, defined and holomorphic on $\mathbf{C}\backslash\{1\}$. Thus $D_1 = \mathbf{C}\backslash\{1\}$, and g is the analytic continuation of f on $D = \mathscr{d}(0, 1)$.

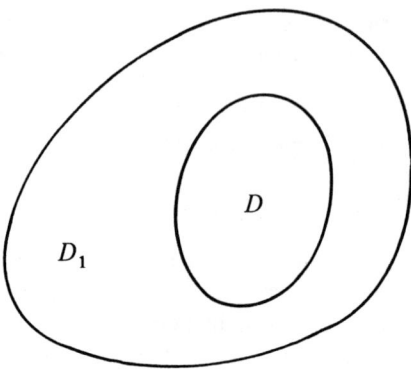

Figure 3.1

THEOREM 3.9 [*Uniqueness of the Analytic Continuation*] *The analytic continuation of a given function f is unique, that is to say, if g_1 and g_2 are analytic continuations in $D_1 \supset D$ of f, then $g_1(z) \equiv g_2(z)$ in D_1.*

PROOF

This follows immediately from Theorem 3.8 using the function $\Delta(z) = g_1(z) - g_2(z)$, which is holomorphic on D and equal to zero on D.

We will now consider the construction of the analytic continuation of a function by the method of Weierstrass: If a function f is holomorphic inside and on a p.w.s. closed curve γ, Cauchy's formula enables us to calculate the values of $f(z)$ inside γ if we know its values on γ. Also, if a function f is defined and holomorphic on a domain D, then the Taylor series enables us to calculate the values of $f(z)$ outside of D, provided f possesses an analytic continuation outside of D.

EXAMPLE

$f(z) = 1 + z + z^2 + \cdots$. This power series defines a holomorphic function in the open disc $d(0, 1)$, namely, the function $1/(1 - z)$. We take now an arbitrary point c of the open disc $d(0, 1)$ as center of a new power series $f_1(z)$. We find that $f_1(z) = \sum_n a_n(z - c)^n$ with $a_n = \dfrac{f^{(n)}(c)}{n!}$. It is not difficult to calculate $f^{(n)}(c)$ with the aid of the power series for f about the origin, $f(z) = 1 + z + z^2 + \cdots$. The disc of convergence $d_1(c, r_c)$ of the new series $f_1(z) = \sum a_n(z - c)^n$ has a radius of convergence $r_c = |c - 1|$. We can pick $c \ni d_1(c, r_c) \backslash d(0, 1) \neq \emptyset$. Thus we have defined the function f outside the disc $d(0, 1)$. If c takes on all the values belonging to the open disc $d(0, 1)$, then we get a set of Taylor series $\sum_n a_n(z - c)^n$ which represent f on $d(0, 1)$. Any of these Taylor series is uniquely determined by its center c, and we denote the Taylor series about the point c by f_c. f_c converges in the disc $d_c = d(c, r_c)$. f_c and f are equal on $d(0, 1) \cap d(c, r_c)$, and both f and f_c are analytic continuations of their common restriction to $d(0, 1) \cap d(c, r_c)$.

When f is defined on a domain D and is expanded about every point of D we are able to define f outside D, provided that $\bigcup_{c \in D} d_c \supset D$. This is not always possible. For instance, if the singular points of f are dense on the boundary of D, then $\bigcup_{c \in D} d_c = D$.

EXAMPLE

$f(z) = \sum_n z^{n!}$. One sees easily that the disc of convergence of f is the open disc $d(0, 1)$. It can be shown that the singularities of f are dense on the boundary of the circle $|z| = 1$.

If $\bigcup_{c \in D} d_c = D_1$ and $D_1 \supset D$, then the analytic continuation of f can be represented in terms of a Taylor series.

Let us call the pair (f_1, d_1), where f_1 is a Taylor series and d_1 is the corresponding disc of convergence, a *function element*. If two function elements (f_1, d_1) and (f_2, d_2) are such that $d_1 \cap d_2 \neq \emptyset$, and in addition $f_1 = f_2$ on $d_1 \cap d_2$, then we say that the two function elements are *contiguous*. Two contiguous function elements are both analytic continuations of their rectriction to $d_1 \cap d_2$. We shall call a finite number of function elements $(f_1, d_1), (f_2, d_2), \ldots, (f_n, d_n)$ a *chain* if each pair of successive function elements are contiguous. In accordance with Weierstrass, an analytic function is defined by the set of all function elements which have the property that any two of them may be joined by a chain.

We saw in 1.4.2 that the set $H(D)$ of all functions which are holomorphic on D is a ring. We can see now that it is a ring without zero divisors. Indeed, if f and $g \in H(D)$, and if $f(z) \cdot g(z) \equiv 0$ in D, and if $f(z_o) \neq 0$, then $f(z)$ is nonzero in a certain neighborhood of z_o, and in this neighborhood $g(z)$ must be zero; that is, $g(z) \equiv 0$ on a subdomain of D. Therefore, $g(z) \equiv 0$ on D.

As a comparison, take the set $C(D)$ where it is easy to find two functions $f \not\equiv 0$ and $g \not\equiv 0 \ni$ their product is zero. Indeed, if $c \in D$, consider two subsets S_1 and S_2 of D,

$$S_1 = D \cap \{z \,|\, \text{Re}(z) \leq \text{Re}(c)\}$$

and

$$S_2 = D \cap \{z \,|\, \text{Re}(z) \geq \text{Re}(c)\}$$

and define:

$$f(z) = \begin{cases} 0 & \text{on } S_1 \\ \text{Re}(z - c) & \text{on } S_2 \end{cases}$$

$$g(z) = \begin{cases} \text{Re}(c - z) & \text{on } S_1 \\ 0 & \text{on } S_2 \end{cases}$$

We see immediately that $fg \equiv 0$ on D, without $f \equiv 0$ or $g \equiv 0$ on D.

In this connection, an important problem is that of constructing an entire function with given zeros. One recognizes immediately that the solution will not be unique because there are entire functions without zeros; for instance, the functions e^z and $e^{g(z)}$, where g is an arbitrary entire function.* So, if f is a function with prescribed zeros, then the product $e^g f$ has the same zeros.

If the zeros are finite in number and different from the origin, say $c_1, c_2, \ldots, c_n$, then we can construct a polynomial

$$P(z) = \left(1 - \frac{z}{c_1}\right)\left(1 - \frac{z}{c_2}\right) \cdots \left(1 - \frac{z}{c_n}\right)$$

whose zeros are just $c_1, c_2, \ldots, c_n$.

If the origin is also a zero then it is sufficient to introduce the factor z if the zero at the origin is a simple one, and z^p if the zero at the origin is of order p. In the latter case,

$$P(z) = z^p \left(1 - \frac{z}{c_1}\right)\left(1 - \frac{z}{c_2}\right) \cdots \left(1 - \frac{z}{c_n}\right)$$

If the zeros of an entire function are infinite in number, we know already that the zeros can't have a limit point; otherwise the function would

---

* e^z has no zeros because $e^z \cdot e^{-z} = 1$ for all z. Similarly, if $g(z)$ is an entire function, it is defined for all z and we have $e^{g(z)} \cdot e^{-g(z)} = 1$, so that $e^{g(z)} \neq 0$.

be identically zero. Thus, $\lim_{n\to\infty} c_n = \infty$, and we can attempt to define an infinite product

$$f(z) = z^p\left(1 - \frac{z}{c_1}\right)\left(1 - \frac{z}{c_2}\right)\cdots\left(1 - \frac{z}{c_n}\right)\cdots$$

In general, this infinite product will not converge. To overcome this difficulty, Weierstrass introduced an entire function without zeros as a convergence factor.

9. Permanency principle of functional relations

In this section we want to show that functional relations involving analytic functions remain true for complex values of the independent variable if they are true for real values of the independent variable.

Take for instance the entire functions $\sin z$ and $\cos z$. We know that for real values x and y

$$\sin(x + y) = \sin x \cos y + \cos x \sin y$$

Does this relation remain true for complex values x and y? The answer is yes.

Let y be fixed and real. Consider the function defined by

$$\phi(z) = \sin(z + y) - \sin z \cos y - \cos z \sin y$$

We know already that the function $\phi(z)$ is entire, and that it is zero for z real. It follows that it is identically zero in $\mathbf{C}$.

Now let z be fixed and consider $\psi(\omega)$ where ω is a complex variable:

$$\psi(\omega) = \sin(z + \omega) - \sin z \cos \omega - \cos z \sin \omega$$

$\psi(\omega)$ is an entire function in ω and zero on the real axis. Therefore $\psi(\omega) \equiv 0$ in $\mathbf{C}$.

As another example, we consider the function $e^{\log z}$. e^z is an entire function and each determination of $\log z$ is holomorphic in $\mathbf{C}\backslash\mathbf{R}_0^-$. Therefore, $e^{\log z}$ has the same domain of holomorphicity as $\log z$. Now $e^{\log z} - z$ is zero for z on the positive real axis. It follows that $e^{\log z} - z \equiv 0$ on $\mathbf{C}\backslash\mathbf{R}_0^-$. In the same way, we may show that the relation $e^{x+y} = e^x e^y$, which is true for real values of x and y, remains true for complex values of these variables.

To generalize from the above examples, let $F(s, t)$ be a continuous complex-valued function of the complex variables s and t, and complex differentiable with respect to s and t. In addition, let $s = f(z)$ and $t = g(z)$

be two function holomorphic on D. Then $F(z) = F(f(z), g(z))$ is differentiable with respect to z, and

$$\frac{d}{dz}\tilde{F}(z) = \frac{\partial F}{\partial s}f'(z) + \frac{\partial F}{\partial t}g'(z)$$

Therefore, if $\tilde{F}(z) = 0$ on a sequence of points $c_1, c_2, \ldots, c_n, \ldots$ which have a limit point in D, then $\tilde{F}(z) \equiv 0$ on D.

In particular, if f is holomorphic on D and $g = f'$, and $F(f, f') = 0$ on a subset of D which contains a sequence of points converging to a point in D, then $F(f, f') = 0$ on D. Moreover, if f possesses an analytic continuation outside D, then the relation $F(f, f') = 0$ will still hold for the analytic continuation.

Problems

3.1.: Prove that the Riemann zeta function ζ defined by

$$\zeta(z) = \sum_{n=1}^{\infty} n^{-z}$$

converges for Re $z > 1$ and converges uniformly for Re $z \geq 1 + \varepsilon$ where $\varepsilon > 0$ is arbitrarily small. Show that ζ is analytic for Re $z \geq 1 + \varepsilon$. (*Hint*: $|n^{-z}| = |e^{-z \log n}| = |e^{-x \log n}||e^{-iy \log n}| = e^{-x \log n} = n^{-x}$)

3.2.: Let f be continuous on $[0, \infty]$. Show that $F_a(z) = \int_0^a e^{-zt}f(t)dt$ is an entire function. If $\int_0^\infty |f(t)|dt < \infty$, show that $F(z) = \int_0^\infty e^{-zt}f(t)dt$ is holomorphic in the half plane Re $z \geq 0$. $\left(\textit{Hint}: F(z) = \sum_{n=0}^{\infty} \int_n^{n+1} e^{-zt}f(t)dt.\right.$
Show that the series converges uniformly.$\Big)$

3.3.: The Fibonacci numbers are defined by $c_0 = 0$, $c_1 = 1$ and $c_n = c_{n-1} + c_{n-2}$. Prove that $\sum_{n=0}^{\infty} c_n z^n$ is a rational function and determine an expression for c_n.
(*Hint*: Set $\phi(z) = c_2 + c_3 z + \cdots$ and show that $z\phi(z) + 1 + z^2\phi(z) + z = \phi(z)$.)

3.4.: Show that if $f(z) = \sum_{n=0}^{\infty} a_n z^n$ and $g(z) = \sum_{n=0}^{\infty} b_n z^n$ in $d(0, r)$, then $f(z) \cdot g(z) = \sum_{n=0}^{\infty} c_n z^n$ with $\{c_n\} = \{a_n\} * \{b_n\}$. Use this relation in order to find the Taylor series for $\dfrac{1}{1-z-z^2}$ about the origin. $\Big($ Hint: If $\dfrac{1}{1-z-z^2} = \sum\limits_{n=0}^{\infty} a_n z^n$, then $(1 - z - z^2)(a_0 + a_1 z + \cdots) = 1.\Big)$

3.5.: Let f be an entire function. Prove the following: If $\exists$ a positive integer n and a positive constant M such that
$$|f(z)| \le M|z|^n, \quad \forall z \in \mathbb{C}$$
then f is a polynomial of degree $\le n$. (Compare with Liouville's theorem.)

3.6.: Define the functions cosh z and sinh z by their Taylor series about the origin:
$$\cosh z = 1 + \frac{z^2}{2!} + \frac{z^4}{4!} + \cdots$$
$$\sinh z = z + \frac{z^3}{3!} + \frac{z^5}{5!} + \cdots$$
Prove the relations $\cos iz = \cosh z$, $\sin iz = i \sinh z$, and $\cosh^2 z - \sinh^2 z = 1$.

3.7.: Determine the radius of convergence of—
  (a) $1 + 4z + 9z^2 + \cdots + (n+1)^2 z^n + \cdots$
  (b) $1 + \alpha z + \binom{\alpha}{2} z^2 + \cdots + \binom{\alpha}{k} z^k + \cdots$, $\quad 0 < \alpha < 1$

3.8.: Investigate the absolute and uniform convergence of the series
$$\sum_{n=0}^{\infty} \frac{1}{2^n} \left(\frac{z-1}{z+1}\right)^n$$

3.9.: Expand $f(z) = \log(2+z)$ in a power series of the form $\sum\limits_{n=0}^{\infty} a_n z^n$ and determine the radius of convergence.

3.10.: Let $f(z) = \sum\limits_{n=0}^{\infty} c_n z^n$ be an entire function $\ni \forall R > 0$ the inequality $|f(z)| \le e^R$ is valid whenever $|z| \le R$. Prove that $|c_n| \le e^n n^{-n} \forall n > 0$.

3.11.: Show that the series
$$\sum_{n=0}^{\infty} \left(\frac{z}{3}\right)^n \quad \text{and} \quad \sum_{n=0}^{\infty} \frac{(\frac{1}{3})^n (z-6i)^n}{(1-2i)^{n+1}}$$
are analytic continuations of each other.

3.12.: Prove the convergence of the series $\sum\limits_{n=1}^{\infty} \dfrac{n^3}{2^n}$.

3.13.: Determine the radius of convergence of the following series:
  a) $\sum\limits_{n=1}^{\infty} \dfrac{z^n}{n^2}$
  b) $\sum\limits_{n=1}^{\infty} \dfrac{n z^n}{2^n}$
  c) $\sum\limits_{n=1}^{\infty} n! z^n$
  d) $\sum\limits_{n=1}^{\infty} z^{2^n}$

3.14.: Show that if r is the radius of convergence of the series $\sum_{n=1}^{\infty} a_n z^n$, then the radius of convergence of the series $\sum_{n=1}^{\infty} a_n^2 z^n$ is r^2.

3.15.: At what points of the boundary of the disc of convergence do the following series converge?

a) $\sum_{n=1}^{\infty} z^n$ b) $\sum_{n=1}^{\infty} \frac{z^n}{n}$ c) $\sum_{n=1}^{\infty} \frac{z^n}{n^2}$

3.16.: Expand the following functions in a Taylor series about the origin:

(a) $\sinh^2 z$ b) $\log \frac{1+z}{1-z}$

Utilize this series to calculate the value of log 3 to two decimal digits.

3.17.: Expand the following functions in a Taylor series about $z = 1$:

a) $\frac{1}{z}$ b) $\frac{1}{z+2}$ c) $\log z$

3.18.: Expand in a Taylor series about the origin:

$$\frac{z}{e^z - 1} = \sum_{n=0}^{\infty} \frac{B_n}{n!} z^n$$

Show that the B_n satisfy the following relations: $B_0 = 1$, and $\binom{n+1}{0} B_0 + \binom{n+1}{1} B_1 + \cdots + \binom{n+1}{n} B_n = 0$ for $n \geq 1$ (symbolically, $(1 + B)^{n+1} = B_{n+1}$).

3.19.: Determine the zeros of the following functions and their order.

a) $e^z - 1$ b) $\sin^2 z$ c) $\sin z^2$

d) $\frac{\sin z}{z}$

---

§2. Laurent series

1. Definition of Laurent series

DEFINITION   A *Laurent series about* c is a series of the following type:

$$\sum_{n=-\infty}^{n=\infty} a_n(z-c)^n = \sum_{n=0}^{\infty} a_n(z-c)^n + \sum_{n=1}^{\infty} \frac{a_{-n}}{(z-c)^n}$$

where c and a_n are complex numbers.

The Laurent series is a generalization of the Taylor series (which does not contain negative powers). We call $\sum_{n=0}^{\infty} a_n(z-c)^n$ the *Taylorian part* and $\sum_{n=1}^{\infty} \frac{a_{-n}}{(z-c)^n}$ the *principle part* of the Laurent series.

The Taylorian part converges, as we know already, inside a certain disc of convergence $d(c, R)$, where R depends on the coefficients a_n, $n = 0, 1, 2, \ldots$. We also know that R can take on all values from 0 to ∞. If $R > 0$, the Taylorian part defines a holomorphic function in $d(c, R)$.

In order to see where the principle part converges, we set

$$u = \frac{1}{z-c}$$

then

$$\sum_{n=1}^{\infty} \frac{a_{-n}}{(z-c)^n} = \sum_{n=1}^{\infty} a_{-n} u^n$$

which is a Taylor series in u. The series $\sum_{n=1}^{\infty} a_{-n} u^n$ converges in a disc $d(0, \rho)$ where ρ depends on the coefficients a_{-n};

$$\rho = \frac{1}{\overline{\lim} \sqrt[n]{|a_{-n}|}}, \quad n = 1, 2, \ldots$$

ρ can also assume values from 0 to ∞. Now, $|u| < \rho$ is equivalent to $\left|\frac{1}{z-c}\right| < \rho$ or $|z-c| > \frac{1}{\rho}$. Set $r = \frac{1}{\rho}$; then we see that the principle part of the Laurent series converges outside the disc $d(c, r)$.

For $\rho > 0$, the principle part defines a holomorphic function in u for $|u| < \rho$. Let $p(u) = \sum_{n=1}^{\infty} a_{-n} u^n$. Then $p(u)$ is holomorphic for $|u| < \rho$. The principle part $\sum_{n=1}^{\infty} \frac{a_{-n}}{(z-c)^n} = p\left(\frac{1}{z-c}\right)$ is holomorphic for $\left|\frac{1}{z-c}\right| < \rho$ or $|z-c| > \frac{1}{\rho}$.

If the intersection of the domains of convergence of the Taylorian part and the principal part is not void, then the Laurent series defines in this intersection a holomorphic function. If $r < R$, then the domains $|z-c| < R$ and $|z-c| > r$ have in common the annulus $r < |z-c| < R$, and this open annulus is the domain of convergence of the Laurent series. If $r > R$, then the Laurent series is divergent for any z. Based on results for Taylor series, we see immediately that on every annulus

$r' \leq |z - c| \leq R'$ with $r < r' < R' < R$ the convergence is uniform, and we may differentiate the series term by term.

We say that a function f defined in the open annulus $r < |z - c| < R$ can be expanded in a Laurent series if $\exists$ a Laurent series $\sum_{n=-\infty}^{\infty} a_n(z - c)^n$ which converges to f in this annulus.

THEOREM 3.10 *[Fundamental Theorem for Functions Holomorphic in an Annulus] A function f which is holomorphic in the annulus $r < |z - c| < R$ possesses a Laurent series expansion about the point c. The Laurent series is unique and converges uniformly on every compact subset of the annulus.*

PROOF

Let r', r'', R'', R' be four numbers satisfying the inequalities $r < r' < r'' < R'' < R' < R$. Denote by γ' the circle $|z - c| = r'$ and by Γ' the circle $|z - c| = R'$, both described in positive sense, and denote by A'' the closed annulus $r'' \leq |z - c| \leq R''$. Using Cauchy's formula for the annulus (Theorem 2.17) for an arbitrary $z \in A''$, we obtain the following representation of $f(z)$:

$$f(z) = \frac{1}{2\pi i} \int_{\Gamma'} \frac{f(u)}{u - z} du + \left[-\frac{1}{2\pi i} \int_{\gamma'} \frac{f(u)}{u - z} du \right]$$

The first term

$$\frac{1}{2\pi i} \int_{\Gamma'} \frac{f(u)}{u - z} du$$

is a function of z holomorphic inside the disc $d(c, R')$ (compare Theorem 2.14) and furnishes the Taylorian part. The second term

$$-\frac{1}{2\pi i} \int_{\gamma'} \frac{f(u)}{u - z} du$$

is holomorphic for $|z - c| > r'$ and furnishes the principle part of the Laurent series of f.

For z in the closed annulus $A'' = \{z \mid r'' \leq |z - c| \leq R''\}$ and for each u on Γ' we have $|z - c| \leq R'' < R' = |u - c|$, which implies $\left|\frac{z - c}{u - c}\right| \leq \frac{R''}{R'} < 1$ and

$$\frac{1}{u - z} = \sum_{n=0}^{\infty} \frac{(z - c)^n}{(u - c)^{n+1}}$$

Now

$$\frac{1}{2\pi i}\int_{\Gamma'} \frac{f(u)}{u-z} du = \frac{1}{2\pi i}\int_{\Gamma'} f(u) \cdot \sum_{n=0}^{\infty} \frac{(z-c)^n}{(u-c)^{n+1}} du$$

$$= \sum_{n=0}^{\infty} (z-c)^n \frac{1}{2\pi i}\int_{\Gamma'} \frac{f(u)}{(u-c)^{n+1}} du$$

$$= \sum_{n=0}^{\infty} a_n(z-c)^n$$

with

$$a_n = \frac{1}{2\pi i}\int_{\Gamma'} \frac{f(u)}{(u-c)^{n+1}} du$$

The rearrangement of the second expression leading to the third one is possible by virtue of uniform convergence. It is worthwhile to remember here that f may not be holomorphic inside Γ', and therefore a_n may not equal $f^{(n)}(c)/n!$ even if $f^{(n)}(c)$ exists.

For z in the closed annulus $A'' = \{z \mid r'' \leq |z-c| \leq R''\}$, and for each u on γ', we have

$$|u-c| = r' < r'' \leq |z-c|$$

so that

$$\left|\frac{u-c}{z-c}\right| \leq \frac{r'}{r''} < 1$$

and

$$\frac{1}{u-z} = \frac{1}{(u-c)-(z-c)} = -\frac{1}{z-c} \cdot \frac{1}{1-\frac{u-c}{z-c}}$$

$$= -\sum_{n=1}^{\infty} \frac{(u-c)^{n-1}}{(z-c)^n}$$

Then

$$-\frac{1}{2\pi i}\int_{\gamma'} \frac{f(u)}{u-z} du = \frac{1}{2\pi i}\int_{\gamma'} f(u) \sum_{n=1}^{\infty} \frac{(u-c)^{n-1}}{(z-c)^n} du$$

If we integrate the last expression termwise we obtain

$$\sum_{n=1}^{\infty} \frac{1}{(z-c)^n} \cdot \frac{1}{2\pi i} \int_{\gamma'} f(u)(u-c)^{n-1} \, du$$

$$= \sum_{u=1}^{\infty} \frac{a_{-n}}{(z-c)^n}$$

with

$$a_{-n} = \frac{1}{2\pi i} \int_{\gamma'} f(u)(u-c)^{n-1} \, du$$

Finally, we get

$$f(z) = \sum_{n=0}^{\infty} a_n (z-c)^n + \sum_{n=1}^{\infty} \frac{a_{-n}}{(z-c)^n}$$

The uniqueness of the Laurent series may be shown in the same way as the uniqueness of the Taylor series was shown: Since the functions $f(u)/(u-c)^{n+1}$, or, equivalently, $f(u)(u-c)^{n-1}$, are holomorphic inside the annulus $r < |z-c| < R$, we may replace in the integral formulas for a_n and a_{-n} the circles Γ' and γ' by the circle γ_ρ with the center at c and the radius ρ, where ρ satisfies the inequality $r < \rho < R$. Then, if $\sum_{n=-\infty}^{\infty} b_n(z-c)^n$ converges uniformly on the circle γ_ρ: $|z-c| = \rho$ to $f(z)$, we divide $f(z)$ by $(z-c)^{k+1}$ and integrate the result over γ_ρ. We find

$$2\pi i \, b_k = \int_{\gamma_\rho} \frac{f(u)}{(u-c)^{k+1}} \, du, \qquad k = 0, \pm 1, \pm 2, \ldots$$

It follows that $b_k = a_k$.

If $a_{-1} = a_{-2} = \cdots = a_{-n} \cdots = 0$, then the principle part vanishes, and the Laurent series reduces to a Taylor series. In this case, f is analytic in the open disc $d(c, R)$.

2. Isolated singularities and their classification

A special case arises when f is holomorphic in the punctured disc $\dot{d}(c, R)$, that is to say when $r = 0$. In this case, it may be possible to define $f(c)$ in such a way that f is holomorphic in the entire disc $d(c, R)$. How do we know that this is possible? Theorem 3.11 answers this question.

Analytic Functions   161

THEOREM 3.11 *If f is holomorphic in the punctured disc $d(c, R)$, then f possesses an analytic extension in the entire disc if and only if $f(z)$ is bounded in a neighborhood of c.*

PROOF

It is clear that, if f is analytic at c, $f(z)$ is bounded in a neighborhood of c. This is already true for continuous functions.

On the other hand, let us suppose that $f(z)$ is bounded—that is, $|f(z)| \leq M$ in a certain neighborhood of c, say in the inside and on the circle γ_ρ: $|z - c| = \rho$. Writing a_n as

$$a_n = \frac{1}{2\pi i} \int_{\gamma_\rho} \frac{f(u)}{(u - c)^{n+1}} \, du, \qquad n = 0, \pm 1, \pm 2, \pm 3, \ldots$$

we find that the Cauchy inequalities

$$|a_n| \leq \frac{M}{\rho^n}$$

remain true for negative n:

$$|a_{-n}| \leq M\rho^n$$

We may take ρ arbitrarily small, which proves that a_{-n} must be zero for $n = 1, 2, \ldots$.

So, if $f(z)$ is bounded in a neighborhood of c, the Laurent series reduces to a Taylor series, and if we set $f(c) = a_0$, f will be holomorphic in the entire disc $d(c, R)$.

If $f(z)$ is not bounded in a neighborhood of c, then at least one of the coefficients a_{-n} of the principal part must be different from zero. In this case, c is an isolated singularity of f. If the principal part of the Laurent series of f contains a finite number of terms, then we shall say that c is a *pole* of f. If $a_{-p-1} = a_{-p-2} = \cdots = 0$—that is, if a_{-p} is the nonzero coefficient with the lowest index—then we shall say that c is a *pole of order p*. In this case, the function $f(z)(z - c)^p$ is holomorphic in a neighborhood of c, and is given there by the Taylor series

$$f(z)(z - c)^p = a_{-p} + a_{-p+1}(z - c) + \cdots$$
$$+ a_{-1}(z - c)^{-p-1} + a_0(z - p)^p + \cdots$$

If, on the other hand, the principal part contains an infinite number of terms, we say that c is an *essential singularity* of f.

EXAMPLES

1 The function $f(z) = \dfrac{z+1}{z-1}$ has its only singularity at $z=1$. Its expansion about that point is given by

$$\frac{z+1}{z-1} = 1 + \frac{2}{z-1}$$

Here the Taylorian part of the Laurent series consists of one term only, namely 1, and the principal part is $\dfrac{2}{z-1}$. Hence $z=1$ is a pole of order 1.

2 The function $e^{1/z}$ has its only singularity at $z=0$, and its Laurent series is given by

$$e^{1/z} = 1 + \frac{1}{z} + \frac{1}{2!z^2} + \frac{1}{3!z^3} + \cdots$$

The Taylorian part of the series of $e^{1/z}$ contains only one term, namely 1. The number of terms of the principal part is infinite. So $z=0$ is an essential singularity.

Let us now consider the values of an analytic function in the neighborhood of an isolated singularity.

If c is a pole of order p of the function f, then g defined by $g(z) = f(z)(z-c)^p$ is holomorphic with $g(c) = a_{-p} \neq 0$ and

$$\lim_{z \to c} f(z) = \lim_{z \to c} \frac{g(z)}{(z-c)^p} = \infty$$

Using the language of geometry, this means that the image under f of a sufficiently small disc $d(c, \varepsilon)$ will be outside of an arbitrarily great disc.

If $f(z)$ has a pole of order p, then $1/f(z)$ has a zero of order p in c, and conversely. Indeed, suppose

$$f(z) = \frac{a_{-p}}{(z-c)^p} + \frac{a_{-p+1}}{(z-c)^{p-1}} + \cdots + a_0 + a_1(z-c) + \cdots$$

with $a_{-p} \neq 0$. Then rewriting $f(z)$ by pulling out the factor $1/(z-c)^p$, we obtain

$$f(z) = \frac{1}{(z-c)^p}[a_{-p} + a_{-p+1}(z-c) + \cdots] = \frac{1}{(z-c)^p} g(z)$$

where $g(z) = a_{-p} + a_{-p+1}(z-c) + \cdots$ is holomorphic in a certain neighborhood of c, with $g(c) = a_{-p} \neq 0$. Then

$$\frac{1}{f(z)} = (z-c)^p \cdot \frac{1}{g(z)}$$

and since $g(c) \neq 0$, $1/g(z)$ is also holomorphic in a certain neighborhood of c. Therefore,

$$\frac{1}{g(z)} = b_0 + b_1(z-c) + \cdots$$

and we see immediately that $1/f(z)$ has a zero of order p in c.

If, on the other hand, $f(z) = (z-c)^p g(z)$ has a zero of order p at c, then $g(c) \neq 0$, and

$$\frac{1}{f(z)} = \frac{1}{(z-c)^p} \cdot \frac{1}{g(z)} = \frac{1}{(z-c)^p}[b_0 + b_1(z-c) + \cdots]$$

so that c is a pole of order p of $1/f(z)$.

EXAMPLE

$\sin z$ has simple zeros at $z = k\pi$, $k = 0, \pm 1, \pm 2, \ldots$ So $\dfrac{1}{\sin z}$ has simple poles at $z = k\pi$.

The behavior of a function in the neighborhood of an essential singularity is quite different.

THEOREM 3.12 [*Weierstrass*] *In the neighborhood of an essential singularity the function comes arbitrarily close to every complex value.*

Geometrically speaking, this means the following: If f is holomorphic in the punctured disc $\dot{d}(c, R)$, and if c is an essential singularity of f, then the image of an arbitrarily small disc $\dot{d}(c, \varepsilon)$ is everywhere dense in the complex plane.

PROOF

Suppose the image of $\dot{d}(c, \varepsilon)$ is not dense in the complex plane; then $f(z)$ for $z \in \dot{d}(c, \varepsilon)$ avoids a certain disc, for instance the disc $d(a, \rho)$. That is, $|f(z) - a| > \rho$ for $z \in \dot{d}(c, \varepsilon)$.

Consider the function $g(z) = \dfrac{1}{f(z) - a}$. g is holomorphic and bounded in $\dot{d}(c, \varepsilon)$, since $\left|\dfrac{1}{f(z) - a}\right| < \dfrac{1}{\rho}$, and possesses therefore a holomorphic extension for the entire disc $d(c, \varepsilon)$. Then $1/g(z)$ is holomorphic in a neighborhood of c, provided that $g(c) \neq 0$, or, if $g(z)$ has a zero of order p at c, then c is a pole of order p of $1/g(z)$. The same is true for $f(z) = a + 1/g(z)$. But this contradicts our hypothesis that c is an essential singularity of $f(z)$.

Picard has shown that, in an arbitrarily small neighborhood of an essential singularity, an analytic function takes on every complex value except possibly one value.

EXAMPLE

However small we take $r > 0$ the image of the punctured disc $\dot{d}(0, r)$ by the function $e^{1/z}$ is the entire complex plane excluding the origin. (We state this fact without proof. See 4.1.3.)

3. Meromorphic functions

DEFINITION We shall say that f is *meromorphic in D* if the only singularities of f in D are poles.

EXAMPLES

1 e^z/z is meromorphic in the entire plane. The origin is a simple pole of e^z/z and is the only singularity of the function.

2 A rational function is meromorphic in the entire plane. Its only singularities are the zeros of the denominator, which are poles of the rational function. (We suppose that the numerator and denominator have no common factor.)

3 The function $\dfrac{1}{\sin z}$ is meromorphic in the entire plane. The only singularities are simple poles at $z = k\pi$, $k = 0, \pm 1, \pm 2, \pm 3, \ldots$.

4 The quotient f/g of two functions holomorphic in D is meromorphic provided the denominator $g \not\equiv 0$. Indeed, the quotient is holomorphic in D except at points where g has a zero. At points where $g(z) = 0$, the quotient f/g has poles, provided $f(z) \neq 0$.

It is easy to show that the set of functions meromorphic in D defines a field. Indeed, if f is meromorphic in D then $1/f$ is holomorphic at points where $f(z) \neq 0$. If $c \in D$ and is a pole of f, then c is a zero of $1/f$, and if $c \in D$ and c is a zero of f, then it is a pole of $1/f$. The field of meromorphic functions is closed with respect to differentiation. In fact, if f is meromorphic, then f' has the same poles as f, but the order of the poles increases by one.

A theorem of G. Mittag-Leffler, which we will not present here, enables us to construct a meromorphic function in the complex plane if its poles and the corresponding principal parts are given. The Mittag-Leffler theorem applies also to meromorphic functions where the number of poles is infinite.

DEFINITION A meromorphic function f is called *periodic* if $f(z + \omega) = f(z) \; \forall \; z \in D$. ω is called the *period* of f.

EXAMPLE

1 The function e^z is periodic with the period $2\pi i$. Indeed, $e^{z+2\pi i} = e^z \cdot e^{2\pi i} = e^z$.

2 The functions $\sin z$ and $\cos z$ are periodic with the period 2π, since $\sin z = \dfrac{1}{2i}(e^{iz} - e^{-iz})$ and $\cos z = \tfrac{1}{2}(e^{iz} + e^{-iz})$.

DEFINITION We call a meromorphic function *elliptic*, or *doubly-periodic*, if the meromorphic function possesses two periods ω_1 and ω_2 whose ratio ω_1/ω_2 is not real.

Clearly, if ω_1 and ω_2 are periods of f, then any combination $\omega = m_1\omega_1 + m_2\omega_2$, where m_1 and m_2 are integers is a period of f. Using Liouville's theorem we can show that an elliptic function is a constant if it has no singularities.

Weierstrass has introduced auxiliary functions which enable us to construct elliptic functions with given periods if we know their zeros and poles.

4. The point at infinity

We shall define the behavior of a function f at infinity by considering the behavior of the function $\phi(t) = f(1/t)$ at the origin. If ϕ is holomorphic, or has a pole of order p or an essential singularity at the origin, we shall say that f shows the same behavior at infinity.

EXAMPLES

1 The function $f(z) = \dfrac{1}{z-1}$ is holomorphic and has a simple zero at infinity since $\phi(t) = \dfrac{t}{1-t}$ exhibits the same behavior at the origin.

2 The function z^2 has a pole of order 2 at infinity since $\phi(t) = 1/t^2$ has a pole of order 2 at $t = 0$.

3 The function e^z has an essential singularity at infinity since $\phi(t) = e^{1/t}$ has an essential singularity at the origin.

4 The polynomial $P(z) = a_0 + a_1 z + \cdots + a_n z^n$ has a pole of order n at infinity since $\phi(t) = a_0 + \dfrac{a_1}{t} + \cdots + \dfrac{a_n}{t^n}$ has a pole of order n at $t = 0$.

5 The rational function $P(z)/Q(z)$ has a pole of order $p - q$ at infinity if the degree p of the polynomial $P(z)$ is greater than the degree q of $Q(z)$; otherwise, it is holomorphic at infinity and has there a zero of order $q - p$.

The rational functions are therefore meromorphic on $\mathbf{C}^\infty$; that is, on the Riemann sphere. As proved below (Theorem 3.14), the converse is true; that is, a function which is meromorphic on the Riemann sphere is rational. Note that there are functions which are meromorphic on the complex plane but not meromorphic on the Riemann sphere.

EXAMPLES

6 The function $e^z + (1/z)$ is meromorphic on the entire plane without being meromorphic on the Riemann sphere, since the point at infinity is an essential singularity of $e^z + (1/z)$. In fact, there is no function—except the constant—which is holomorphic on the Riemann sphere, a consequence of Liouville's theorem.

7 The function $\dfrac{1}{\sin z}$ has as its only singularities the poles $z = k\pi$, $k = 0, \pm 1, \pm 2, \ldots$. Hence $\dfrac{1}{\sin z}$ is meromorphic in $\mathbf{C}$. It is not meromorphic on the Riemann sphere since the point at infinity is a limit point of poles, and thus not an isolated singularity. Thus the point at infinity cannot be a pole, since a pole is an isolated singularity.

THEOREM 3.13 *An entire function f which has a pole of order n at infinity is a polynomial of degree n.*

PROOF

Let f be developed in a Taylor series with center 0:

$$f(z) = a_0 + a_1 z + a_2 z^2 + \cdots$$

The convergence radius of that series is infinite. In order to study the behavior of f at infinity, set

$$\phi(t) = f\left(\frac{1}{t}\right) = a_0 + \frac{a_1}{t} + \frac{a_2}{t^2} + \cdots + \frac{a_n}{t^n} + \cdots$$

Since f has a pole of order n at infinity, $a_n \neq 0$ and $a_k = 0$ for $k > n$. It follows that f is a polynomial of degree n.

THEOREM 3.14 *A function f which is meromorphic on the Riemann sphere is a rational function.*

PROOF

f has only a finite number of poles, since otherwise the limit point of the poles would be a nonisolated singularity, that is, a singularity which is not a pole.

Let $s_1, s_2, \ldots, s_r$ be the poles of f with orders respectively $n_1, n_2, \ldots, n_r$. Then about each singularity s_i the function f possesses a Laurent expansion with principal part P_i:

$$P_i(z) = \frac{a^i_{-1}}{z - s_i} + \frac{a^i_{-2}}{(z - s_i)^2} + \cdots + \frac{a^i_{-n_i}}{(z - s_i)^{n_i}}$$

Clearly P_i is a rational function. On the other hand, $f - P_i$ is holomorphic in a neighborhood of s_i. It follows that the function $f - \sum_{i=1}^{r} P_i$ is an entire function which is either holomorphic at infinity or has a pole at infinity. Hence $f - \sum_{i=1}^{r} P_i$ is either a constant or a polynomial P:

$$f(z) - \sum_{i=1}^{r} P_i(z) = a_0 + a_1 z + \cdots + a_n z^n = P(z)$$

Thus f is a rational function, since P and all the P_i are rational functions.

Problems

3.20.: The Bessel functions are defined by the generating function

$$\exp\left[\frac{x}{2}\left(u - \frac{1}{u}\right)\right] = \sum_{n=-\infty}^{\infty} J_n(x)u^n$$

Show that

$$J_n(x) = \frac{1}{\pi}\int_0^\pi \cos(n\phi - x\sin\phi)\,d\phi$$

or

$$J_n(x) = \sum_{v=0}^{\infty} \frac{(-1)^v \left(\frac{x}{2}\right)^{n+2v}}{v!(n+v)!}$$

3.21.: Determine the Laurent series of the following functions.

a) $\dfrac{e^z - e^{-z}}{z^3}$ about $z = 0$, valid for $|z| > 0$

b) $\dfrac{z^2 + 1}{z - 1}$ about $z = 1$, valid for $|z - 1| > 0$

3.22: Find the Laurent expansion of $\dfrac{1}{(z-1)(z-2)}$ valid for $1 < |z| < 2$.

Hint: Use partial fractions.

3.23.: Find the Laurent series of the following functions:

a) $\dfrac{e^{3z}}{(z-a)^3}$ about $z = a$ b) $(z+3)\sin\dfrac{1}{z-3}$ about $z = 3$

3.24.: Find the Laurent series expansion of:

a) $f(z) = \dfrac{z+a}{z-a}$ about a

b) $f(z) = z\cosh z^{-2}$ about the origin:

3.25.: Expand in a Laurent series:

a) $\dfrac{1 - e^{2z}}{z^4}$ about the origin   b) $\dfrac{1}{z^2(z-1)}$ about $z = 0$ and $z = 1$

c) $\dfrac{e^{2z}}{(z-1)^2}$ about $z = 1$ d) $\dfrac{1 + e^z}{\sin z + z\cos z}$ about $z = 0$

Compute the residue and determine the order of the poles in each case.
3.26.: Determine the nature of the singular points of

$$f(z) = \frac{e^{1/z}}{z(1+z)^2}.$$

3.27.: Expand the function $e^{1/(z-1)}$ in a Laurent series about $z = 1$.
3.28.: Determine the singularities at the point $z = \infty$ of the following functions:

a) $\dfrac{1}{1+e^z}$ b) $e^{-1/z}$

---

§3. Residues and their applications

1. The residue theorem

We defined in 2.2.2 the residue of a function at an isolated singularity c by

$$\text{Res}(f, c) = \frac{1}{2\pi i} \int_\Gamma f(z)\, dz$$

where Γ is a circle with center at c and with a radius sufficiently small to insure that the closed disc $\bar{d}(c, r)$ does not contain any other singularity besides c. If we compare the expression $\text{Res}(f, c)$ with the expression we found for the coefficients of the Laurent series,

$$a_n = \frac{1}{2\pi i} \int_\Gamma \frac{f(u)}{(u-c)^{n+1}}\, du$$

we see immediately that $\text{Res}(f, c) = a_{-1}$.

EXAMPLES

1 $\text{Res}\left(\dfrac{1}{1-z}, 1\right) = 1$

2 $\text{Res}\left(\dfrac{1}{(1-z)^2}, 1\right) = 0$

3 $\text{Res}(e^{1/z}, 0) = 1$
4 $\text{Res}(\cos(1/z), 0) = 0$

There are some practical methods for calculating the residue of a function at a pole. For example, if c is a simple pole of f, then

$$f(z) = \frac{a_{-1}}{z-c} + a_0 + a_1(z-c) + \cdots$$

in some neighborhood of c and

$$\operatorname{Res}(f, c) = a_{-1} = \lim_{z \to c} f(z)(z-c)$$

EXAMPLE

The residue of $f(z) = \dfrac{e^{iz}}{z^2 + a^2}$ (a real) at $z = ai$, which is a simple pole of f, may be obtained by

$$\operatorname{Res}(f, ai) = \lim_{z \to ai} \frac{e^{iz}}{z^2 + a^2}(z - ai) = \lim_{z \to ai} \frac{e^{iz}}{z + ai} = \frac{e^{-a}}{2ai}$$

In general, if f is the quotient of two holomorphic functions, $f(z) = \dfrac{g(z)}{h(z)}$, where h has a simple zero at c with $g(c) \neq 0$, the procedure discussed above leads to the residue given by

$$\operatorname{Res}\left(\frac{g}{h}, c\right) = \lim_{z \to c} (z-c) \frac{g(z)}{h(z)} = \lim_{z \to c} \frac{g(z)}{\frac{h(z) - h(c)}{z - c}} \quad [\text{since } h(c) = 0]$$

$$= \frac{g(c)}{h'(c)}$$

If h has a zero of order p, then $\dfrac{g(z)}{h(z)}(z-c)^p$ is holomorphic at c, and if $\dfrac{g(z)}{h(z)}(z-c)^p = b_0 + b_1(z-c) + \cdots$, then b_{p-1} will be the residue of $f(z) = \dfrac{g(z)}{h(z)}$ at c since it is equal to a_{-1} in the Laurent series for f.

Now we want to show how we may apply the knowledge of the residues at singular points in order to evaluate integrals over closed curves and, later, generalized integrals.

We shall give here a weak version of Cauchy's residue theorem which is very useful in many applications.

THEOREM 3.15 [*Residue Theorem*] *Let D be a simply connected domain and $s_1, s_2, \ldots, s_r$ r points belonging to D. Let γ be a p.w.s. closed curve in D which does not pass through any of the points s_j. Let f be holomorphic in $D\setminus\{s_1, s_2, \ldots, s_r\}$. Then*

$$\int_\gamma f(z)\, dz = 2\pi i \sum_{i=1}^{r} I(\gamma, s_i) \operatorname{Res}(f, s_i)$$

PROOF

The points s_j, $j = 1, 2, \ldots, r$, are isolated singularities of f. Let us write the Laurent series of f around s_j: We find $f(z) = T_j(z) + P_j(z)$ where $T_j(z)$ is the Taylorian part and $P_j(z)$ the principal part of the Laurent series. Clearly, T_j is holomorphic at s_j.

The series for $P_j(z)$,

$$P_j(z) = \frac{a^j_{-1}}{(z - s_j)} + \frac{a^j_{-2}}{(z - s_j)^2} + \cdots$$

converges uniformly on any compact $K \subset \mathbb{C}\setminus\{s_j\}$. It follows that

$$\int_\gamma P_j(z)\, dz = \int_\gamma \sum_{k=1}^{\infty} \frac{a^j_{-k}}{(z-s_j)^k}\, dz = \sum_{k=1}^{\infty} \int_\gamma \frac{a^j_{-k}}{(z-s_j)^k}\, dz$$

$$= a^j_{-1} \int_\gamma \frac{dz}{z - s_j} = \operatorname{Res}(f, s_j) \cdot 2\pi i \cdot I(\gamma, s_j)$$

Now it is easy to see that

$$f - \sum_{j=1}^{r} P_j$$

is holomorphic throughout D. Indeed, f and $\sum_{j=1}^{r} P_j$ have the only singularities $s_1, s_2, \ldots, s_r$ in D. Therefore, $s_1, s_2, \ldots, s_r$ are the only possible singularities in D of $f - \sum_{j=1}^{r} P_j$. Now $f - P_j = T_j$ is holomorphic at s_j, but $\sum_{i \neq j} P_i$ is also holomorphic at s_j. Hence $f - \sum_{j=1}^{r} P_j$ is holomorphic throughout D.

Since D is simply connected, γ is homotopic to a point with respect to D. Therefore,

$$\int_\gamma \left(f(z) - \sum_{j=1}^{r} P_j(z) \right) dz = 0$$

or

$$\int_\gamma f(z)\,dz = \sum_{j=1}^r \int_\gamma P_j(z)\,dz = 2\pi i \sum_{j=1}^r \mathrm{Res}(f, s_j) I(\gamma, s_j)$$

It follows immediately that singularities s_j which are not inside γ need not be taken into account. And if we have $I(\gamma, s_j) = 1$ for all s_j, as happens in most applications, then

$$\int_\gamma f(z)\,dz = 2\pi i \sum_{j=1}^r \mathrm{Res}(f, s_j)$$

2. Applications of the residue theorem

We present some applications of the residue theorem which will allow us to evaluate definite real integrals using complex integrals over closed curves. The techniques used will be illustrated by some typical examples.

CATEGORY 1   Integrals of the form

$$I = \int_0^{2\pi} R(\cos t, \sin t)\,dt$$

where $R(x, y)$ is a rational function of x and y which is continuous on the unit circle $x^2 + y^2 = 1$.

Set $z = e^{it}$, $0 \le t \le 2\pi$. Then

$$I = \int_{\partial(0,1)} R\left[\frac{1}{2}\left(z + \frac{1}{z}\right), \frac{1}{2i}\left(z - \frac{1}{z}\right)\right] \frac{dz}{iz}$$

$$= 2\pi i \sum_{s_j \in d(0,1)} \mathrm{Res}\left[\frac{R[\ ]}{iz}, s_j\right]$$

where s_j ($j = 1, 2, \ldots$) are the singular points of $\dfrac{R[\ ]}{iz}$ inside the unit circle.

$$R[\ ] = R\left[\frac{1}{2}\left(z + \frac{1}{z}\right), \frac{1}{2i}\left(z - \frac{1}{z}\right)\right].$$

EXAMPLE

$$I = \int_0^{2\pi} \frac{dt}{a + \cos t} \quad \text{with } a > 1$$

Setting $z = e^{it}$, we find

$$I = -2i \int_{\partial(0,\,1)} \frac{dz}{z^2 + 2az + 1} = 4\pi \sum_{s_j \in d(0,\,1)} \text{Res}\left(\frac{1}{z^2 + 2az + 1}, s_j\right)$$

The singularities of $f = \dfrac{1}{z^2 + 2az + 1}$ are

$$-a + \sqrt{a^2 - 1} \quad \text{and} \quad -a - \sqrt{a^2 - 1}$$

The only singularity inside $\partial(0, 1)$ is a simple pole at $s = -a + \sqrt{a^2 - 1}$.

In order to find the residue, we use

$$\text{Res}(f, s) = \lim_{z \to s} (z - s) f(z) = \lim_{z \to s} \frac{1}{(z + a + \sqrt{a^2 - 1})} = \frac{1}{2\sqrt{a^2 - 1}}$$

It follows that $I = 2\pi/\sqrt{a^2 - 1}$

CATEGORY 2 Integrals of the form

$$I = \int_{-\infty}^{\infty} R(x)\, dx$$

where $R(z) = P(z)/Q(z)$ is a rational function with no poles on the real axis, and where the degree of $Q(z)$ is at least two units greater than the degree of $P(z)$.

In order to calculate I, we integrate the complex function R over the boundary γ of the semidisc in the upper half plane with center 0 and radius r. (Fig. 3.2).

The boundary γ consists of the segment $(-r, r)$ on the real axis and of the semicircle $\partial(r)$ centered at the origin. If r is large enough, all the poles of $R(z)$ in the upper half plane will lie inside γ, and we have:

$$\int_\gamma R(z)\, dz = \int_{-r}^{r} R(x)\, dx + \int_{\partial(r)} R(z)\, dz = 2\pi i \sum_{\text{Im } s_j > 0} \text{Res}(R(z), s_j)$$

where s_j, $j = 1, 2, \ldots$, are the poles of $R(z)$.

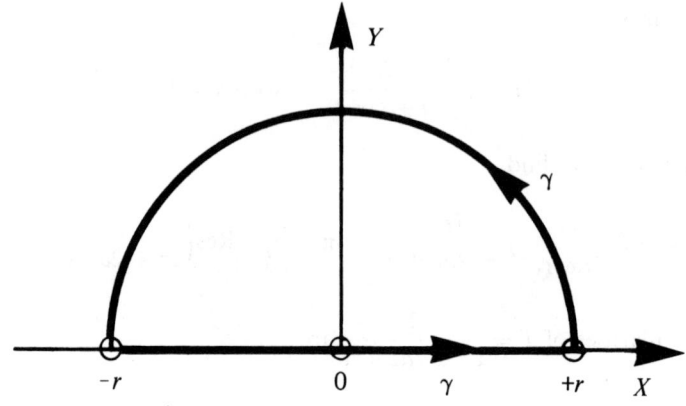

Figure 3.2

Let $r \to \infty$; then

$$\lim_{r \to \infty} \int_{-r}^{r} R(x)\, dx = \int_{-\infty}^{\infty} R(x)\, dx$$

since I is a convergent integral

$$\lim_{r \to \infty} \int_{-r}^{r} R(x)\, dx = \lim_{\substack{a \to \infty \\ b \to \infty}} \int_{-a}^{b} R(x)\, dx$$

Since the degree of $Q(z)$ surpasses the degree of $P(z)$ by at least two units, $\exists$ for sufficiently large r a constant $A \ni |R(z)| < A/r^2$ for all $z \in \partial(r)$. It follows that

$$\left| \int_{\partial(r)} R(z)\, dz \right| \leq \pi r \frac{A}{r^2} = \frac{\pi A}{r}$$

hence $\lim_{r \to \infty} \int_{\partial(r)} R(z)\, dz = 0$ and

$$\int_{-\infty}^{\infty} R(x)\, dx = 2\pi i \sum_{\operatorname{Im} s_j > 0} \operatorname{Res}(R(z), s_j)$$

EXAMPLE

$$I = \int_{-\infty}^{\infty} \frac{dx}{1 + x^2}$$

The only singularity of $R(z) = \dfrac{1}{1+z^2}$ in the upper half plane is a simple pole at $z = i$. We find the residue at $z = i$ by

$$\operatorname{Res}(R(z), i) = \lim_{z \to i} (z - i)R(z) = \lim_{z \to i} \frac{1}{z + i} = \frac{1}{2i}$$

It follows that $I = 2\pi i(1/2i) = \pi$.

CATEGORY 3   Integrals of the form

$$I = \int_{-\infty}^{\infty} f(x)e^{ix}\, dx$$

where f is real valued on the real axis, whose real and imaginary parts are the integrals

$$\int_{-\infty}^{\infty} f(x) \cos x\, dx \quad \text{and} \quad \int_{-\infty}^{\infty} f(x) \sin x\, dx$$

respectively, which are important in applications.

We suppose that f is holomorphic in the upper half plane $\operatorname{Im} z \geq 0$ except at a finite number of points $s_1, s_2, \ldots, s_r$ which do not lie on the real axis. Moreover, we suppose that $\lim_{z \to \infty} f(z) = 0$ for $\operatorname{Im} z \geq 0$.

Now take the rectangle γ with the vertices $b, b + ic, -a + ic, -a$. (Fig. 3.3). Choose a, b, and c sufficiently large to insure that all the singularities $s_1, s_2, \ldots, s_r$ lie inside the rectangle. Then

$$\int_{\gamma} f(z)e^{iz}\, dz = 2\pi i \sum_{j=1}^{r} \operatorname{Res}(f(z)e^{iz}, s_j)$$

It follows that

$$\int_{-a}^{b} f(x)e^{ix}\, dx = 2\pi i \sum_{j=1}^{r} \operatorname{Res}(f(z)e^{iz}, s_j) - \int_{b}^{b+ic} \cdots$$
$$+ \int_{-a+ic}^{b+ic} \cdots + \int_{-a}^{-a+ic} f(z)e^{iz}\, dz$$

We want to show that if $a \to \infty$, $b \to \infty$ independently and if $c \to \infty$, then the three integrals on the right hand converge to 0. For

$$I_1 = \int_{b}^{b+ic} f(z)e^{iz}\, dz$$

176   Chapter 3

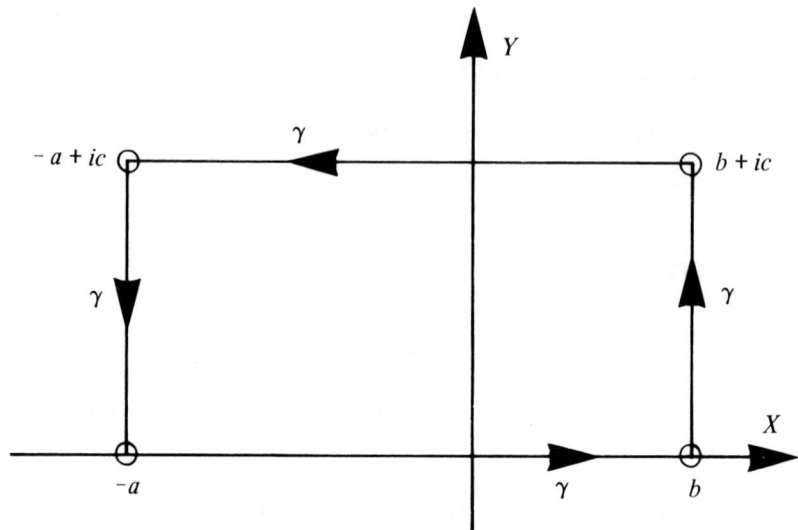

Figure 3.3

set $z = b + iy$; then

$$I_1 = \int_0^c f(b + iy)e^{i(b+iy)} \cdot i \, dy$$

Hence

$$|I_1| \leq M(b) \int_0^c e^{-y} \, dy \leq M(b)$$

where $M(b) = \max_{0 \leq y \leq c} |f(b + iy)|$. In the same way, we find

$$|I_3| = \left| \int_{-a}^{-a+ic} f(z)e^{iz} \, dz \right| \leq M(a)$$

where $M(a) = \max_{0 \leq y \leq c} |f(-a + iy)|$.

For

$$I_2 = \int_{-a+ic}^{b+ic} f(z)e^{iz} \, dz$$

we find, setting $z = x + ic$,

$$|I_2| = \left| \int_{-a}^{b} f(x + ic)e^{i(x+ic)} \, dx \right| \leq M(c) \int_{-a}^{b} e^{-c} \, dx = e^{-c} M(c)(a + b)$$

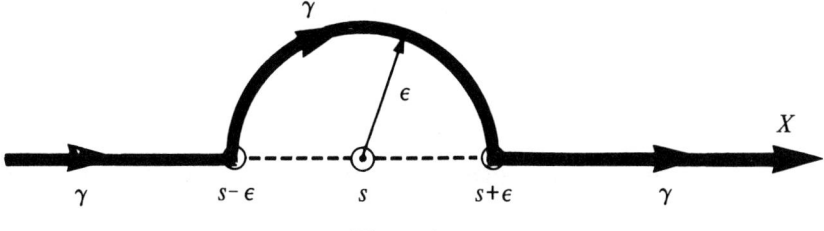

Figure 3.4

Since $\lim_{z \to \infty} f(z) = 0$ for Im $z \geq 0$, it is possible to choose a and b so large that $M(a) \leq \varepsilon$, $M(b) \leq \varepsilon$, and c so large that $e^{-c}M(c)(a+b) < \varepsilon$. It follows then that $|I_1| + |I_2| + |I_3| < 3\varepsilon$. Since ε is arbitrarily small, we have $\lim_{a,b,c \to \infty} (I_1 + I_2 + I_3) = 0$. Consequently, we find

$$\int_{-\infty}^{\infty} f(x)e^{ix}\, dx = 2\pi i \sum_j \operatorname{Res}(f(z)e^{iz}, s_j)$$

EXAMPLE

$$I = \int_{-\infty}^{\infty} \frac{\cos x}{1+x^2}\, dx = \operatorname{Re}\left(\int_{-\infty}^{\infty} \frac{e^{ix}}{1+x^2}\, dx\right)$$

$$= \operatorname{Re}\left[2\pi i \operatorname{Res}\left(\frac{e^{iz}}{1+z^2}, i\right)\right]$$

$$= \operatorname{Re}\left[2\pi i \lim_{z \to i} \frac{e^{iz}}{z+i}\right] = \operatorname{Re}\left[2\pi i \cdot \frac{1}{2ie}\right] = \frac{\pi}{e}$$

We have supposed that f has no singular points on the real axis. If f has a simple pole at s on the real axis, then we modify γ slightly by replacing the interval $[s - \varepsilon, s + \varepsilon]$ on the real axis by the semicircle $\partial(\varepsilon)$ lying in the upper half plane with center at s and radius ε. (Fig. 3.4).

Clearly if f has a simple pole at s, then so does $f(z)e^{iz}$.

Let

$$f(z)e^{iz} = T(z) + \frac{a_{-1}}{z-s}$$

be the Laurent expansion of $f(z)e^{iz}$ about s and $T(z)$ its Taylorian part. Then

$$\int_{\partial(\varepsilon)} f(z)e^{iz}\,dz = \int_{\partial(\varepsilon)} \frac{a_{-1}}{z-s}\,dz + \int_{\partial(\varepsilon)} T(z)\,dz$$

In order to find the value of the first integral on the right-hand side, we set $z - s = e^{it}$ and obtain:

$$\int_{\partial(\varepsilon)} \frac{a_{-1}}{z-s}\,dz = \int_{\pi}^{0} \frac{a_{-1}}{e^{it}} \cdot ie^{it}\,dt = a_{-1}i\int_{\pi}^{0} dt = -a_{-1}\pi i$$

For the second integral we use the inequality

$$\left|\int_{\partial(\varepsilon)} T(z)\,dz\right| \leq M(\varepsilon) \cdot \pi\varepsilon$$

where $M(\varepsilon) = \max\{|T(z)| \,|\, z \in \partial(\varepsilon)\}$. Since $T(z)$ is continuous at s, it follows that

$$\lim_{\varepsilon \to 0} \int_{\partial(\varepsilon)} T(z)\,dz = 0$$

and

$$\lim_{\varepsilon \to 0} \int_{\partial(\varepsilon)} f(z)e^{iz}\,dz = a_{-1}\pi i$$

so that

$$\lim_{\varepsilon \to 0}\left(\int_{-\infty}^{s-\varepsilon}\cdots + \int_{s+\varepsilon}^{\infty} f(x)e^{ix}\,dx\right) = 2\pi i\left[\sum_{\operatorname{Im} s_j > 0} \operatorname{Res}(f(z)e^{iz}, s_j) + \frac{a_{-1}}{2}\right]$$

EXAMPLE

$$I = \int_0^\infty \frac{\sin x}{x}\,dx = \frac{1}{2}\int_{-\infty}^{\infty} \frac{\sin x}{x}\,dx$$

$$= \frac{1}{2}\operatorname{Im}\left(\int_{-\infty}^{\infty} \frac{e^{ix}}{x}\,dx\right)$$

since $\dfrac{\sin x}{x}$ is an even function. Now e^{iz}/z has no singular points in the upper half plane, so

$$\lim_{\varepsilon \to 0}\left(\int_{-\infty}^{-\varepsilon}\cdots + \int_{\varepsilon}^{\infty} \frac{\sin x}{x}\,dx\right) = \frac{1}{2}\operatorname{Im}(\pi i a_{-1})$$

where $a_{-1} = \text{Res}(e^{iz}/z, 0) = 1$. Hence

$$\lim_{\varepsilon \to 0} \int_\varepsilon^\infty \frac{\sin x}{x} dx = \frac{\pi}{2}$$

Now $\dfrac{\sin x}{x}$ is analytic at the origin, so $\int_\varepsilon^\infty \dfrac{\sin x}{x} dx$ is a continuous function of ε; hence we have finally:

$$\int_0^\infty \frac{\sin x}{x} dx = \frac{\pi}{2}$$

For the fourth category, we need the definition of the function z^α for real positive values of α.

Suppose α is a real positive number. We define z^α by $z^\alpha = e^{\alpha \log z}$.
We have

$$|z^\alpha| = |e^{\alpha \log z}| = |e^{\alpha(\log|z| + i \arg z)}| = e^{\alpha \log|z|} = |z|^\alpha$$

and

$$\arg(z^\alpha) = \arg(e^{\alpha \log|z|}) + \arg(e^{i\alpha \arg z}) = \alpha \arg z$$

z^α is holomorphic in $\mathbf{C} \backslash \mathbf{R}_0^-$ as a composite function $z^\alpha = e^z \circ \alpha \log z$ of two holomorphic functions. Clearly, z^α is, like $\log z$, a multivalued function. When z describes a circle about the origin in the positive sense, then $\log z$ increases by $2\pi i$ and z^α is multiplied by $e^{2\pi i \alpha}$.

CATEGORY 4   Integrals of the form

$$I = \int_0^\infty x^\alpha R(x) \, dx$$

with $0 < \alpha < 1$, where $R(z) = P(z)/Q(z)$ is a rational function without poles on the positive x-axis ($z = 0$ may be a simple pole) and with a zero of at least order 2 at infinity. (That is, the degree of $Q(z)$ is greater than the degree of $P(z)$ by at least two units.) The integral converges since, for large values of x, we have

$$x^\alpha R(x) \le \frac{A}{x^{2-\alpha}}$$

To compute the integral we use the closed curve γ which consists of the segment $[\varepsilon, r]$ and the counterclockwise oriented circle $\partial(0, r)$, then

180   Chapter 3

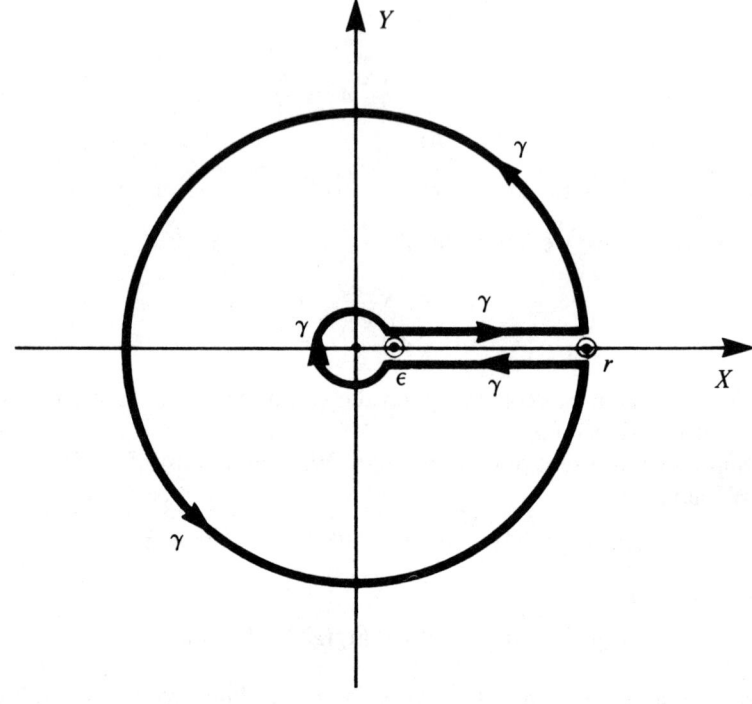

Figure 3.5

the segment $[r, \varepsilon]$ and the clockwise oriented circle $\partial(0, \varepsilon)$. (See Figure 3.5) Let $s_1, s_2, \ldots, s_r$ be the poles of R inside γ. Then

$$\int_\gamma z^\alpha R(z)\, dz = 2\pi i \sum_{j=1}^r \operatorname{Res}(z^\alpha R(z), s_j)$$

We suppose that ε is sufficiently small and r sufficiently large ∋ all singular points of R except the origin lie inside γ.

We start integrating on the segment $[\varepsilon, r]$ with a determination of $z^\alpha R(z)$ which is real on the x-axis. As we have seen above, after turning around the origin on $\partial(0, r)$, $z^\alpha R(z)$ is multiplied by $e^{2\pi i \alpha}$. We have:

$$\int_\gamma z^\alpha R(z)\, dz = \int_\varepsilon^r x^\alpha R(x)\, dx + \int_{\partial(0, r)} z^\alpha R(z)\, dz + \int_r^\varepsilon e^{2\pi i \alpha} x^\alpha R(x)\, dx$$

$$- \int_{\partial(0, \varepsilon)} z^\alpha R(z)\, dz = 2\pi i \sum_{j=1}^r \operatorname{Res}(z^\alpha R(z), s_j)$$

or

$$(1 - e^{2\pi i\alpha})\int_\varepsilon^r x^\alpha R(x)\, dx = 2\pi i \sum_{j=1}^r \text{Res}(z^\alpha R(z), s_j) - \int_{\partial(0,r)} z^\alpha R(z)\, dz$$
$$+ \int_{\partial(0,\varepsilon)} z^\alpha R(z)\, dz$$

Now let $\varepsilon \to 0$ and $r \to \infty$; then

$$\lim_{\substack{\varepsilon \to 0 \\ r \to \infty}} \int_\varepsilon^r x^\alpha R(x)\, dx = \int_0^\infty x^\alpha R(x)\, dx$$

Now

$$\lim_{\varepsilon \to 0} \int_{\partial(0,\varepsilon)} z^\alpha R(z)\, dz = 0$$

since $|z^\alpha R(z)| \leq A/|z|^{1-\alpha}$ so that

$$\left| \int_{\partial(0,\varepsilon)} z^\alpha R(z)\, dz \right| \leq \frac{A}{\varepsilon^{1-\alpha}} \cdot 2\varepsilon\pi = 2A\pi\varepsilon^\alpha$$

From the inequality

$$\left| \int_{\partial(0,r)} z^\alpha R(z)\, dz \right| \leq \frac{A}{r^{2-\alpha}} \cdot 2r\pi = \frac{2A\pi}{r^{1-\alpha}}$$

we find also

$$\lim_{r \to \infty} \int_{\partial(0,r)} z^\alpha R(z)\, dz = 0$$

Thus we have

$$(1 - e^{2\pi i\alpha})I = 2\pi i \sum_{j=1}^r \text{Res}(z^\alpha R(z), s_j)$$

EXAMPLE

Evaluate the integral

$$I = \int_0^\infty \frac{x^\alpha}{x(1+x)}\, dx, \qquad 0 < \alpha < 1$$

Here we find

$$(1 - e^{2\pi i\alpha})I = 2\pi i \, \text{Res}\left(\frac{z^\alpha}{z(1+z)}, -1\right)$$
$$= 2\pi i \lim_{z \to -1} z^{\alpha-1} = -2\pi i (-1)^\alpha$$

or, since $e^{\pi i} = -1$,

$$I = \frac{2\pi i e^{\pi i \alpha}}{e^{2\pi i \alpha} - 1} = \frac{2\pi i}{e^{\pi i \alpha} - e^{-\pi i \alpha}} = \frac{\pi}{\sin \pi \alpha}$$

CATEGORY 5   Integrals of the form

$$I = \int_0^\infty R(x) \log x \, dx$$

where $R(z) = P(z)/Q(z)$ is a rational function without poles at the origin or on the positive x-axis, and where degree $Q(x) \geq$ degree $P(z) + 2$ so that I is convergent.

Consider $\int_\gamma R(z) \log z \, dz$, where γ is the same closed curve as in category 4. On the segment $[\varepsilon, r]$ we take the determination of $\log z$ which is real on the positive x-axis. Then if $s_1, s_2, \ldots, s_r$ are the nonzero singular points of $R(z) \log^2 z$, we find, by letting $\varepsilon \to 0$ and $r \to \infty$, that

$$\int_0^\infty R(x) \log^2 x \, dx - \int_0^\infty R(x)(\log x + 2\pi i)^2 \, dx$$

$$= 2\pi i \sum_{j=1}^r \text{Res}(R(z) \log^2 z, s_j)$$

or

$$4\pi i \int_0^\infty R(x) \log x \, dx - 4\pi^2 \int_0^\infty R(x) \, dx$$

$$= -2\pi i \sum_{j=1}^r \text{Res}[R(z) \log^2 z, s_j]$$

If, in addition, R is real-valued on the real axis, then

$$\int_0^\infty R(x) \log x \, dx = -\frac{1}{2} \text{Re}\left[\sum_{j=1}^r \text{Res}(R(z) \log^2 z, s_j)\right]$$

and

$$\int_0^\infty R(x) \, dx = -\frac{1}{2\pi} \text{Im}\left[\sum_{j=1}^r \text{Res}(R(z) \log^2 z, s_j)\right]$$

EXAMPLE

$$I = \int_0^\infty \frac{\log x}{1+x^2} dx$$

The singularities of $\frac{1}{1+z^2}$ are i and $-i$. Using the relations $\arg i = \pi/2$, $\arg(-i) = (3\pi)/2$, we find

$$\text{Res}\left(\frac{\log^2 z}{1+z^2}, i\right) = \frac{\log^2 i}{2i} = i\frac{\pi^2}{8}$$

and

$$\text{Res}\left(\frac{\log^2 z}{1+z^2}, -i\right) = -\frac{\log^2(-i)}{2i} = -i\frac{9\pi^2}{8}$$

From the above we obtain

$$I = \int_0^\infty \frac{\log x}{1+x^2} dx = 0, \quad \int_0^\infty \frac{dx}{1+x^2} = -\frac{1}{2\pi}(-\pi^2) = \frac{\pi}{2}$$

REMARK

In the example above, we take the branch of log z which is holomorphic in $\mathbf{C}\backslash(\mathbf{R}^+ \cup \{0\})$ and which is real on the x-axis to obtain $\log i = (\pi i)/2$ and $\log(-i) = (3\pi i)/2$. This branch of log z is continuous on the closed curve of Figure 3.5 (Page 180).

It should be pointed out that the five categories of applications of the residue theorem we have shown are only a sample of the many possible applications of that theorem.

3. The logarithmic residue

DEFINITION Let f be meromorphic in D and $c \in D$. The *logarithmic residue of f at c* is the residue of

$$\frac{f'}{f} = \frac{d}{dz}\{\log f(z)\}$$

evaluated at c.

The singularities of f'/f are the zeros of f, since the holomorphicity of f in the neighborhood of a point implies the holomorphicity of f' in the same neighborhood.

If c is a zero of order α of f,
$$f(z) = (z-c)^\alpha g(z)$$
with g holomorphic in a neighborhood of c and $g(c) \neq 0$. Then
$$f' = (z-c)^{\alpha-1} g(z) + (z-c)^\alpha g'(z)$$
and
$$\frac{f'}{f} = \frac{\alpha}{z-c} + \frac{g'}{g}$$
so that we get
$$\operatorname{Res}\left(\frac{f'}{f}, c\right) = \operatorname{Res}\left(\frac{\alpha}{z-c}, c\right) + \operatorname{Res}\left(\frac{g'}{g}, c\right)$$

But $\operatorname{Res}(g'/g, c) = 0$, since g'/g is holomorphic in c, and therefore $\operatorname{Res}(f'/f, c) = \alpha$.

On the other hand, if c is a pole of order β of the function f, then
$$f(z) = (z-c)^{-\beta} g(z)$$
with g holomorphic in a neighborhood of c and $g(c) \neq 0$. It is left to the reader to show that, in this case,
$$\operatorname{Res}\left(\frac{f'}{f}, c\right) = -\beta$$

From the preceding discussion we obtain:

THEOREM 3.16 Let γ be a piecewise smooth closed curve and $I(\gamma, z) = 1$ for all z interior to γ. If f is meromorphic in the interior of γ, and holomorphic and nonzero on γ, we have
$$\frac{1}{2\pi i} \int_\gamma \frac{f'(u)}{f(u)} = Z - P$$
where Z is the number of zeros and P the number of poles of f in the interior of γ.

Note that a zero of order α is counted as α zeros, and a pole of order β is counted as β poles. Remember that
$$\frac{1}{2\pi i} \int_\gamma \frac{f'(u)}{f(u)} du = \frac{1}{2\pi i} \sum_s \operatorname{Res}\left(\frac{f'}{f}, s\right)$$
where the sum is to be taken over all singularities of f'/f in the interior of γ. (Recall Theorem 3.15.)

It is clear that if f is holomorphic on γ and in its interior, then $P = 0$ and we obtain

$$\frac{1}{2\pi i}\int_\gamma \frac{f'(u)}{f(u)}\,du = Z$$

If we set $v = f(u)$ in the integral $\dfrac{1}{2\pi i}\int_\gamma \dfrac{f'(u)}{f(u)}\,du$, then

$$\frac{1}{2\pi i}\int_\gamma \frac{f'(u)}{f(u)}\,du = \frac{1}{2\pi i}\int_\Gamma \frac{dv}{v} = I(\Gamma, 0)$$

where Γ is the image of γ under f. This result may be formulated as follows:

THEOREM 3.17 [*Principle of Variation of the Argument*] *Let f and γ be as in Theorem 3.16., and let Γ be the image of γ under f. Then $I(\Gamma, 0)$, the index of Γ with respect to the origin, is equal to the number of zeros of f inside γ, counted with their multiplicity, minus the number of poles of f inside γ, also counted with their multiplicity.*

THEOREM 3.18 [*Rouché's Theorem*] *Let f and g be holomorphic in the interior of and on the closed piecewise smooth curve γ, which has $I(\gamma, z) = 1$ for all points of int γ. Let us suppose that on γ $|f(z)| > |g(z)|$. Then inside γ the functions f and $f + g$ have the same number of zeros.*

PROOF

The number of zeros of $f + g$ on the interior of γ is given by:

$$Z(f + g) = \frac{1}{2\pi i}\int_\gamma \frac{(f + g)'}{f + g}\,dz$$

If we write $f + g = f\left(1 + \dfrac{g}{f}\right)$ and

$$(f + g)' = f'\left(1 + \frac{g}{f}\right) + f\left(1 + \frac{g}{f}\right)'$$

then

$$Z(f + g) = \frac{1}{2\pi i}\int_\gamma \frac{f'}{f}\,dz + \frac{1}{2\pi i}\int_\gamma \frac{\left(1 + \dfrac{g}{f}\right)'}{1 + \dfrac{g}{f}}\,dz$$

Now setting

$$v = 1 + \frac{g}{f}$$

we obtain:

$$Z(f+g) = \frac{1}{2\pi i}\int_\gamma \frac{f'}{f}\,dz + \frac{1}{2\pi i}\int_\Gamma \frac{dv}{v} = Z_f + I(\Gamma, 0)$$

where Γ is the image of γ by the mapping $1 + g/f$.

On γ we have $\left|\frac{g}{f}\right| < 1$. It follows, for z on γ, that $\left(1 + \frac{g}{f}\right)(z)$ is inside the disc $d(1, 1)$. Therefore $I(\Gamma, 0) = 0$ and $Z(f+g) = Z(f)$.

REMARK

In Rouché's theorem, we must have $|f(z)| > |g(z)|$ on γ. It is not sufficient to suppose $|f(z)| \geq |g(z)|$. Indeed, take $f(z) = z^2(z+2)$ and $g(z) = z$, and γ the circle $|z| = 1$. Then we have

$$|z^2(z+2)| \geq |z|$$

because for $z = -1$ we have $|f(z)| = |g(z)|$, and the number of zeros of f inside γ is 2, while that of $f + g$ is 1.

Applications of Rouché's theorem:

1 Another proof of the Gauss-d'Alembert theorem:
Let $P(z) = a_n z^n + a_{n-1} z^{n-1} + \cdots + a_0$ be a polynomial of degree $n > 0$. Set $f(z) = a_n z^n$ and $g(z) = a_{n-1} z^{n-1} + \cdots + a_0$. Clearly,

$$\frac{|a_n z^n|}{|a_{n-1} z^{n-1} + \cdots + a_0|} = \left|\frac{f(z)}{g(z)}\right|$$

converges to infinity with $z \to \infty$. So, if we choose R sufficiently large, then for $|z| \geq R$ we have $|f(z)| > |g(z)|$, and therefore in the interior of the circle $|z| = R$ the number of zeros of f is equal to the number of zeros of $f + g = P$. $f(z) = a_n z^n$ has a zero of order n inside the circle $|z| = R$. Thus $f + g$ has n zeros inside the same circle.

2 The following theorem is a direct consequence of Rouché's theorem:

THEOREM 3.19 *Let $f_1, f_2, \ldots, f_n, \ldots$ be a sequence of functions holomorphic in D which converge uniformly on every compact domain $K \subset D$ to a nonconstant function f. If $f - W_o$ has a zero of order p at $c \in D$, then in a disc $d(c, r)$, with r arbitrarily small, the function f_n assumes the value W_o p times for all $n > N(r)$.*

PROOF

Let r be so small that $\bar{d}(c, r) \subset D$ and c is the only zero in $d(c, r)$. Then, since $f - W_o$ is continuous and nonzero on the circle $\partial(c, r)$, ∃ a number $m > 0 \ni |f(z) - W_o| > m$ on that circle. On the other hand, the sequence f_n converges uniformly on this circle to f, so that ∃ $N(r) \ni$ for $n > N(r)$ we have $|f_n(z) - f(z)| < m\ \forall\ z \in \partial(c, r)$. So, finally, for $n > N(r)$ we have

$$|f_n(z) - f(z)| < m < |f(z) - W_o|, \qquad \forall\ z \in \partial(c, r)$$

It follows from Rouché's theorem that $f - W_o + f_n - f = f_n - W_o$ and $f - W_o$ have the same number of zeros in the open disc $d(c, r)$, and, for $n > N(r)$, f_n assumes the value W_o p times in the disc $d(c, r)$.

3 Another direct result of Rouché's theorem is:

THEOREM 3.20 *If the functions $f_1, f_2, \ldots, f_n, \ldots$ are holomorphic and univalent in D and converge uniformly on each compact $K \subset D$ to a nonconstant function f, then f is also univalent.*

PROOF

Suppose $f(z_1) = f(z_2)$, $z_1, z_2 \in D$ and $z_1 \neq z_2$. Consider the functions $g_n(z) = f_n(z) - f_n(z_1)$. Clearly $g = \lim_{n \to \infty} g_n$ has a zero at z_1 and at z_2. So for $n > N$, g_n will have a zero in the neighborhood of z_1 and in the neighborhood of z_2. If we pick these neighborhoods so that they have no points in common, then this means that f_n is not univalent, contrary to our hypothesis.

Problems

3.29.: Prove that $\text{Res}(f + g, c) = \text{Res}(f, c) + \text{Res}(g, c)$.
3.30.: Let $g(z)$ be analytic in c and let $f(z)$ have a simple pole at c. Show that $\text{Res}(fg, c) = g(c)\text{Res}(f, c)$.
3.31.: Let $g(z)$ be analytic at c and let $h(z)$ have a pole of order 2 at c. Show that $\text{Res}(g(z)h(z), c) = g'(c)\text{Res}((z - c)h(z), c) + g(c)\text{Res}(h(z), c)$.
3.32.: Verify by direct computation that the sum of the residues of

$$f(z) = \frac{z + a}{z(z^2 + 1)}$$

in the complex plane is zero.

More generally, show that the sum of the residues of the rational function $f(z) = P(z)/Q(z)$ is equal to zero if the degree of $Q(z)$ exceeds the degree of $P(z)$ by at least two units.

3.33.: a) Show that

i) $\operatorname{Res}\left(\dfrac{z}{\sin z}, 0\right) = 0$

ii) $\operatorname{Res}\left(\tan z, \dfrac{\pi}{2}\right) = -1$

iii) $\operatorname{Res}\left(\dfrac{\sin z}{z^2}, 0\right) = 1$

iv) $\operatorname{Res}\left(\dfrac{1 - \cos z}{z^3}, 0\right) = \dfrac{1}{2}$

b) Determine the residues of $\dfrac{z}{(z-1)(z-2)}$ at $z = 1$ and at $z = 2$.

3.34.: Evaluate:

a) $\displaystyle\int_0^\infty \dfrac{x^{\alpha-1}}{1+x^3}\,dx, \qquad 0 < \alpha < 1$

b) $\displaystyle\int_0^\infty \dfrac{x^\alpha}{x^2+a^2}\,dx, \qquad -1 < \alpha < 1,\ a > 0$

3.35.: Evaluate:

a) $\displaystyle\int_0^\infty \dfrac{dx}{x^4+1}$

b) $\displaystyle\int_0^\infty \dfrac{x^2\,dx}{x^4+1}$

c) $\displaystyle\int_0^\infty \dfrac{x^2\,dx}{(x^2+1)^2}$

3.36.: Evaluate:

a) $\displaystyle\int_{-\infty}^\infty \dfrac{\cos x\,dx}{x^4+1}$

b) $\displaystyle\int_{-\infty}^\infty \dfrac{\sin x\,dx}{x^3+1}$

c) $\displaystyle\int_0^\infty \dfrac{\cos x\,dx}{(x^2+1)^2}$

3.37.: Show that:

a) $\int_0^{2\pi} \dfrac{d\phi}{1+\sin^2 \phi} = \pi\sqrt{2}$

b) $\int_0^{2\pi} \dfrac{\sin^2 \phi}{1+\frac{1}{2}\cos \phi}\, d\phi = 4\pi(2-\sqrt{3})$

c) $\int_0^{\pi} \dfrac{d\phi}{(1+\frac{1}{2}\cos \phi)^2} = \dfrac{8\pi}{3\sqrt{3}}$

d) $\int_0^{\pi/2} \dfrac{d\phi}{1+\sin^2 \phi} = \dfrac{\pi}{2\sqrt{2}}$

3.38.: Show that:

a) $\int_0^{\infty} \dfrac{\log x}{(1+x^2)^2}\, dx = -\dfrac{\pi}{4}$

b) $\int_{-\infty}^{\infty} \dfrac{e^{ax}}{1+e^x}\, dx = \dfrac{\pi}{\sin a\pi}, \quad 0 < a < 1$

c) $\int_0^{\infty} \dfrac{(\log x)^2}{1+x^2} = \dfrac{\pi^3}{8}$

d) $\int_0^{\infty} \dfrac{\log(1+x)}{1+x^2}\, dx = \dfrac{\pi \log 2}{2}$

3.39.: Calculate the residues of

$$z\dfrac{f'(z)}{f(z)} \quad \text{and} \quad g(z)\dfrac{f'(z)}{f(z)}$$

at a zero or a pole of $f(z)$ assuming that $g(z)$ is holomorphic at these points.

3.40.: Evaluate:

$$I = \int_0^{\infty} \dfrac{x^{2m}}{1+x^{2n}}\, dx \quad \text{for} \quad n > m \geq 0$$

3.41.: Show that:

$$I = \int_{-\infty}^{\infty} \dfrac{dx}{(1+x^2)^n} = \dfrac{\pi}{2^{n-2}} \cdot \dfrac{(2n-2)!}{[(n-1)!]^2}$$

Problems using Rouché's theorem:

3.42.: Let $a > 0$ and $b > e^a$. Show that the equation $e^{az} = bz^n$ has n roots in $d(0, 1)$. (*Hint*: Show that $|e^{az}| < |bz^n|$ for $|z| = 1$.)

3.43.: Show that two roots of the equation

$$z^4 - 3z^2 + 1 = 0$$

lie in the annulus $1 < |z| < 2$, and that the other two roots lie in the disc $d(0, 1)$. (*Hint*: Show that $|z^4| > |-3z^2 + 1|$ for $|z| = 2$ and $|z^4| < |-3z^2 + 1|$ for $|z| = 1$.)

3.44.: Show that the five roots of the equation $z^5 + 15z + 1 = 0$ lie in the interior of the circle $|z| = 2$.

chapter 4
Conformal Mappings

In this chapter we shall show the equivalence of the two notions "complex differentiable" and "conformal." Moreover, we shall prove Riemann's mapping theorem.

§1. Conformal mappings

1. Definitions and relations

Let $f = U + iV$ be a continuously differentiable mapping $D \to \mathbf{C}$. This means U_x, U_y, V_x, V_y, as well as $\dfrac{\partial f}{\partial x} = U_x + iV_x$, $\dfrac{\partial f}{\partial y} = U_y + iV_y$, and $\dfrac{\partial f}{\partial z} = \dfrac{1}{2}\left(\dfrac{\partial f}{\partial x} - i\dfrac{\partial f}{\partial y}\right)$, $\dfrac{\partial f}{\partial \bar{z}} = \dfrac{1}{2}\left(\dfrac{\partial f}{\partial x} + i\dfrac{\partial f}{\partial y}\right)$, exist and are continuous throughout D. Moreover we, shall suppose that the Jacobian

$$\frac{\partial(U, V)}{\partial(x, y)} \neq 0$$

in D.

Let $\gamma = \gamma(t) = x(t) + iy(t)$, $0 \le t \le 1$, be a smooth arc in D, and let $\Gamma = \Gamma(t) = f(\gamma(t))$, $0 \le t \le 1$, be its image under the mapping f. Then $\Gamma = \Gamma(t)$ is a smooth arc in $f(D)$. Indeed,

$$\Gamma'(t) = \frac{d}{dt}\Gamma(t) = \frac{\partial f}{\partial x}x'(t) + \frac{\partial f}{\partial y}y'(t)$$

$$= (U_x + iV_x)x'(t) + (U_y + iV_y)y'(t) \neq 0$$

since $y'(t) \neq 0$ and $\dfrac{\partial(U, V)}{\partial(x, y)} \neq 0$. So we see that the image of a smooth arc γ by f is again smooth. $\Gamma'(t)$ may also be written: $\Gamma'(t) = \dfrac{\partial f}{\partial z}\gamma'(t) + \dfrac{\partial f}{\partial \bar{z}}\bar{\gamma}'(t)$.

If $\gamma_1 = \gamma_1(t)$ and $\gamma_2 = \gamma_2(t)$ are two smooth arcs in D passing through $c = \gamma_1(t_o) = \gamma_2(t_o)$, then the angle between the two arcs is defined by:

$$\text{angle}(\gamma_1, \gamma_2, c) = \arg \gamma_2'(t_o) - \arg \gamma_1'(t_o)$$

Let Γ_1 and Γ_2 be the images under f of γ_1 and γ_2, respectively. Then the angle$(\Gamma_1, \Gamma_2, f(c))$ between the two arcs Γ_1 and Γ_2 at $f(c)$ is determined by

$$\text{angle}(\Gamma_1, \Gamma_2, f(c)) = \arg(\Gamma_2'(t_o)) - \arg(\Gamma_1'(t_o))$$

DEFINITION 1 The continuously differentiable mapping f is conformal at c if, for each pair (γ_1, γ_2) of smooth arcs passing through c, we have

$$\text{angle}(\Gamma_1, \Gamma_2, f(c)) = \text{angle}(\gamma_1, \gamma_2, c)$$

DEFINITION 2 The continuously differentiable mapping f is conformal at c if

$$\arg \Gamma'(t_o) - \arg \gamma'(t_o)$$

does not depend on γ.

It is easy to show that the two definitions are equivalent: angle$(\Gamma_1, \Gamma_2, f(c)) =$ angle$(\gamma_1, \gamma_2, c) \Leftrightarrow \arg \Gamma_2'(t_o) - \arg \Gamma_1'(t_o) = \arg \gamma_2'(t_o) - \arg \gamma_1'(t_o) \Leftrightarrow \arg \Gamma_2'(t_o) - \arg \gamma_2'(t_o) = \arg \Gamma_1'(t_o) - \arg \gamma_1'(t_o)$.

DEFINITION The continuously differentiable mapping $f: D \to \mathbf{C}$ is conformal in D if it is conformal at all but isolated points of D.

THEOREM 4.1 *The notions "complex differentiable in D" and "conformal in D" are equivalent.*

PROOF

We use definition 2 of conformality, and suppose f is not a constant. (If f is a constant, then it may be considered to be conformal.)

(i) f *is complex differentiable in* $D \Rightarrow f$ *is conformal in* D: If f is complex differentiable and not constant in D, then $\dfrac{\partial(U, V)}{\partial(x, y)} = |f'|^2 = 0$ only at isolated points (see 3.1.7).

Conformal Mappings

Now

$$\Gamma'(t_o) = \frac{\partial f}{\partial z}\gamma'(t_o) + \frac{\partial f}{\partial \bar{z}}\bar{\gamma}'(t_o)$$

and

$$\arg \Gamma'(t_o) = \arg \gamma'(t_o) + \arg\left[\frac{\partial f}{\partial z} + \frac{\partial f}{\partial \bar{z}}\frac{\bar{\gamma}'(t_o)}{\gamma'(t_o)}\right]$$

and, since f is complex differentiable, we have $\frac{\partial f}{\partial \bar{z}} = 0$ and $\frac{\partial f}{\partial z} = f'$. Hence

$$\arg \Gamma'(t_o) - \arg \gamma'(t_o) = \arg f'(c)$$

does not depend on γ.

(ii) *f is conformal in $D \Rightarrow f$ is complex differentiable in D:* Clearly, if $\frac{\partial f}{\partial \bar{z}} \neq 0$, then

$$\arg \Gamma'(t_o) - \arg \gamma'(t_o) = \arg\left[\frac{\partial f}{\partial z} + \frac{\partial f}{\partial \bar{z}}\frac{\bar{\gamma}'(t_o)}{\gamma'(t_o)}\right]$$

depends on γ, since for $t = t_o$ and γ varying, but passing through c, we have

$$\frac{\bar{\gamma}'(t)}{\gamma'(t)} = e^{-2i \arg \gamma'(t)}$$

since

$$\left|\frac{\bar{\gamma}'(t)}{\gamma'(t)}\right| = 1 \quad \text{and} \quad \arg \frac{\bar{\gamma}'(t)}{\gamma'(t)} = -2 \arg \gamma'(t)$$

Hence the complex number $\frac{\partial f}{\partial z} + \frac{\partial f}{\partial \bar{z}}\frac{\bar{\gamma}'(t)}{\gamma'(t)}$ describes a circle with center $\frac{\partial f}{\partial z}$ and radius $\left|\frac{\partial f}{\partial \bar{z}}\right|$. It follows that $\arg \Gamma'(t_o) - \arg \gamma'(t_o)$ is not constant, that is, it depends on $\gamma'(t)$, and f is not conformal.

If f is continuously differentiable with respect to x and y, but not complex differentiable, and if the quotient $\frac{f_z + f_{\bar{z}}}{f_z - f_{\bar{z}}}$ remains bounded with modulus $\leq k$, we say that the mapping f is *k-quasi-conformal*. When f is complex differentiable, then $f_{\bar{z}} \equiv 0$ and f is conformal, and therefore 1-quasi-conformal.

2. Mapping at points where $f' = 0$

Let the nonconstant mapping f be complex differentiable in D and let $f'(c) = 0$ for some $c \in D$. We want to show that angles are not preserved at c, but are multiplied by some positive integer.

Since f is analytic in D, we have

$$f(z) = f(c) + f'(c)(z - c) + \cdots$$

Suppose that c is a zero of order n of $f(z) - f(c)$; that is,

$$f'(c) = f''(c) = \cdots = f^{(n-1)}(c) = 0 \quad \text{and} \quad f^{(n)}(c) \neq 0$$

Then

$$f(z) - f(c) = (z - c)^n [a_n + a_{n+1}(z - c) + \cdots] = (z - c)^n g(z)$$

where g is complex differentiable at c with $g(c) = a_n \neq 0$.

Now let γ_1, γ_2 be two smooth arcs passing through c, and let Γ_1 and Γ_2 be their respective images under f. Take $z_1 \neq c$ on γ_1 and $z_2 \neq c$ on γ_2, and let $z_1 \to c$ on γ_1 and $z_2 \to c$ on γ_2. It follows that

$$\text{angle}(\Gamma_1, \Gamma_2, f(c)) = \lim_{\substack{z_1 \to c \\ z_2 \to c}} \frac{f(z_2) - f(c)}{f(z_1) - f(c)}$$

$$= \lim_{\substack{z_1 \to c \\ z_2 \to c}} \frac{(z_2 - c)^n g(z_2)}{(z_1 - c)^n g(z_1)}$$

$$= \lim_{\substack{z_1 \to c \\ z_2 \to c}} n \arg \frac{z_2 - c}{z_1 - c} + \lim_{\substack{z_1 \to c \\ z_2 \to c}} \arg \frac{g(z_2)}{g(z_1)}$$

$$= n \lim_{\substack{z_1 \to c \\ z_2 \to c}} \arg \frac{z_2 - c}{z_1 - c} + \arg \lim_{\substack{z_1 \to c \\ z_2 \to c}} \frac{g(z_2)}{g(z_1)}$$

$$= n \, \text{angle}(\gamma_1, \gamma_2, c)$$

since

$$\lim_{\substack{z_1 \to c \\ z_2 \to c}} \frac{g(z_2)}{g(z_1)} = 1 \quad \text{and} \quad \arg 1 = 0$$

In conclusion, we see that at points c where $f'(c) = 0$ the angles are multiplied by a positive integer which is equal to the order of the zero of $f(z) - f(c)$ at c, and so the inverse function is many-valued.

EXAMPLE

$f(z) = z^n$. The origin is a zero of order n of $f(z) - f(0)$. It follows that the angles at the origin are multiplied by n. This can be seen also by the relation

$$\arg z^n = n \arg z$$

We have seen in 1.2.3 that the inverse function $\sqrt[n]{z}$ is a many valued function: it has n determinations which are holomorphic in $\mathbf{C}\backslash(\mathbf{R}^- \cup \{0\})$, namely:

$$(\sqrt[n]{z})_k = |\sqrt[n]{z}| \cdot \exp\left(\frac{i}{n} \operatorname{Arg} z + \frac{2k\pi i}{n}\right), \qquad k = 1, 2, \ldots, n$$

We can also write

$$(\sqrt[n]{z})_k = \exp\left(\frac{1}{n} \operatorname{Log} z + \frac{2k\pi i}{n}\right), \qquad k = 1, 2, \ldots$$

which shows the holomorphicity of all the determinations in $\mathbf{C}\backslash(\mathbf{R}^- \cup \{0\})$. (Note in the second representation that although k can take all values $1, 2, \ldots$, we have only n determinations since $\exp(2\pi i) = 1$.)

In any simply connected domain which does not contain the origin, it is possible to define a branch of $\sqrt[n]{z}$ similar to that for $\log z$.

If f is multivalued, then at points where $f' = 0$ angles may be multiplied by any positive real number.

EXAMPLE

$f(z) = z^\alpha$, with α real and positive. As we have seen in 3.3.2, f is defined by

$$f(z) = e^{\alpha \log z}$$

which implies $\arg f(z) = \alpha \arg z$ so that at $z = 0$ angles are multiplied by α.

In any simply connected domain D which does not contain the origin, it is possible to define a branch of $f = z^\alpha$, as was done for $\log z$, so that D and $f(D)$ are isomorphic.

3. The open mapping theorem

We want to show that an analytic function maps open sets onto open sets; that is, that f is an open mapping (see 0.3.2).

THEOREM 4.2 *Let f be a nonconstant analytic function on D. Then f is an open mapping; that is, if O is an open subset of D, then $f(O)$ is open. (Since continuous functions preserve connectivity, we may also state this as: If $D_1 \subset D$ is a domain, then $f(D_1)$ is a domain.)*

PROOF

Let $c \in D_1$; we wish to show that $\exists\, \rho > 0 \ni d(f(c), \rho) \subset f(D_1)$.

Clearly, $f(z) - f(c)$ is analytic in D and has a zero at c; let $p \geq 1$ be its order. Then $f(z) - f(c) = (z - c)^p g(z)$ with $g(c) \neq 0$.

Choose r sufficiently small so that $\bar{d}(c, r) \subset D_1$ and c is the only zero of $f(z) - f(c)$ in $\bar{d}(c, r)$. Denote by ρ the minimum of $|f(z) - f(c)|$ on the circle $\partial(c, r)$. Clearly $\rho > 0$; otherwise, $f(z) = f(c)$ for some z on $\partial(c, r)$.

Now, if ρ_1 is arbitrary but less than ρ, then for any ϕ, $0 \leq \phi \leq 2\pi$, we have

$$|\rho_1 e^{i\phi}| < |f(z) - f(c)|, \quad \forall z \in \partial(c, r)$$

Applying the Rouché's theorem it follows that $f(z) - f(c)$ and $f(z) - f(c) - \rho_1 e^{i\phi}$ have the same number of zeros—that is, p zeros in $d(c, r)$. Therefore, the values of f cover the disc $d(f(c), \rho)$ p times. This proves that f is an open mapping.

If $f'(c) \neq 0$, then $p = 1$, and the values of f cover $d(f(c), \rho)$ only once. Also, since f is continuous, $\exists\, \varepsilon > 0 \ni f(d(c, \varepsilon)) \subset d(f(c), \rho)$. f then defines a one-to-one correspondence between $d(c, \varepsilon)$ and $f(d(c, \varepsilon))$. In conclusion, we state that at points where $f' \neq 0$, f possesses locally—that is, in some neighborhood—an inverse f^{-1}.

If $f'(c) = 0$, then $p \geq 2$, and the values of f cover $d(f(c), \rho)$ at least twice. f is therefore certainly not one-to-one on $d(c, r)$.

Let $w = f(z)$. Then at points where $f'(c) = 0$, $w - f(c) = f(z) - f(c) = (z - c)^p g(z)$, with $p \geq 2$ and $g(c) \neq 0$. f behaves locally much like the function $(z - c)^p$, whose inverse possesses p holomorphic determinations.

To see this, we choose one of the p determinations of $\sqrt[p]{g(z)}$. We find that the derivative of $(z - c)\sqrt[p]{g(z)}$ at $z = c$ is nonzero:

$$[(z - c)\sqrt[p]{g(z)}]'_{z=c} = [(z - c)(\sqrt[p]{g(z)})' + \sqrt[p]{g(z)}]_{z=c} = \sqrt[p]{g(c)} \neq 0$$

Hence $t = (z - c)\sqrt[p]{g(z)}$ possesses locally a holomorphic inverse function $z = \phi(t) = \phi(\sqrt[p]{w - f(c)})$, so that at points c where $f - f(c)$ has a zero of order p, f possesses locally p inverse functions $\phi(\sqrt[p]{w - f(c)})$, according to the choice of the determination of $\sqrt[p]{w - f(c)}$. We say then that c is a *critical point of order p*.

THEOREM 4.3 Let $f \in H(D)$ and let $\gamma \subset D$ be a simple p.w.s. closed curve, with int $\gamma \subset D$. If $\Gamma = f(\gamma)$ is simple, and if $f^{-1}(\Gamma) = \gamma$, or at least $f^{-1}(\Gamma) \cap D = \gamma$, then f defines an isomorphism int $\gamma \to$ int Γ.

PROOF

The only thing we have to show is that f is one-to-one from int γ onto int Γ.

(i) $c \in$ int $\gamma \Rightarrow f(c) \in$ int Γ: Let $c \in$ int γ and denote by Z the number of zeros of $f - f(c)$ in int γ. Then we have:

$$Z = \frac{1}{2\pi i} \int_\gamma \frac{f'(z)}{f(z) - f(c)} dz = \frac{1}{2\pi i} \int_\Gamma \frac{du}{u - f(c)} = I(\Gamma, f(c))$$

Now the number of zeros of $f - f(c)$ in int γ is at least one; hence $Z \geq 1$. But, since $f(c) \notin \Gamma$ and Γ is simple, we have that $I(\Gamma, f(c))$ is either ± 1 if $f(c) \in$ int Γ, or 0 if $f(c) \notin$ int Γ. Hence $I(\Gamma, f(c)) = 1$ and $f(c) \in$ int Γ.

(ii) $w_0 \in$ int $\Gamma \Rightarrow \exists$ one and only one $z_0 \in$ int $\gamma \ni f(z_0) = w_0$: The number of times f attains the value w_0 is nonnegative and is given by

$$Z = \frac{1}{2\pi i} \int_\gamma \frac{f'(z) \, dz}{f(z) - w_0} = \frac{1}{2\pi i} \int_\Gamma \frac{du}{u - w_0} = I(\Gamma, w_0) = 1$$

since $w_0 \in$ int Γ and Γ is simple.

EXAMPLE

Take $f(z) = e^z$ and let the domain $D = \{z \mid -a < \text{Re } z < a, -c < \text{Im } z < c < \pi\}$. D is an open rectangle with boundary γ. (Figure 4.1)

$e^z = \exp z = \exp x \cdot \exp iy$ maps a straight line parallel to the x-axis into a ray from the origin with argument y. A parallel to the y-axis is mapped onto a circle $\partial(0, r)$ with $r = |\exp z| = \exp x$. Thus, D is transformed by f into $f(D)$:

$$f(D) = \{w \mid -c < \arg w < c, e^{-a} < |w| < e^a\}$$

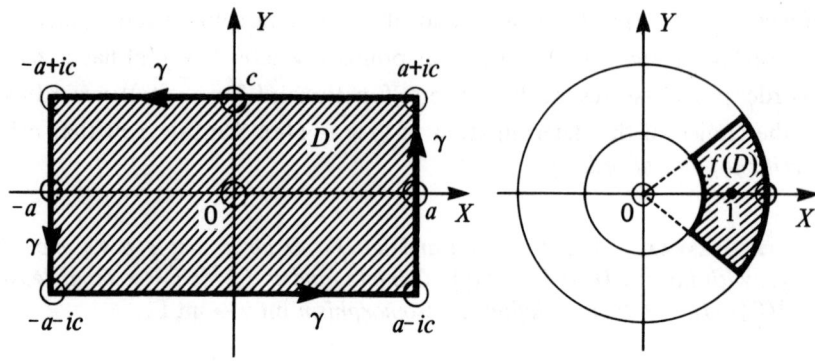

Figure 4.1

If we denote by γ the boundary of D and by Γ the boundary of the image $f(D)$, then obviously $f(\gamma) = \Gamma$ and $f^{-1}(\Gamma) \cap D = \gamma$. Both boundaries γ and Γ are simple; hence, using Theorem 4.3, D and $f(D)$ are isomorphic. Since this is true for all positive values of $c < \pi$ and for all values of a, we conclude that the strip $St = \{z \mid -\infty < \operatorname{Re} z < \infty, \ -\pi < \operatorname{Im} z < \pi\}$ is isomorphic to $\mathbf{C}\backslash(\mathbf{R}^- \cup \{0\})$.

A set which is isomorphic to $\mathbf{C}$ or to $\mathbf{C}$ minus one or several rays is called *a fundamental set*.

The strip $St_0 = \{z \mid -\infty < \operatorname{Re} z < \infty, \ -\pi < \operatorname{Im} z < \pi\}$ is a fundamental set for $f(z) = e^z$. Indeed, St_0 is isomorphic to $\mathbf{C}$ minus the negative x-axis. The inverse mapping $f^{-1}: f(D) \to D$ is obviously the principle value of $\log z$.

Clearly, any strip $St_k = \{z \mid -\infty < \operatorname{Re} z < \infty, \ (k-1)\pi < \operatorname{Im} z < (k+1)\pi\}$, where k is an integer, is a fundamental set.

By this example, we grasp the geometric behavior of $\exp z$ at the point at infinity, which is an essential singularity. We see that, however large we take R, outside of $d(0, R)$ $\exp z$ assumes all values except the value 0 an infinite number of times. Also, since the function $1/z$ transforms the outside of $d(0, R)$ into the inside of $d(0, 1/R)$, it follows that, however small we take ε, the function $\exp(1/z)$ assumes inside of $d(0, \varepsilon)$ all complex values except 0 an infinite number of times.

Problems

4.1.: Consider the mapping $w = z^2$. Find the image of the following curves:
 a) $z = re^{it}, \quad 0 \le t \le 2\pi$
 b) $y = x + b$

c) $z = 1 + e^{it}$, $0 \le t \le 2\pi$
d) $x = a$
e) $y = b$

4.2.: Consider the mapping $w = \sqrt{1-z^2}$. Determine $a \ni$ the hyperbola $x^2 - y^2 = a^2$ is mapped into itself.

4.3.: Consider the mapping $w = \sin z$. Find the images of the following domains:

a) $-\dfrac{\pi}{2} \le \operatorname{Re} z \le \dfrac{\pi}{2}$

b) $-\dfrac{\pi}{2} + k\pi \le \operatorname{Re} z \le \dfrac{\pi}{2} + k\pi$

4.4.: Consider the mapping $w = z^2$. Calculate the area of the image of the square $0 \le x \le 1, 0 \le y \le 1$ by using the relation $J = |f'|^2$ where J is the Jacobian of the mapping, or by determining the image of the square.

4.5.: Consider the mapping $w = 1 + z^2$. Find the image C' of the circle $C: |z - 1| = \sqrt{2}$, and show that the angle between the tangent to C at z_0 and the tangent to C' at $1 + z_0^2$ is equal to $\arg(w')_{z_0} = \arg 2z_0 = \arg z_0$

4.6.: Determine the locus of points for which the dilatation ratio is c $(c > 0)$ under the transformation

$$w = az^2 + bz$$

4.7.: Determine the locus of points in which the tangent to a curve does not change its direction under the transformation

$$w = az^2 + bz$$

4.8.: Show that $w = e^z$ maps the lines $y = \alpha x$ into the spirals $\rho = \exp(\phi/\alpha)$.

4.9.: Show that $w = z + 1/z$ maps the circles $|z| = c$ into ellipses and the rays $\arg z = \phi$ into hyperbolas. Determine the foci.

4.10.: Find the image of $\operatorname{Re} z > 0$ and $\operatorname{Im} z > 0$ under the mapping $w = \log z$.

4.11.: Show that the image of the strip $-\pi < y \le \pi$ under the mapping $z \to 1 + e^z$ is $\mathbf{C} \backslash \{1\}$.

---

§2. Homographies

1. Definitions and basic properties

A *homography*, or *Moebius transformation*, is a transformation of the form:

$$w = \frac{az + b}{cz + d}$$

with $ad - bc \ne 0$.

If $c = 0$ and $d = 1$ then the transformation reduces to

$$w = az + b$$

and is called an *affine transformation*.

Let us recall the definitions of 1.4.6: If f is an isomorphism on D, then we say that D and $f(D)$ are isomorphic. If $f(D) = D$, then we say that f is an *automorphism*. The set of all automorphisms of D will be denoted by $A(D)$.

It is easy to see that $A(D)$ is a group (see 0.2.1).

We denote the set of the affine transformations by L. We see easily that an affine transformation defines an automorphism of the complex plane, that is, $L \subset A(\mathbf{C})$, and that the set L of all affine transformations forms a group $(L, \circ)$. But more can be shown:

THEOREM 4.4 *The group of automorphisms of the complex plane is made up of affine transformations*: $A(\mathbf{C}) = L$.

PROOF

Let $w = f(z)$ be an automorphism of the complex plane. Then f is holomorphic on $\mathbf{C}$. By Liouville's theorem, it would be a constant if it were also holomorphic at infinity. The point at infinity is an isolated singularity which cannot be an essential singularity, because otherwise the image of a disc $d(0, \varepsilon)$ under $\phi(z) = f(1/z)$ would be dense in the plane, or—equivalently—the image of the exterior of a disc $d(0, 1/\varepsilon)$ would be dense in the complex plane; this is impossible, because the image of the exterior of $d(0, 1/\varepsilon)$ does not intersect the image of $d(0, 1/\varepsilon)$, which is an open set.

Thus f has a pole at ∞, and is therefore a polynomial. But a polynomial is not univalent unless its degree is one, since, by the theorem of Gauss-d'Alembert, the equation $f(z) - c = 0$ which is of degree n assumes n solutions, which are all different except at isolated values of c.

The group L of affine transformations is transitive in the complex plane, which means that if we select arbitrarily two points z_1 and z_2 in the plane, we can always associate them by an affine transformation $w = az + b$ which transforms z_1 to z_2.

The subgroup of affine transformations which leave the point z_0 fixed is given by

$$w - z_0 = a(z - z_0)$$

where a is an arbitrary complex number.

Any homographic transformation

$$w = \frac{az+b}{cz+d}, \qquad ad - bc \neq 0$$

defines a mapping of the Riemann sphere S^2 onto itself which transforms $z = \infty$ to $w = a/c$ and $z = -d/c$ ($c \neq 0$) to $w = \infty$. If $c = 0$, then the transformation is affine and leaves the point at infinity fixed.

Obviously, the representation of a homography by

$$h(z) = \frac{az+b}{cz+d}$$

is not unique since we can divide numerator and denominator of the fraction by the same number. We choose this number ∋ the determinant $ad - bc = 1$. It is then easy to check that the product of composition $h_1 \circ h_2$ of two homographies h_1 and h_2 is again a homography and that the inverse of the homography h exists and is given by

$$h^{-1}(z) = \frac{dz-b}{-cz+a}$$

This proves that the homographies form a group. We denote it by H.

To each homography $h = \dfrac{az+b}{cz+d}$ with determinant 1, there corresponds a matrix

$$M(h) = \begin{pmatrix} a & b \\ c & d \end{pmatrix}$$

with the following properties:

a) $\quad M(h_1 \circ h_2) = M(h_1) \cdot M(h_2)$
b) $\quad M(h^{-1}) = M(h)^{-1}$

So ∃ a group isomorphism between H and the 2×2 complex matrices with determinant 1.

In the topology induced by $\mathbf{R}^3$ on the Riemann sphere a homography h is a homeomorphism. Since we can define a derivative $h'(z)$ at any point of S^2 ($S^2 = \mathbf{C}^\infty$), we shall call h an automorphism of S^2.

Let $A(S^2)$ be the group of all automorphisms of S^2.

THEOREM 4.5 *The group of automorphisms of the sphere equals H:* $A(S^2) = H$.

PROOF

H is transitive on the sphere S^2. Indeed, it contains the subgroup of the affine transformations L, which is transitive in the plane, and it contains a transformation by which the point $-d/c$ is transformed to ∞ and the point ∞ to a/c.

On the other hand, the subgroup which leaves the point at infinity fixed is the group of automorphisms of the plane. Thus it is identical with the group L of the affine transformations $w = az + b$.

Let $T \in A(S^2)$. Then either T leaves the point at ∞ fixed and $T \in L \subset H$, or $\exists$ a $z_0 \ni T(\infty) = z_0$. By virtue of the transitivity of H $\exists$ also a homographic transformation $h_1 \ni h_1(\infty) = z_0$. $h_1^{-1} \circ T$ is an automorphism of S^2 which leaves the point at infinity fixed. Hence $h_1^{-1} \circ T$ is an element L_1 of L. It follows that $T = h_1 \circ L_1 \in H$, or $A(S^2) = H$.

Among the homographic transformations there is one that is particularly simple, namely, the inversion I defined by $w = 1/z$. We have $|w| = 1/|z|$ and $\arg w = -\arg z$.

This transformation can be carried out in two steps:

1. Change z to z_1 by leaving the argument unchanged and by altering the modulus:

$$|z_1| = \frac{1}{|z|} \quad \text{and} \quad \arg z_1 = \arg z$$

This amounts to a symmetry with respect to the unit circle (see Figure 4.2).

2. Change z_1 to $\bar{z}_1$, which is a symmetry with respect to the x-axis; this yields finally $w = 1/z = \bar{z}_1$.

THEOREM 4.6 *Every homographic transformation* $h(z) = \dfrac{az + b}{cz + d}$ *can be generated by two affine transformations* L_1, L_2, *and an inversion* I.

PROOF

Set

$$z' = cz + d \quad \text{(affine transformation } L_1\text{)}$$

$$z'' = \frac{1}{z'} \quad \text{(inversion } I\text{)}$$

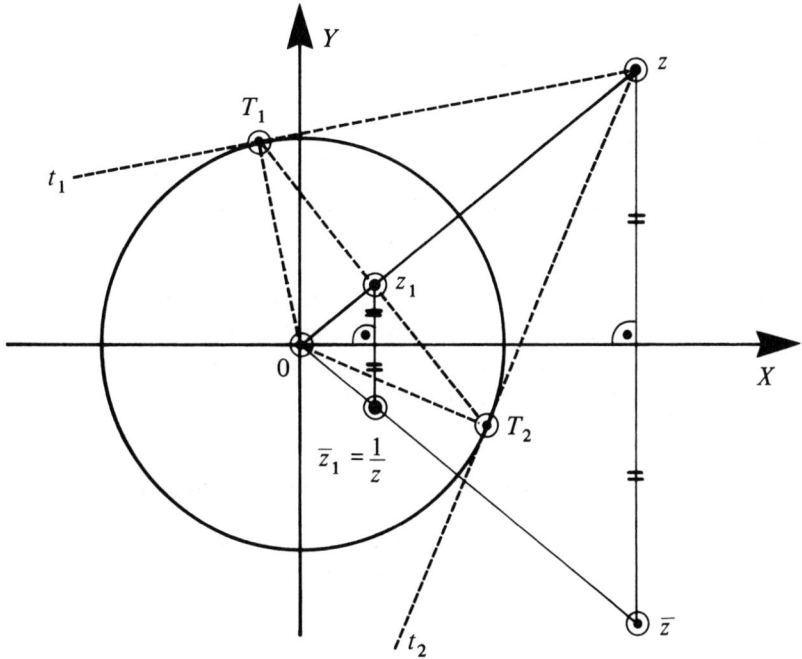

Figure 4.2

and
$$w = \frac{a}{c} + \frac{bc - ad}{d} z'' \quad \text{(affine transformation } L_2\text{)}$$

which leads to
$$w = \frac{az + b}{cz + d}$$

and $h = L_2 \circ I \circ L_1$.

THEOREM 4.7 *Homographic transformations preserve the set of circles and straight lines: a straight line or a circle is transformed either into a straight line or a circle.*

PROOF

This follows immediately for homographic transformations of the type $L: w = az + b$ because they consist of translations, rotations, and homothetic transformations (i.e., dilatations). It can be shown by a simple calculation that Theorem 4.7 also holds for the inversion:

The family of circles and straight lines is represented by the equation

$$\alpha z \bar{z} + \beta z + \overline{\beta z} + \gamma = 0$$

where α and γ are real. (If $\alpha = 0$ then the above equation reduces to a straight line). If an inversion $I : w = 1/z$ is carried out, we find that $\alpha z \bar{z} + \beta z + \overline{\beta z} + \gamma = 0$ is carried into $\alpha + \beta w + \overline{\beta w} + \gamma w \bar{w} = 0$, which is again a circle or a straight line, depending whether $\gamma \neq 0$ or $\gamma = 0$.

Theorem 4.7 is then a consequence of Theorem 4.6.

It can easily be verified that $w = 1/z$ transforms the interior of the unit circle into its exterior, from which we conclude that the interior of the unit circle is isomorphic to its exterior, or, more generally, the interior of every circle is isomorphic to its exterior.

DEFINITION The cross ratio of four points is defined by

$$(z_1, z_2, z_3, z_4) = \frac{(z_1 - z_2)(z_3 - z_4)}{(z_1 - z_3)(z_2 - z_4)}$$

THEOREM 4.8 *Homographies leave the cross ratio invariant.*

The verification is easily done, first for $w = az + b$, and then for $w = 1/z$.

This property permits us to write down immediately the homographic transformation which carries the three points z_1, z_2, z_3 into the points w_1, w_2, w_3:

$$\frac{w - w_1}{w - w_2} \cdot \frac{w_2 - w_3}{w_1 - w_3} = \frac{z - z_1}{z - z_2} \cdot \frac{z_2 - z_3}{z_1 - z_3}$$

2. Symmetry

DEFINITION We say that the points z and z^* are symmetric with respect to the circle (straight line) $C(\alpha, R)$ if all circles Γ passing through z and z^* cut C at a right angle.

This implies (a) that the power of the center α with respect to each of the circles Γ equals R^2, (b) that the points z and z^* are associated with α by the relation $|z - \alpha||z^* - \alpha| = R^2$.

THEOREM 4.9 [*The Principle of Symmetry*] *The homographies preserve the symmetry; that is: if C is a circle and C' is its image, and z and z* are symmetric with respect to C, then their images w and w* are symmetric with respect to C'.*

Indeed, it follows immediately from Theorems 4.1 and 4.7 that the circles and lines Γ which pass through z and z^* are transformed into circles and straight lines Γ' which pass through w and w^*. Since homographies are conformal, Γ' cuts C' at a right angle. So w and w^* are symmetric with respect to C'.

The principle of symmetry can be utilized in the study of a homography which carries a given circle (straight line) C into another circle (straight line) C'. We pick a point z_1 on C and its image w_1 on C', and a pair z_2, z_2^* symmetric with respect to C and their images, a pair w_2, w_2^* points symmetric with respect to C'. The transformation is then determined by

$$(w, w_1, w_2, w_2^*) = (z, z_1, z_2, z_2^*)$$

The circle C' is determined by the three points w_1, w_2, w_2^*; its center α' lies on the straight line through w_2 and w_2^* and is determined by

$$|\alpha' - w_1|^2 = |\alpha' - w_2||\alpha' - w_2^*|.$$

EXAMPLE

Determine the most general homographic transformation which transforms the unit circle $|z| = 1$ into itself:

If the point $z = \alpha$ inside the unit circle is carried into $w = 0$, then its image $z^* = 1/\bar{\alpha}$ obtained by inversion is carried into the inverse of $w = 0$, $w^* = \infty$. Thus we can write immediately:

$$w = k \cdot \frac{z - \alpha}{z - \frac{1}{\bar{\alpha}}} = k\bar{\alpha} \frac{z - \alpha}{\bar{\alpha}z - 1}$$

For $z = 1$ we wish to have $|w| = 1$, which implies, since

$$|1 - \alpha| = |1 - \bar{\alpha}|$$

that

$$|k\bar{\alpha}| = 1 \quad \text{or} \quad k\bar{\alpha} = e^{it}$$

It follows that

$$w = e^{it} \frac{z - \alpha}{\bar{\alpha}z - 1}$$

Note that for a given α we can determine $t \ni w'(0)$ has an arbitrarily given argument.

Denote the subgroup of homographies which transform the unit disc into itself by S. Then S consists of all homographies of the form:

$$w = e^{it} \frac{z - \alpha}{\bar{\alpha}z - 1} \quad \text{with} \quad |\alpha| < 1$$

THEOREM 4.10 *The group of automorphisms $A(\mathscr{d}(0, 1))$ of the unit disc $\mathscr{d}(0, 1)$ is equal to S.*

PROOF

S is transitive in $\mathscr{d}(0, 1)$. Indeed, α is carried into 0, where α is some number with modulus < 1. Moreover, an automorphism which leaves the origin fixed is a rotation:

$$w = az \quad \text{with} \quad a = e^{i\phi}$$

Indeed, by Schwarz's lemma (Theorem 3.7), we have $|f(z)| \leq |z|$ for $z \in \mathscr{d}(0, 1)$; but also $|z| \leq |f(z)|$, applying this lemma on the inverse function. It follows that $|f(z)| = |z|$, or $f(z) = e^{i\phi}z$.

Let $T \in A(\mathscr{d}(0, 1))$ be an automorphism of the unit disc $\mathscr{d}(0, 1)$ which transforms 0 into α ($|\alpha| < 1$). $\exists$ a homography $h_1 \in S$ which transforms α into 0, from which it follows that the transformation $h_1 \circ T = h_2 \in A(\mathscr{d}(0, 1))$ leaves the origin fixed. Thus h_2 is a homography of the form $e^{i\phi}z$. So we find $h_1 \circ T = h_2$, or $T = h_1^{-1} \circ h_2$, and T is a homography $\in S$. Thus $S = A(\mathscr{d}(0, 1))$.

The symmetry principle of Schwarz:

If the boundary of the domain D contains γ, an arc of a circle or a segment of a straight line, whose image γ' by the mapping $f \in H(D)$ and $C(\bar{D})$ is also an arc of a circle or a segment of a straight line, then f has an extension outside of D which can be obtained by the principle of symmetry by associating a point z^*, exterior to D and symmetric with respect to γ to an interior point z of D, with the value which is symmetric to $f(z)$ with respect to γ'.

Problems

4.12.: Find the affine transformation which leaves fixed the point $z = c$. Represent it as a composite of a rotation and a dilatation. Particular cases:
   a) Find the affine transformation with the fixed point i which carries the point $2i$ into the point $-2i$. Find the angle of rotation and the ratio of dilatation.
   b) Determine the fixed point, the rotation and dilatation of the following affine mappings:
   i) $w = iz + 1$
   ii) $w = 2iz + 1 + i$
   iii) $w = (2 + 4i)z + 2$

4.13.: Compute the cross ratio of the fourth roots of i.

4.14.: Show that if the complex numbers z_1, z_2, z_3 and z_4 lie on a circle, then their cross ratio is real.

4.15.: Find the image of a circle and of a straight line by the mapping $w = 1/z$.
   Which circles are transformed into straight lines and which straight lines remain straight lines? In particular, find the images of:
   a) $|z - 2 - 2i| = \sqrt{5}$
   b) $\operatorname{Re} z = \pm 1, \pm 2$
   c) $\operatorname{Im} z = \pm 1, \pm 2$

4.16.: Represent $w = \dfrac{5z - 2}{z + 3}$ as a composite of rotations, dilatations, translations and of the inversion mapping.

4.17.: Find a homography which maps:
   a) the points a, b, c into $0, 1, \infty$
   b) the points $0, 1, \infty$ into a, b, c
   c) the points $-1, 0, 2$ into $0, 3, 6$

4.18.: If $h_1(z) = \dfrac{z + 1}{z - 1}$ and $h_2(z) = \dfrac{z + 2}{z - 3}$, find $h_1 \circ h_2$, $h_2 \circ h_1$, $h_2^{-1} \circ h_1$.

4.19.: Let $a, b, a \neq b$, be the two fixed points of a homography h. Show that h may be written as:

$$\frac{w - a}{w - b} = k \frac{z - a}{z - b}$$

where k is a complex number.
   Show that the cross ratio of a, b, z, w is constant.
   A particular case: Determine a, b and k for the homography

$$w = \frac{5z - 2}{z + 3}$$

Show that w maps the upper half plane $\operatorname{Im} z \geq 0$ onto itself.

4.20.: Which homographies transform:
  a) the half plane Re $z \geq 0$ onto the half plane Im $z \geq 0$
  b) the upper half plane Im $z \geq 0$ onto itself
  c) the upper half plane Im $z \geq 0$ onto the lower half plane Im $z \leq 0$
  d) the half plane Im $z \geq 0$ onto the closed disc $\overline{d}(0, 1)$?

4.21.: Find the homographies which carry the circle $|z - c| = r$ onto the circle $|z| = 1$.

4.22.: Show that if the two fixed points a, b of a homography h coincide, then h may be written as follows:

$$w = h(z) = \frac{k}{z - a} + b$$

4.23.: Show that the six homographies:

$$z, \quad \frac{1}{z}, \quad 1 - z, \quad \frac{1}{1 - z}, \quad \frac{z - 1}{z}, \quad \frac{z}{z - 1}$$

form a subgroup of H.

4.24.: Find an isomorphism of the strip $0 < \text{Im } z < \pi$ and the unit disc $d(0, 1)$. (*Hint:* e^z maps the strip onto the upper half plane Im $z > 0$; then take a homography which maps the upper half plane onto the unit disc.)

4.25.: Find a mapping which transforms the domain D defined by $0 < \arg z < \pi/4$ onto the unit disc. (*Hint:* z^4 maps D onto the upper half plane. Combine it with a homography which maps the upper half plane onto $d(0, 1)$.)

4.26.: Determine an isomorphism of $d(0, 1) \cap d(1, 1)$ with the upper half plane Im $z > 0$. (*Hint:* Find a homography which maps the points of intersection between the circles $|z| = 1$ and $|z - 1| = 1$ into the origin and infinity, respectively, and combine it with a rotation and z^α.)

4.27.: Show that $\text{Log } \frac{1 + z}{1 - z}$ is an isomorphism of the disc $d(0, 1)$ with the strip $-\pi/2 < \text{Im } z < \pi/2$. (*Hint:* Let $z_1 = 1 - z$, $z_2 = 1/z_1$, $z_3 = -1 + 2z_2$, $z_4 = \text{Log } z_3$.)

4.28.: Show that

$$w = \coth\left(\frac{z}{2}\right) = \frac{e^z + 1}{e^z - 1}$$

maps the semistrip Re $z \geq 0$, $-\pi \leq \text{Im } z \leq \pi$, onto Re $w \geq 0$. (*Hint:* Let $z_1 = e^z$, $z_2 = z_1 - 1$, $z_3 = 1/z_2$, $z_4 = 1 + 2z_3$.)

---

§3. Riemann's mapping theorem

Riemann's mapping theorem is the fundamental theorem of conformal representation.

THEOREM 4.11 *Every simply connected domain $D_1 \neq \mathbf{C}$ is isomorphic to the open disc $d(0, 1)$.*

Since isomorphicity is an equivalence relation it follows that any two simply connected domains $\neq \mathbf{C}$ are isomorphic.

PROOF

First step: We show that D_1 is isomorphic to a simply connected domain $D \subset d(0, 1)$ with $0 \in D$.

Let $D_1 \neq \mathbf{C}$ and $a \in \mathbf{C}\backslash D_1$. Since D_1 is simply connected and does not contain the point a, by 2.3.1 we can define on D_1 a branch of $\log(z - a)$ which is an isomorphism $D_1 \to D_2$ where the domain D_2 is the image of D_1 by $\log(z - a)$.

Let w_0 be a point in D_2. Then $\exists$ an open disc $d(w_0, r) \subset D_2$. Since $z \in D_2 \Rightarrow z + 2\pi i \notin D_2$, we have $d(w_0 + 2\pi i, r) \cap D_2 = \emptyset$. Consider the homography $h: z \to \dfrac{1}{z - w_0 - 2\pi i}$ and denote by D_3 the image of D_2 by h. D_3 is a domain since h is an isomorphism. D_3 is bounded because

$$z \in D_2 \Rightarrow |z - w_0 - 2\pi i| > r \Rightarrow \left|\dfrac{1}{z - w_0 - 2\pi i}\right| < \dfrac{1}{r}.$$

By means of a translation and a homography, we then transform the domain D_3 into a domain D lying in the unit disc $d(0, 1)$ with $0 \in D$.

Second step: We show that the set $\mathscr{F}(D) = \{f \mid f \text{ is an isomorphism } D \to d(0, 1) \text{ with } f(0) = 0 \text{ and } f'(0) \geq 1\}$ is compact.

$\mathscr{F}(D)$ is not empty, because the identity mapping $z \to z$ is in $\mathscr{F}(D)$. Note that if h is any isomorphism $D \to d(0, 1)$, then $\exists$ an automorphism g of $d(0, 1) \ni f = g \circ h$ satisfies the conditions $f(0) = 0, f'(0) \geq 1$.

If we provide $\mathscr{F}(D)$ with the topology of Section 2.4—that is, the topology of uniform convergence on every compact $K \subset D$—then $\mathscr{F}(D)$ is precompact by virtue of Theorem 2.21. Indeed, $\mathscr{F}(D)$ is uniformly bounded since we have $|f(z)| < 1 \; \forall \; f \in \mathscr{F}(D)$ and $z \in D$.

The set $\mathscr{F}(D)$ is also closed, because if $f_1, f_2, \ldots, f_n \ldots$ is a sequence of functions belonging to $\mathscr{F}(D)$ and uniformly convergent to f on each compact $K \subset D$, we have $f(0) = \lim_{n \to \infty} f_n(0) = 0$ and $f'(0) = \lim_{n \to \infty} f_n'(0) \geq 1$. f is not a constant since $f'(0) \geq 1$, and the univalence of the f_n, $n = 1, 2, \ldots$, implies the univalence of f (Theorem 3.20). Moreover, $|f(z)| < 1 \; \forall \; z \in D$. Indeed, if $|f(z)| = 1$ for $z \in D$ then z is a relative maximum which implies f is a constant in contradiction with $f'(0) \geq 1$. Thus $f \in \mathscr{F}(D)$, and $\mathscr{F}(D)$ is compact.

Third step: We show that the function $g \in \mathscr{F}(D)$ with the maximal $g'(0)$ defines the isomorphism $D \leftrightarrow d(0, 1)$.

For all $f \in \mathscr{F}(D)$, the value $f'(0)$ is a linear, continuous function of f. Since $\mathscr{F}(D)$ is compact, the set $\{f'(0) | f \in \mathscr{F}(D)\}$ has a maximum. Thus $\exists$ an isomorphism $g \in \mathscr{F}(D)$ for which $g'(0)$ is a maximum, that is, $g'(0) \geq f'(0) \; \forall \; f \in \mathscr{F}(D)$.

$g(z)$ is an isomorphism $D \leftrightarrow d(0, 1)$:

Indeed, if there existed an $a \in d(0, 1) \ni a \notin g(D)$, then we could construct an isomorphism $h \in \mathscr{F}(D)$ as follows, $\ni h'(0) > g'(0)$, contrary to the hypothesis that $g'(0)$ is maximal:

Let h_1 be an automorphism of $d(0, 1)$ with $h_1(a) = 0$. Then $h_1(g(D))$ is a simply connected domain which does not contain the origin. Let r be a branch of $\sqrt{z}$ on $h_1(g(D))$ and h_2 another automorphism of $d(0, 1)$ such that $h_2 \circ r \circ h_1(0) = 0$ and $(h_2 \circ r \circ h_1)'(0) > 0$. (See the example in 4.2.2.) Set $\phi = h_2 \circ r \circ h_1$. Then $\phi(0) = 0$, $\phi'(0) > 0$ and $|\phi(z)| < 1$ for $z \in g(D)$.

Let ψ be the inverse of ϕ: $\psi = \phi^{-1} = h_1^{-1} \circ r^{-1} \circ h_2^{-1}$. Then ψ is holomorphic on the unit disc, since h_1 and h_2 are automorphisms of $d(0, 1)$ and r^{-1} is z^2, which is holomorphic throughout $\mathbf{C}$. Moreover, $\psi(0) = 0$, and $|\psi(z)| < 1$ for $|z| < 1$. It follows by the Schwarz lemma (Theorem 3.7) that $|\psi'(0)| < 1$ since ψ cannot be a rotation because of $r^{-1} = z^2$. Thus $\phi'(0) = \dfrac{1}{\psi'(0)} > 1$ (note that $\phi'(0) > 0$).

Set $h = \phi \circ g$. Then $h \in \mathscr{F}(D)$ and $h'(0) = \phi'(g(0)) \cdot g'(0) = \phi'(0) \cdot g'(0) > g'(0)$, in contradiction to the hypothesis that $g'(0)$ is maximal on $\mathscr{F}(D)$.

The function g is therefore an isomorphism of D onto $d(0, 1)$.

Problems

4.29.: Determine the image of the given domain D by the following conformal mappings:
  a) $w = e^z$; D: rectangle with vertices at $a, a + ib, c, c + ib$, (a, b, c real, $b < 2\pi$)
  b) $w = \dfrac{1}{z}$; D: $\operatorname{Re} z > 0$

c) $w = \dfrac{i-z}{i+z}$; D: Im $z > 0$

d) $w = \dfrac{z-1}{z+1}$; D: Re $z > 0$

e) $w = z + \dfrac{1}{z}$; D: $|z| < 1$, Im $z > 0$

f) $w = \dfrac{1-\cos z}{1+\cos z}$; D: $0 < \text{Re } z < \dfrac{\pi}{2}$

g) $w = \left(\dfrac{z+1}{z-1}\right)^2$; D: $|z| < 1$, Im $z > 0$

h) $w = \sqrt{z}$; D: $y^2 > 4ax$

4.30.: Show that the function $h(x, y) = \arctan(y/x) = \arg z$ is harmonic in Re $z > 0$ with

$$h(0, y) = \begin{cases} \pi & \text{for } y > 0 \\ -\pi & \text{for } y < 0 \end{cases}$$

4.31.: Determine the function which is harmonic in the annulus $r \leq |z - c| \leq R$ and which attains the value a on the circle $\partial(c, r)$ and the value A on the circle $\partial(c, R)$. (*Hint*: Consider Re $\log(z - c)$.)

Solutions to the Odd-Numbered Problems

P. = Problem; A. = Answer

0.1.: P. Show that equivalence in the sense of set theory is an equivalence relation.
A. We shall show that the equivalence in the sense of set theory satisfies conditions (*i*), (*ii*) and (*iii*) of an equivalence relation:
  (*i*) The equivalence is reflexive; that is, a set A is always equivalent to itself. Indeed, take for the one-to-one correspondence the identity mapping.
  (*ii*) The equivalence is symmetric; that is, if the set A is equivalent to the set B, then the set B is equivalent to the set A. If f is the one-to-one correspondence $A \to B$, then f^{-1} is the one-to-one correspondence $B \to A$.
  (*iii*) The equivalence is transitive. If f is one-to-one $A \to B$ and g is one-to-one $B \to C$, then $g \circ f$ is one-to-one $A \to C$.

0.3.: P. Show that the set of real numbers is not countable.
A. Let **P** be the set of positive integers and **R** the set of real numbers.
We want to show that, whatever mapping f we choose, f can never be **P** onto **R**, i.e., $f(\mathbf{P})$ cannot cover **R**. $f(\mathbf{P})$ cannot even cover the open interval $]0, 1[$: We shall write the numbers of $]0, 1[$ in decimal form:

$$0.a_1 a_2 \ldots a_n \ldots$$

where $a_n \in \{0, 1, 2, \ldots, 8, 9\}$, $n = 1, 2, \ldots$.
Since all numbers $\in]0, 1[$ with a finite number of $a_n \neq 0$ can be written in two different ways—that is, if a_k is the last nonzero number, then

$$0.a_1 a_2 \ldots a_k 0\,0\,0 \ldots = 0.a_1 a_2 \ldots (a_k - 1)\,9\,9\,9 \ldots$$

we shall choose the first way in order to avoid ambiguity.

Now let f be any mapping $\mathbf{P} \to]0, 1[$. Then

$$f(1) = 0.a_{11}a_{12}a_{13}\cdots$$
$$f(2) = 0.a_{21}a_{22}a_{23}\cdots$$
$$\vdots$$
$$f(k) = 0.a_{k1}a_{k2}a_{k3}\cdots$$
$$\vdots$$

with $a_{kn} \in \{0, 1, 2, ..., 9\}$ $\forall k$ and $\forall n$

Then choose $r = 0.b_1 b_2 \ldots b_n \ldots$ with $b_k \in \{1, 2, ..., 8\}$ such that $b_1 \neq a_{11}, b_2 \neq a_{22}, ..., b_k \neq a_{kk}, \ldots$. Then $r \in]0, 1[$ and $r \neq f(k)$ $\forall k$; i.e., $r \notin f(\mathbf{P})$. (We have chosen $b_k \in \{1, 2, ..., 8\}$ instead of $b_k \in \{0, 1, ..., 9\}$ in order to avoid that $b_k = 0 \, \forall k$ or $b_k = 9 \, \forall k$ which would mean $r \notin]0, 1[$.)

0.5.: P. Show that if $(R, +, \cdot)$ is a ring then $(R, \cdot)$ need not be a group.

A. Consider the ring of the integers. Obviously the integers are not a group with respect to multiplication. Indeed, 2^{-1} is not an integer; thus 2 has no inverse in the ring of integers.

0.7.: P. Show that the polynomials of degree $\leq n$ in one variable, with real coefficients, form a real vector space.

A. Let $P(x) = a_0 + a_1 x + \cdots + a_n x^n$ be such a polynomial. Then with each polynomial we can associate a point P of $\mathbf{R}^{n+1}$: $P = (a_0, a_1, ..., a_n)$. This is a one-to-one correspondence which preserves the algebraic structure:

If
$$Q(x) = b_0 + b_1 x + \cdots + b_n x^n$$

corresponds with the point $Q \in \mathbf{R}^{n+1}$,

$$Q = (b_0, b_1, ..., b_n)$$

then $P(x) + Q(x)$ corresponds to

$$P + Q = (a_0 + b_0, a_1 + b_1, ..., a_n + b_n) \in \mathbf{R}^{n+1}$$

and

$$cP(x) = c(a_0 + a_1 x + \cdots + a_n x^n)$$
$$= ca_0 + ca_1 x + \cdots + ca_n x^n$$

corresponds to the point

$$cP = (ca_0, ca_1, ..., ca_n) \in \mathbf{R}^{n+1}$$

It follows that the set of all polynomials in one variable of degree $\leq n$, with real coefficients, forms a real vector space isomorphic to $\mathbf{R}^{n+1}$.

0.9.: P. Let S_1 and S_2 be two connected sets with $S_1 \cap S_2 \neq \emptyset$. Show that $S = S_1 \cup S_2$ is connected. How can one generalize this result?

A. If S is not connected then $\exists$ two open sets O_1 and $O_2 \ni S \cap O_1 \neq \emptyset$, $S \cap O_2 \neq \emptyset$, $S \subset O_1 \cup O_2$ and $S \cap O_1 \cap O_2 = \emptyset$.

Now S_1 is connected. Hence either $S_1 \subset O_1$ or $S_1 \subset O_2$; otherwise— i.e., if $S_1 \cap O_1 \neq \emptyset$ and $S_1 \cap O_2 \neq \emptyset$—$S_1$ could not be connected, since $S_1 \cap O_1 \cap O_2 = \emptyset$ and $S_1 \subset O_1 \cup O_2$. Suppose $S_1 \subset O_1$. Then $S_2 \subset O_1$, since $S_1 \cap S_2 \neq \emptyset$ and either $S_2 \subset O_1$ or $S_2 \subset O_2$ by virtue of the connectedness of S_2. It follows $S_1 \cup S_2 \subset O_1$ and $S \cap O_2 = (S_1 \cup S_2) \cap O_2 = \emptyset$, in contradiction with the hypothesis that S is not connected.

One could generalize this result for a collection of sets S_i where the index i varies in some finite or infinite set I of indices.

0.11.: P. Show that in a Hausdorff space the boundary of a compact set is compact.

A. Let K be the compact set. We have seen that the boundary of a set is closed (see 0.3.1, definition of boundary), and that a compact set in a Hausdorff space is closed. It follows $\overline{K} = K$ and that the boundary of K, as a closed subset of a compact set, is compact.

0.13.: P. Show that the function space $C[a, b]$ is complete with the norm $\|f\| = $ l.u.b.$\{|f(x)| \,|\, x \in [a, b]\}$, but is not complete with the Hilbert norm $\|\cdot\|_{L^2}$.

A. (i) $C[a, b]$ is complete with the norm $\|f\| = $ l.u.b.$\{|f(x)| \,|\, x \in [a, b]\}$. Let $f_1, f_2, \ldots, f_n, \ldots$ be a Cauchy sequence of functions $\in C[a, b]$, i.e., f_n is continuous on $[a, b]$, $n = 1, 2, \ldots$. Since for every $x \in [a, b]$ we have $|f_{n+p}(x) - f_n(x)| < \varepsilon$ for $n > N(\varepsilon)$ and p an arbitrary positive integer, we see that, for a fixed x, $f_1(x), f_2(x), \ldots, f_n(x), \ldots$ form a Cauchy sequence of real numbers and converges therefore to a real number, say $f(x)$. So $f(x)$ is a real function defined for every $x \in [a, b]$. Let us show that f is continuous on $[a, b]$:

$$|f(x) - f(x')| \leq |f(x) - f_n(x)| + |f_n(x) - f_n(x')| + |f_n(x') - f(x')|$$

Since $|f_{n+p}(x) - f_n(x)| < \varepsilon \;\forall\; x$, for $n > N(\varepsilon)$ and $\forall\, p \in \mathbf{P}$, it follows that in the limit $|f(x) - f_n(x)| \leq \varepsilon$ as well as $|f(x') - f_n(x')| \leq \varepsilon$, and since $f_n \in C[a, b]$, we have

$$|x - x'| < \delta(\varepsilon) \Rightarrow |f_n(x) - f_n(x')| < \varepsilon$$

Hence

$$|f(x) - f(x')| \leq \varepsilon + \varepsilon + \varepsilon = 3\varepsilon \text{ for } |x - x'| < \delta(\varepsilon)$$

which proves that $f(x) \in C[a, b]$.

(ii) $C[a, b]$ is not complete with the Hilbert norm $\|\cdot\|_{L^2}$. Let f_n be

$$f_n(x) = \begin{cases} 0 & \text{for } a \leq x \leq \dfrac{a+b}{2} \\ nx - n\dfrac{a+b}{2} & \text{for } \dfrac{a+b}{2} \leq x \leq \dfrac{a+b}{2} + \dfrac{1}{n} \\ 1 & \text{for } \dfrac{a+b}{2} + \dfrac{1}{n} \leq x \leq b \end{cases}$$

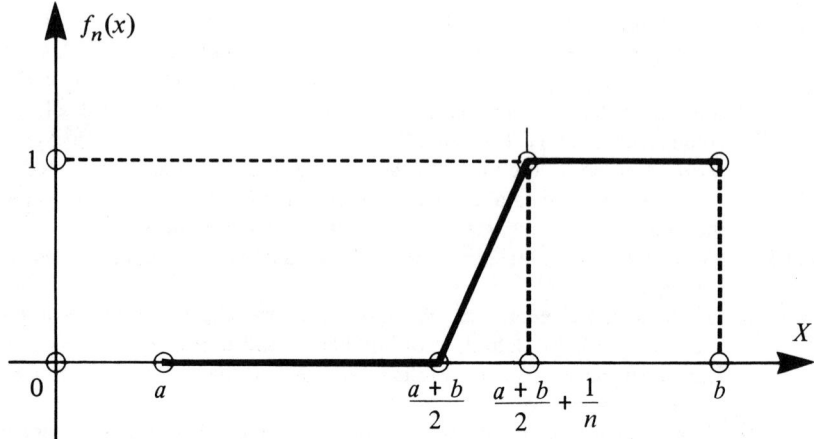

Figure A.1

It is easy to check that f_n is continuous throughout $[a, b]$, i.e., $f_n \in C[a, b]$. We shall show that f_n converges in the Hilbert norm to

$$f(x) = \begin{cases} 0 & \text{for} \quad a \leq x \leq \dfrac{a+b}{2} \\ 1 & \text{for} \quad \dfrac{a+b}{2} < x \leq b \end{cases}$$

Indeed:

$$\|f_n - f\|_{L^2}^2 = \int_a^b (f_n - f)^2 \, dx = \int_{(a+b)/2}^{(a+b)/2 + (1/n)} \left(nx - n\frac{a+b}{2} - 1 \right)^2 dx$$

$$\left(\text{setting } x' = x - \frac{a+b}{2} \right)$$

$$= \int_0^{1/n} (nx' - 1)^2 \, dx \leq \frac{1}{n}$$

since $(nx' - 1)^2 \leq 1$ for $0 \leq x' \leq 1/n$.

It follows that $\lim_{n \to \infty} \|f_n - f\|_{L^2}^2 = 0$. Since f is not continuous for $x = \dfrac{a+b}{2}$, $f \notin C[a, b]$ we see that $C[a, b]$ is not complete with the Hilbert norm.

0.15.: P. Let $S \subset \mathbf{R}$ be bounded. Does l.u.b.S or g.l.b.S belong to S? Give sufficient conditions.

A. Consider
$$S = \left\{\frac{n-1}{n}\right\}, \quad n = 1, 2, \ldots$$

l.u.b.$S = 1$ and does not belong to S.

If $S =]a, b[$, neither l.u.b.$S = b$ nor g.l.b.$S = a$ belong to S.

Clearly, if S is closed, then l.u.b.S and g.l.b.S belong to S. Let us show that S is closed $\Rightarrow$ l.u.b.$S \in S$: Let λ be l.u.b.S. Then $\lambda - 1/n$ is not an upper bound of S. $\exists\, s_n \in S \ni s_n > \lambda - 1/n$ for $n = 1, 2, \ldots$. Since $\lambda - 1/n < s_n \leq \lambda$, $\lim s_n = \lambda$, and since S is closed $\lambda \in S$ (Theorem 0.12).

0.17.: P. Show that if (E, d) is a metric space, then so is $\left(E, \dfrac{d}{1+d}\right)$.

A. $\dfrac{d}{1+d}$ is a distance if d is. Indeed:

(i) $\dfrac{d(x, y)}{1 + d(x, y)} = 0 \Leftrightarrow d(x, y) = 0 \Leftrightarrow x = y$

(ii) $\dfrac{d(y, x)}{1 + d(y, x)} = \dfrac{d(x, y)}{1 + d(x, y)}$, since $d(x, y) = d(y, x)$

(iii) $\dfrac{d(x, z)}{1 + d(x, z)} \leq \dfrac{d(x, y)}{1 + d(x, y)} + \dfrac{d(y, z)}{1 + d(y, z)}$

The last inequality can be verified as follows: Multiplying both sides of the inequality by
$$(1 + d(x, y))(1 + d(y, z))(1 + d(z, x))$$
we find, after simplification, that
$$d(x, z) \leq d(x, y) + d(y, z) + 2d(x, y)d(y, z) + d(x, y)d(y, z)d(x, z)$$
Since d is a distance and is ≥ 0, the last inequality follows immediately.

0.19.: P. Let (E, d) be a metric space. Show that the identity is an homeomorphism between (E, d) and $\left(E, \dfrac{d}{1+d}\right)$.

A. Clearly the identity is a one-to-one correspondence. It remains to show that it, as well as its inverse, is continuous.

Let $x_1, x_2, \ldots, x_n, \ldots$ be a sequence of elements of E converging to c in the metric d. We want to show that it converges to c in the metric $\dfrac{d}{1+d}$ and conversely. We have $\lim\limits_{n \to \infty} d(x_n, c) = 0$. Clearly,

$$0 \leq \frac{d(x_n, c)}{1 + d(x_n, c)} \leq d(x_n, c)$$

It follows that

$$\lim_{n\to\infty} \frac{d(x_n, c)}{1 + d(x_n, c)} = 0$$

Conversely, let

$$\lim_{n\to\infty} \frac{d(x_n, c)}{1 + d(x_n, c)} = 0$$

Then for $n \geq N(\varepsilon)$ we have:

$$\frac{d(x_n, c)}{1 + d(x_n, c)} \leq \varepsilon$$

or $d(x_n, c)(1 - \varepsilon) \leq \varepsilon$, so

$$d(x_n, c) \leq \frac{\varepsilon}{1 - \varepsilon}$$

Choose $\varepsilon < \frac{1}{2}$. Then $d(x_n, c) \leq 2\varepsilon$, and it follows that $\lim_{n\to\infty} d(x_n, c) = 0$.

0.21.: P. Let K and L be two compact subsets of the metric space (E, d). Define the distance $\text{dist}(K, L)$ by

$$\text{dist}(K, L) = \text{g.l.b.}\{d(x, y) | x \in K, y \in L\}$$

Show that if $K \cap L = \emptyset$, then $\text{dist}(K, L) > 0$.

A. Clearly, $\text{dist}(K, L) \geq 0$. It remains to show that $\text{dist}(K, L) = 0$ is impossible if $K \cap L = \emptyset$. Suppose $\text{dist}(K, L) = 0$. Then for a given $n \; \exists \; x_n \in K$ and $y_n \in L \ni d(x_n, y_n) \leq 1/n$. If we take $n = 1, 2, \ldots$, we get a sequence $\{x_n\}$ and a sequence $\{y_n\}$. And since $x_n \in K$, $n = 1, 2, \ldots$, and K is compact, $\exists$ a subsequence $\{x_n'\}$ of $\{x_n\}$ converging to $x_0 \in K$ and, similarly, a subsequence $\{y_n'\}$ of $\{y_n\}$ converging to $y_0 \in L$. Obviously, $d(x_0, y_0) = \lim_{n\to\infty} d(x_n, y_n) = 0$, so $x_0 = y_0$ and $K \cap L \neq \emptyset$.

0.23.: P. Show that the mapping $x \to 1/x$, $x \in \,]0, 1[$, is continuous but not uniformly continuous.

A. The continuity of $1/x$ follows from the fact that the quotient of two continuous functions is continuous as long as the denominator does not vanish, and $x \neq 0$ in $]0, 1[$.

In order to see that $1/x$ is not uniformly continuous, let ε be given. Then, however small we might choose $\delta(\varepsilon)$, $\exists \; x'$, x'' with $|x' - x''| < \delta(\varepsilon) \ni \left|\frac{1}{x'} - \frac{1}{x''}\right| > \varepsilon$. Indeed, if $x'' = x' + \frac{1}{2}\delta(\varepsilon)$, then clearly $|x'' - x'| < \delta(\varepsilon)$ and

$$\left|\frac{1}{x'} - \frac{1}{x''}\right| = \left|\frac{x' - x''}{x'x''}\right| = \frac{\frac{1}{2}\delta(\varepsilon)}{x'(x' + \frac{1}{2}\delta(\varepsilon))}$$

Obviously,
$$\lim_{x' \to 0} \frac{\frac{1}{2}\delta(\varepsilon)}{x'(x' + \frac{1}{2}\delta(\varepsilon))} = \infty$$

Therefore, it is possible to choose $x' \in {]}0, 1{[} \ni \frac{\frac{1}{2}\delta(\varepsilon)}{x'(x' + \frac{1}{2}\delta(\varepsilon))} > \varepsilon$.

1.1.: P. Reduce to the form $a + bi$.

    a) $(4 + i) - (3 - 2i)$

    b) $(4 + i)(2 - 3i)$

    c) $\dfrac{1}{1 + i}$

A. a) $1 + 3i$

    b) $11 - 10i$

    c) $\dfrac{1}{1 + i} = \dfrac{1 - i}{1 - i^2} = \dfrac{1}{2} - \dfrac{i}{2}$

1.3.: P. Prove that

    a) $(1 - i)^m = 2^{m/2}\left(\cos\dfrac{m\pi}{4} - i \sin\dfrac{m\pi}{4}\right)$

    b) $(1 + i\sqrt{3})^m = 2^m\left(\cos\dfrac{m\pi}{3} + i \sin\dfrac{m\pi}{3}\right)$

    c) $(1 - \cos\alpha + i \sin\alpha)^n = 2^n(i)^n \sin^n\left(\dfrac{\alpha}{2}\right)\left(\cos\dfrac{n\alpha}{2} - i \sin\dfrac{n\alpha}{2}\right)$

A. a) $1 - i = \sqrt{2}\left(\cos\dfrac{\pi}{4} - i \sin\dfrac{\pi}{4}\right) = \sqrt{2}\cdot\text{cis}\left(-\dfrac{\pi}{4}\right)$

It follows that
$$(1 - i)^m = 2^{m/2}\,\text{cis}\left(-\dfrac{m\pi}{4}\right) = 2^{m/2}\left(\cos\dfrac{m\pi}{4} - i \sin\dfrac{m\pi}{4}\right)$$

    b) $1 + i\sqrt{3} = 2\left(\cos\dfrac{\pi}{3} + i \sin\dfrac{\pi}{3}\right) = 2\,\text{cis}\dfrac{\pi}{3}$

It follows that
$$(1 + i\sqrt{3})^m = 2^m \operatorname{cis}\frac{m\pi}{3} = 2^m\left(\cos\frac{m\pi}{3} + i\sin\frac{m\pi}{3}\right)$$

c) $1 - \cos\alpha + i\sin\alpha = \cdot 2i\sin\frac{\alpha}{2}\left(\cos\frac{\alpha}{2} - i\sin\frac{\alpha}{2}\right)$

It follows that
$$(1 - \cos\alpha + i\sin\alpha)^n = 2^n i^n \sin^n\left(\frac{\alpha}{2}\right)\left(\cos\frac{n\alpha}{2} - i\sin\frac{n\alpha}{2}\right)$$

1.5.: *P.* Show that $|z| \geq \max(|x|, |y|)$.
A. $|z| = \sqrt{x^2 + y^2}$ and $\sqrt{x^2 + y^2} \geq \sqrt{x^2} = |x|$. Also, $\sqrt{x^2 + y^2} \geq \sqrt{y^2} = |y|$. It follows that $|z| \geq \max(|x|, |y|)$.

1.7.: *P.* Show that $|z_1 - z_2| \geq ||z_1| - |z_2||$.
A. We have
$$|z_1| = |z_1 - z_2 + z_2| \leq |z_1 - z_2| + |z_2|$$

as well as
$$|z_2| = |z_1 - z_2 - z_1| \leq |z_1 - z_2| + |z_1|$$

which implies
$$|z_1| - |z_2| \leq |z_1 - z_2|$$

and
$$|z_2| - |z_1| \leq |z_1 - z_2|$$

or
$$||z_1| - |z_2|| \leq |z_1 - z_2|$$

1.9.: *P.* a) Show by induction that $\left|\sum_{i=1}^{n} c_i\right| \leq \sum_{i=1}^{n} |c_i|$ and that the equal sign holds only if $\arg c_1 = \arg c_2 = \cdots = \arg c_n$.

b) Show that
$$\left|\prod_{i=1}^{n} c_i\right| = \prod_{i=1}^{n} |c_i| \quad \text{and} \quad \arg\left(\prod_{i=1}^{n} c_i\right) = \sum_{i=1}^{n} \arg c_i$$

c) Show for integer n that $|c|^n = |c^n|$ and $\arg c^n = n \arg c$.

d) Show also that $|\sqrt[n]{c}| = \sqrt[n]{|c|}$ and
$$\arg\sqrt[n]{c} = \frac{1}{n}\arg c + \frac{2k\pi}{n} \quad (k = 0, 1, 2, \ldots, n-1)$$

e) Show that if the segment joining z_1 to z_2 does not cut R_0^-, where $R_0^- = R^- \cup \{0\}$, then
$$\operatorname{Arg}\left(\frac{z_1}{z_2}\right) = \operatorname{Arg} z_1 - \operatorname{Arg} z_2$$

A. a) We have $|c_1 + c_2| \leq |c_1| + |c_2|$ with $|c_1 + c_2| = |c_1| + |c_2|$ iff $\arg c_1 = \arg c_2$.

Suppose we have proved $\left|\sum_{i=1}^{k} c_i\right| \leq \sum_{i=1}^{k} |c_i|$ and $\left|\sum_{i=1}^{k} c_i\right| = \sum_{i=1}^{k} |c_i|$ iff $\arg c_1 = \arg c_2 = \cdots = \arg c_k$ for $k = 1, 2, \ldots, n-1$. Now consider $\left|\sum_{i=1}^{n} c_i\right|$. Set $c_1 + c_2 + \cdots + c_{n-1} = s$ then

$$\left|\sum_{i=1}^{n} c_i\right| = |s + c_n| \leq |s| + |c_n|$$

$$= \left|\sum_{i=1}^{n-1} c_i\right| + |c_n| \leq \sum_{i=1}^{n-1} |c_i| + |c_n| = \sum_{i=1}^{n} |c_i|$$

and $|s + c_n| = |s| + |c_n|$ iff $\arg s = \arg c_n$. Now $|s| = \left|\sum_{i=1}^{n-1} c_i\right| = \sum_{i=1}^{n-1} |c_i|$ iff $\arg c_1 = \arg c_2 = \cdots = \arg c_{n-1} = \arg s$, so, finally, $\left|\sum_{i=1}^{n} c_i\right| = \sum_{i=1}^{n} |c_i|$ iff $\arg c_1 = \arg c_2 = \cdots = \arg c_n$.

b) For two factors c_1 and c_2 we have shown that $|c_1 c_2| = |c_1||c_2|$ and $\arg c_1 c_2 = \arg c_1 + \arg c_2$. Suppose we have proved

$$\left|\prod_{i=1}^{n-1} c_i\right| = \prod_{i=1}^{n-1} |c_i| \quad \text{and} \quad \arg \prod_{i=1}^{n-1} c_i = \sum_{i=1}^{n-1} \arg c_i$$

Then

$$\left|\prod_{i=1}^{n} c_i\right| = \left|\prod_{i=1}^{n-1} c_i \cdot c_n\right| = \left|\prod_{i=1}^{n-1} c_i\right| |c_n|$$

$$= \prod_{i=1}^{n-1} |c_i| |c_n| = \prod_{i=1}^{n} |c_i|$$

and

$$\arg \prod_{i=1}^{n} c_i = \arg\left(\prod_{i=1}^{n-1} c_i \cdot c_n\right) = \arg \prod_{i=1}^{n-1} c_i + \arg c_n$$

$$= \sum_{i=1}^{n-1} \arg c_i + \arg c_n = \sum_{i=1}^{n} \arg c_i$$

c) Follows immediately from b) by taking $c_1 = c_2 = \cdots = c_n$.

d) Set $r = \sqrt[n]{c}$; Then obviously $r^n = c$. Using (c), we find $|r^n| = |r|^n$ or $\sqrt[n]{|r^n|} = |r|$ and $\sqrt[n]{|c|} = |\sqrt[n]{c}|$. Then $\arg r^n = n \arg r$ or $\arg r = 1/n \arg r^n$ and $\arg \sqrt[n]{c} = 1/n \arg c$.

e) One of the values of $\arg\left(\dfrac{z_1}{z_2}\right)$ is given by $\phi = \operatorname{Arg} z_1 - \operatorname{Arg} z_2$. Now, if the line from z_1 to z_2 does not intersect the negative x-axis or the origin, the angle at 0 of the triangle $z_1 0 z_2$ is easily seen to be equal to $|\phi|$. Therefore, we have $|\phi| < \pi$ and consequently $\phi = \operatorname{Arg} \dfrac{z_1}{z_2}$, which proves the statement.

1.11.: P. Determine the set of points satisfying each of the following:
 a) $|z - a| + |z - b| = c$
 b) $|z - a| \cdot |z - b| = c$
 c) $\left|\dfrac{z - a}{z - b}\right| = c$

A. Let us recall that $|z - z'|$ is the distance from z to z'.
 a) We suppose that a and b are given complex numbers, z is variable, and that $c \geq |a - b|$. Then $|z - a| + |z - b| = c$ means that the sum of the distances from z to a and from z to b is constant $= c$. This means that z moves in an ellipse whose foci are a and b.
 b) The product of the distances from z to a and from z to b is constant $= c$. This implies that z moves in a lemniscate.
 c) The quotient of the distances from z to a and from z to b is constant $= c$, indicating that z moves on a circle except when $c = 1$; for $c = 1$, z moves on a straight line. (Can be easily seen by computation.)

1.13.: P. a) If $w = \sqrt{z_1 z_2}$, show that

$$|z_1| + |z_2| = \left|\dfrac{z_1 + z_2}{2} + w\right| + \left|\dfrac{z_1 + z_2}{2} - w\right|$$

b) Prove the identity

$$|z_1 + z_2| + |z_1 - z_2| = \left|z_1 + \sqrt{z_1^2 - z_2^2}\right| + \left|z_2 + \sqrt{z_1^2 - z_2^2}\right|$$

A. a) Obviously,

$$\left|\dfrac{z_1 + z_2}{2} + \sqrt{z_1 z_2}\right| = \left|\dfrac{(\sqrt{z_1} + \sqrt{z_2})^2}{2}\right|$$

and

$$\left|\dfrac{z_1 + z_2}{2} - \sqrt{z_1 z_2}\right| = \left|\dfrac{(\sqrt{z_1} - \sqrt{z_2})^2}{2}\right|$$

so we have to prove

$$2|\sqrt{z_1}|^2 + 2|\sqrt{z_2}|^2 = |\sqrt{z_1} + \sqrt{z_2}|^2 + |\sqrt{z_1} - \sqrt{z_2}|^2$$

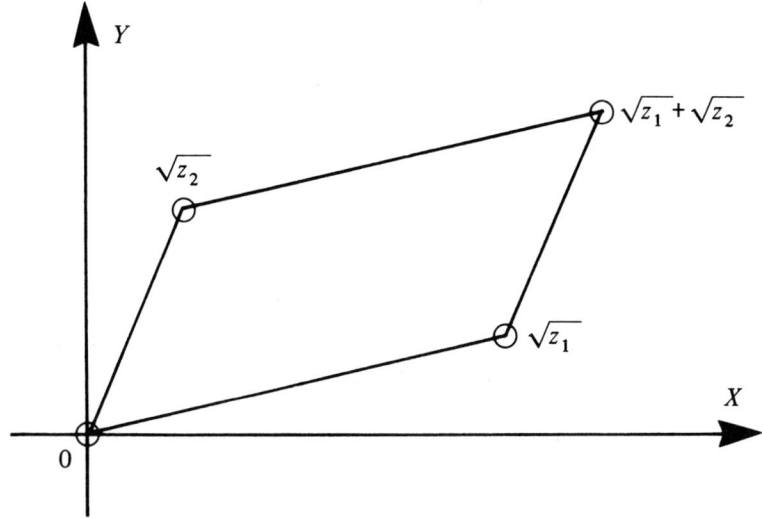

Figure A.2

But this is the well-known parallelogram relation for the parallelogram whose vertices are $(0, \sqrt{z_1}, \sqrt{z_2}, \sqrt{z_1} + \sqrt{z_2})$.

The parallelogram relation says: The sum of the squares of the sides of a parallelogram is equal to the sum of the squares of its diagonals.

b) Set $z_1 = \dfrac{u_1 + u_2}{2}$, $z_2 = \dfrac{u_1 - u_2}{2}$, and b) reduces to a).

1.15.: P. Show that the mapping $z \to \bar{z}$ is an automorphism of $\mathbf{C}$.

A. To show that the mapping $z \to \bar{z}$ is an algebraic automorphism of $\mathbf{C}$, we must show that the algebraic structure of $\mathbf{C}$ is preserved by it.

Now we know that

$$\overline{z_1 + z_2} = \bar{z}_1 + \bar{z}_2 \quad \text{and} \quad \overline{z_1 z_2} = \bar{z}_1 \bar{z}_2$$

so that the algebraic structure is preserved. It can be shown that the mapping $z \to \bar{z}$ is the only automorphism of $\mathbf{C}$ different from the identity.

1.17.: P. a) Describe the relative positions of z and the following points in the complex plane, relative to the Riemann sphere:

1) $-z$ 2) $\bar{z}$ 3) $1/z$ 4) $(z + \bar{z})/2$
5) $(z - \bar{z})/2$ 6) $\pm iz$ 7) $z/\bar{z}$ 8) $z/|z|$

b) Show that the equation of the stereographic projection is:
$$z = \frac{\xi + i\eta}{1 - \zeta}, \quad \zeta \neq 1$$
and, conversely, that
$$\xi = \frac{x}{|z|^2 + 1}, \quad \eta = \frac{y}{|z|^2 + 1}, \quad \zeta = \frac{|z|^2}{|z|^2 + 1}$$

c) Show that the stereographic projection is a homeomorphism between the complex plane and $S^2 \backslash \{N\}$.

d) Prove that the stereographic projection of a straight line in **C** is a circle on the Riemann sphere passing through N.

A. a) 1) The points z and $-z$ are symmetric with respect to the origin. Their projections on the Riemann sphere lie on a plane through N and O symmetric with respect to the diameter NO.

2) The projections of z and $\bar{z}$ lie on the sphere symmetric with respect to the $\xi\zeta$-plane.

3) On the sphere the points z and $\dfrac{1}{\bar{z}}$ are symmetric with respect to the equatorial plane, whereas $\dfrac{1}{\bar{z}}$ and $\dfrac{1}{z}$ are symmetric with respect to the $\xi\zeta$-plane. Consequently, z and $\dfrac{1}{z}$ are symmetric with respect to that diameter of the sphere which is parallel to the ξ-axis, i.e. the diameter which connects the points $+1$ and -1 on the sphere.

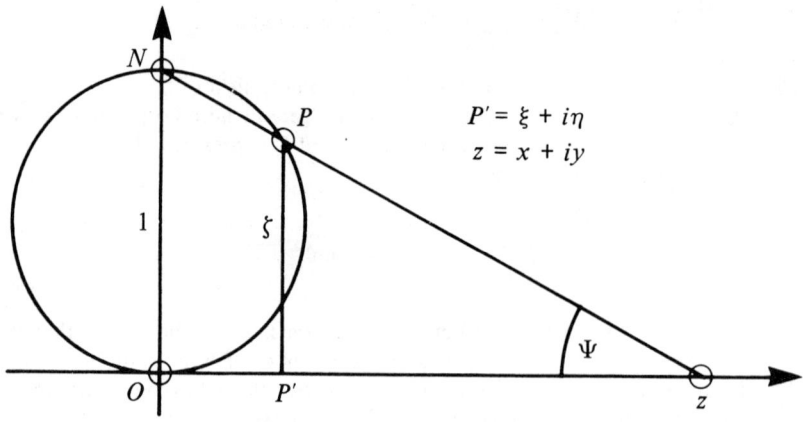

Figure A.3

4) The point $\dfrac{z + \bar{z}}{2}$ is the center of the segment joining z and $\bar{z}$ and lies on the real axis; its projection on the Riemann sphere lies on the meridian of the $\xi\zeta$-plane.

5) The point $\dfrac{z - \bar{z}}{2} = iy$ lies on the imaginary axis and its projections on S^2 on the meridian of $\eta\zeta$ – plane.

6) We find the points iz and $-iz$ by rotating z about the origin by $\pi/2$ and $-\pi/2$, respectively. A rotation of $\mathbf{C}$ about the origin corresponds to a rotation of S^2 about the ζ-axis by the same angle.

7) The point $z/\bar{z}$ lies on the unit circle $\partial(0, 1)$ since $|z/\bar{z}| = 1$ and $\arg(z/\bar{z}) = \arg z - \arg \bar{z} = 2 \arg z$. It follows that the projection of $z/\bar{z}$ onto S^2 lies on the equator on the meridian corresponding to $2 \arg z$.

8) $z/|z|$ lines on the unit circle $\partial(0, 1)$ and has the same argument as z. Its projection lies on the equator as befits all points of $\partial(0, 1)$.

b) The relation $z = \dfrac{\xi + i\eta}{1 - \zeta}$ has been established in the text, where it was also shown that

$$\xi = \frac{x}{1 + x^2 + y^2}, \quad \eta = \frac{y}{1 + x^2 + y^2}, \quad \zeta = \frac{x^2 + y^2}{1 + x^2 + y^2}$$

Since $x^2 + y^2 = |z|^2$, the desired result follows immediately.

c) We must show that the projection π is bicontinuous, i.e., π and π^{-1} are both continuous. That follows immediately, since

$$z = \frac{\xi + i\eta}{1 - \zeta}$$

is continuous as long as $\zeta \neq 1$, but $\zeta = 1$ only for N. Conversely,

$$\xi + i\eta = \frac{z}{|z|^2 + 1} \quad \text{and} \quad \zeta = \frac{|z|^2}{|z|^2 + 1}$$

are continuous since numerator and denominator are continuous and $|z|^2 + 1 \neq 0$.

d) We find that the stereographic projection of a straight line S in $\mathbf{C}$ is the intersection of the plane passing through N and S with the sphere S^2, and that is a circle passing through N. Note that every straight line passes through the point at infinity whose projection on S^2 is N. An analytic proof is given in the text.

1.19.: P. Give an example of two nonopen sets whose union is open.
A. Cut the open circle $d(0,1)$ into two sets S_1 and S_2, $S_1 = d(0,1) \cap (\text{Im } z \geq 0)$ and $S_2 = d(0,1) \cap (\text{Im } z \leq 0)$. More generally, take any domain D. If $c = a + ib \in D$, then consider

$$S_1 = D \cap (\text{Re } z \geq a) \quad \text{and} \quad S_2 = D \geq (\text{Re } z \leq a)$$

Then S_1 and S_2 are not open, since the points $D \cap (\text{Re } z = a)$ are interior points of neither S_1 nor S_2, and $D = S_1 \cup S_2$.

1.21.: P. Give examples of two starlike domains D_1 and $D_2 \ni$
  a) $D_1 \cup D_2$ is not starlike
  b) $D_1 \cap D_2$ is not starlike
  c) $D_1 \cap D_2$ is convex.

A. a) Figure A.4 shows a nonstarlike domain which is the union of two starlike domains. One can also use two starlike domains with $D_1 \cap D_2 = \emptyset$.

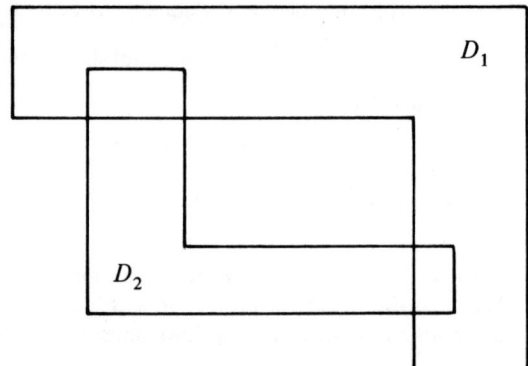

Figure A.4

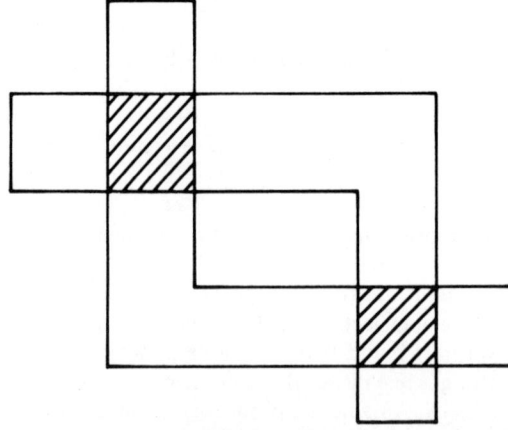

Figure A.5

b) Take two starlike domains whose intersection is the union of two sets whose intersection is ∅, as indicated in Figure A.5.

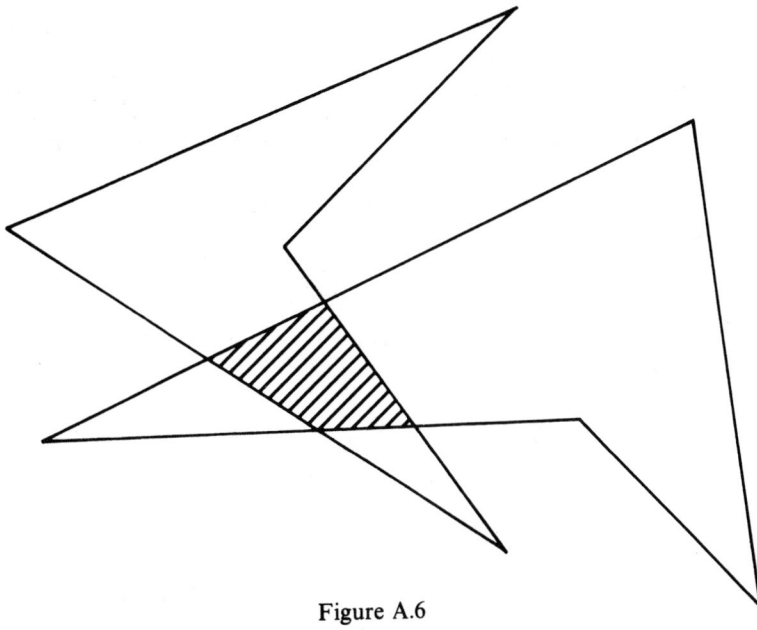

Figure A.6

c) Figure A.6 gives a convex domain which is the intersection of two starlike domains.

1.23.: *P.* Show that a ring (annulus) is not starlike.
A. Using Figure A.7

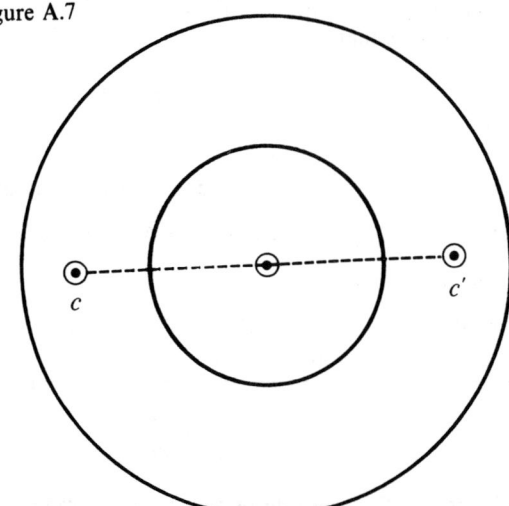

Figure A.7

let c be any point of the ring R; then take its symmetric c' with respect to the center of R. Clearly, the segment joining c to c' does not belong to R, so R is not starlike.

1.25.: **P.** Let S_1, S_2 be two open sets of **R**. Show that the Cartesian product $S_1 \times S_2$ is open in $\mathbf{R}^2$.

A. Let $(a, b) \in S_1 \times S_2$. Then $a \in S_1$ and $b \in S_2$ and $\exists \varepsilon \ni]a - \varepsilon, a + \varepsilon[\subset S_1$ and $]b - \varepsilon, b + \varepsilon[\subset S_2$. It follows that $]a - \varepsilon, a + \varepsilon[\times]b - \varepsilon, b + \varepsilon[\subset S_1 \times S_2$. But $]a - \varepsilon, a + \varepsilon[\times]b - \varepsilon, b + \varepsilon[$ is open in the standard topology of $\mathbf{R}^2$.

1.27.: **P.** Show that if the sequence $\{a_n\}$ is convergent, then

$$\lim_{n \to \infty} \frac{\sum_{i=1}^{n} a_i}{n} = \lim_{n \to \infty} a_n$$

A. Set

$$b_n = \frac{\sum_{i=1}^{n} a_i}{n} \quad \text{and} \quad \lim_{n \to \infty} a_n = a$$

then

$$|b_n - a| = \left| \frac{1}{n} \sum_{i=1}^{n} a_i - a \right| = \frac{1}{n} \left| \sum_{i=1}^{n} (a_i - a) \right|$$

$$\leq \frac{1}{n} \sum_{i=1}^{r} |a_i - a| + \frac{1}{n} \sum_{i=r+1}^{n} |a_i - a|$$

and choose $r \ni |a_i - a| < \varepsilon$ for $i > r$.

It follows that

$$|b_n - a| \leq \frac{1}{n} \sum_{i=1}^{r} |a_i - a| + \frac{(n-r)\varepsilon}{n} \leq \varepsilon + \frac{1}{n} \sum_{i=1}^{r} |a_i - a|$$

Since r is fixed, we can choose n sufficiently large, say $n > N(\varepsilon)$, so that $\frac{1}{n} \sum_{i=1}^{r} |a_i - a| < \varepsilon$. Then for $n > N(\varepsilon)$ we have $|b_n - a| \leq 2\varepsilon$, which proves $\lim_{n \to \infty} b_n = a$.

1.29.: **P.** Show that $f(z) = |z|$ is nowhere differentiable.

A. Set $f(z) = w = u + i0 = \sqrt{x^2 + y^2} = r$. Now

$$\frac{\partial w}{\partial r} = \frac{\partial r}{\partial r} = 1$$

which means that the rate of change in the radial direction is 1. However, along the boundary of the circle, the radius r is constant. Therefore, the derivative in the direction of the tangent,

$$\frac{\partial f}{\partial r} \cdot \frac{\partial r}{\partial \theta} = \frac{\partial w}{\partial r} \cdot \frac{\partial r}{\partial \theta} = 0 \frac{\partial r}{\partial \theta} = 0$$

So we see that the derivative in the radial direction is different from the derivative in the direction of the tangent. Since for a differentiable function, the derivatives in all directions must be equal, we conclude that $|z|$ is nowhere differentiable.

1.31.: P. Show that the function $f(x, y) = \dfrac{x + iy}{x - iy}$ is not continuous at the origin.

A. Consider the sequences

$$\left.\begin{array}{l} P_n = \left(\dfrac{1}{n}, 0\right) \\[2mm] P_n' = \left(0, \dfrac{1}{n}\right) \end{array}\right\} \quad n = 1, 2, \ldots$$

Both sequences are converging to the origin. Now

$$f(P_n) = \frac{1/n}{1/n} = 1 \quad \text{and} \quad f(P_n') = \frac{i(1/n)}{-i(1/n)} = -1$$

so that $\lim_{n \to \infty} P_n \neq \lim_{n \to \infty} P_n'$, which shows that f is not continuous at the origin.

1.33.: P. Show that the function $f(x, y) = x^2 + iy^2$ is uniformly continuous in the open disc $d(0, 1)$, but not uniformly continuous in **C**.

A. Consider $|f(x, y) - f(x', y')| = |x^2 + iy^2 - x'^2 - iy'^2| = |(x + x')(x - x') + i(y + y')(y - y')| \leq |(x + x')(x - x')| + |(y + y')(y - y')| \leq 2(|x - x'| + |y - y'|)$, since in $d(0, 1)$ we have $|x + x'| \leq 2$ as well as $|y + y'| \leq 2$. It follows that

$$|f(x, y) - f(x', y')| \leq \varepsilon \quad \text{for} \quad \sqrt{(x - x')^2 + (y - y')^2} \leq \frac{\varepsilon}{4}$$

that is, $f(x, y)$ is uniformly continuous in $d(0, 1)$.

If (x, y) is not restricted to $d(0, 1)$ but can vary over **C**, then if we choose $y + y' = x + x' = M > 0$,

$$|f(x, y) - f(x', y')| = M|(x - x') + i(y - y')|$$
$$= M\sqrt{(x - x')^2 + (y - y')^2}$$

So, however small we choose $\sqrt{(x - x')^2 + (y - y')^2}$, the expression $|f(x, y) - f(x', y')| > \varepsilon$ for sufficiently large M. Hence $x^2 + iy^2$ is not uniformly continuous throughout **C**.

230   Solutions to the Odd-Numbered Problems

1.35.: P. Show that the function $f(x, y) = x + 2iy$ is continuous in **C**. Determine for given c and $\varepsilon > 0$ the corresponding $\delta = \delta(c, \varepsilon) \ni f(d(c, \delta)) \subset d(f(c), \varepsilon)$.

A. Set $c = a + ib$ where $a + ib \in \mathbf{C}$. Then
$$|f(x, y) - f(a, b)| = |x + 2iy - a - 2ib|$$
$$= |(x - a) + 2i(y - b)| \leq |2(x - a) + 2i(y - b)|$$
$$= 2\sqrt{(x - a)^2 + (y - b)^2}$$

Hence, for $\delta = \sqrt{(x - a)^2 + (y - b)^2} = \varepsilon/2$, we have $|f(x, y) - f(a, b)| \leq \varepsilon$, or $f(d(c, \delta)) \subset d(f(a, b), \varepsilon)$.

1.37.: P. Show that the origin is the only point where $f(z) = z \operatorname{Re} z$ is complex differentiable.

A. $f(z) = zx = (x + iy)x = x^2 + ixy = U(x, y) + iV(x, y)$, or $U(x, y) = x^2$, $V(x, y) = xy$. Hence, using the Cauchy-Riemann equations,

$$U_x = 2x \quad \text{and} \quad V_y = x$$
$$2x = x \Rightarrow x = 0$$
$$U_y = 0, \quad V_x = y \Rightarrow y = 0$$

So the only point where f satisfies the Cauchy-Riemann equations is the origin.

1.39.: P. Show that

$$\frac{\partial}{\partial x} = \frac{\partial}{\partial z} + \frac{\partial}{\partial \bar{z}}, \quad \frac{\partial}{\partial y} = i\left(\frac{\partial}{\partial z} - \frac{\partial}{\partial \bar{z}}\right)$$

and

$$\nabla^2 = \frac{\partial^2}{\partial x^2} + \frac{\partial^2}{\partial y^2} = 4\frac{\partial^2}{\partial z \, \partial \bar{z}}$$

A. We have defined (1.4.3):

$$\frac{\partial}{\partial z} = \frac{1}{2}\left(\frac{\partial}{\partial x} - i\frac{\partial}{\partial y}\right), \quad \frac{\partial}{\partial \bar{z}} = \frac{1}{2}\left(\frac{\partial}{\partial x} + i\frac{\partial}{\partial y}\right)$$

By adding, we find $\dfrac{\partial}{\partial z} + \dfrac{\partial}{\partial \bar{z}} = \dfrac{\partial}{\partial x}$ and, by subtracting,

$$\frac{\partial}{\partial z} - \frac{\partial}{\partial \bar{z}} = -i\frac{\partial}{\partial y} \quad \text{or} \quad \frac{\partial}{\partial y} = i\left(\frac{\partial}{\partial z} - \frac{\partial}{\partial \bar{z}}\right)$$

For ∇^2 we find:

$$\nabla^2 = \frac{\partial}{\partial x}\left(\frac{\partial}{\partial x}\right) + \frac{\partial}{\partial y}\left(\frac{\partial}{\partial y}\right) = \left(\frac{\partial}{\partial z} + \frac{\partial}{\partial \bar{z}}\right)\left(\frac{\partial}{\partial z} + \frac{\partial}{\partial \bar{z}}\right) + i^2\left(\frac{\partial}{\partial z} - \frac{\partial}{\partial \bar{z}}\right)\left(\frac{\partial}{\partial z} - \frac{\partial}{\partial \bar{z}}\right)$$
$$= \frac{\partial^2}{\partial z^2} + \frac{\partial^2}{\partial \bar{z}^2} + 2\frac{\partial^2}{\partial z \, \partial \bar{z}} - \frac{\partial^2}{\partial z^2} - \frac{\partial^2}{\partial \bar{z}^2} + 2\frac{\partial^2}{\partial z \, \partial \bar{z}} = 4\frac{\partial^2}{\partial z \, \partial \bar{z}}$$

1.41.: *P.* Consider the mapping $f(z) = z^2$. What are the transforms of lines parallel to the x-axis and lines parallel to the y-axis? Show by some examples that $\arg f'(z)$ determines the angle of rotation of the tangent.

A. $f(z) = U(x, y) + iV(x, y) = z^2 = x^2 - y^2 + 2ixy$
Hence
$$U(x, y) = x^2 - y^2 \quad \text{and} \quad V(x, y) = 2xy$$

If x is constant, then $\dfrac{V}{2x} = y$ and $-U + x^2 = y^2$; hence $x^2 - U = \left(\dfrac{V}{2x}\right)^2$. This relationship between U and V defines a parabola in the (U, V) plane, so parallels to the y-axis are transformed into parabolas. If y is constant, then the same way we find $\left(\dfrac{V}{2y}\right)^2 = U + y^2$, which is again a parabola.

Now $f'(z) = 2z$ and $\arg f'(z) = \arg z$. Take for instance $y = r$, where r is a real constant. It is a parallel to the x-axis and coincides with its tangent. Its image satisfies the equation $V^2 = 4r^2 U + 4r^4$. Hence, if $V' = \dfrac{dV}{dU}$, then $2VV' = 4r^2$ or $V' = 2r^2/V$.

If x is given, then $V = 2xr$ and $V' = \dfrac{2r^2}{2xr} = \dfrac{r}{x}$ and if ϕ is the angle of rotation of the tangent to the parabola at $U = x^2 - y^2$ and $V = 2xy$, then $\tan \phi = r/x = y/x$, or $\phi = \arctan(y/x) = \arg z = \arg f'(z)$.

2.1.: *P.* Show that homotopy with respect to a domain D is a transitive property.

A. Let $\gamma_1, \gamma_2, \gamma_3$ be three closed curves in a domain D and assume that γ_1 and γ_2 are homotopic with respect to D. Let the family of curves $\gamma_{12}(t, u)$, $0 \le t \le 1, 0 \le u \le 1$ generate that homotopy, so that $\gamma_{12}(t, 0) = \gamma_1$ and $\gamma_{12}(t, 1) = \gamma_2$. Assume furthermore that γ_2 and γ_3 are homotopic with respect to D, with the family of curves $\gamma_{23}(t, u)$, $0 \le t \le 1, 0 \le u \le 1$ generating the homotopy, so that $\gamma_{23}(t, 0) = \gamma_2$ and $\gamma_{23}(t, 1) = \gamma_3$. Now, by a proper choice of the parameter t it can be arranged that the functions $\gamma_{12}(t, 1)$ and $\gamma_{23}(t, 0)$ will be identical. The function $\gamma_{13}(t, u)$ defined by

$$\gamma_{13}(t, u) = \begin{cases} \gamma_{12}(t, 2u) & \text{for } 0 \le u \le \tfrac{1}{2} \\ \gamma_{23}(t, 2u - 1) & \text{for } \tfrac{1}{2} \le u \le 1 \end{cases}$$

will then be continuous in $0 \le t \le 1, 0 \le u \le 1$ and yield

$$\gamma_{13}(t, 0) = \gamma_{12}(t, 0) = \gamma_1,$$
$$\gamma_{13}(t, \tfrac{1}{2}) = \gamma_{12}(t, 1) = \gamma_{23}(t, 0) = \gamma_2,$$
$$\gamma_{13}(t, 1) = \gamma_{23}(t, 1) = \gamma_3,$$

all within the domain D. Thus it establishes the homotopy of γ_1 and γ_3 with respect to D.

2.3.: P. Show that the circle $\partial(0, r)$ and the ellipse centered at the origin with axes a and b parallel to the x-axis and y-axis, respectively, are homotopic with respect to the open square $\{(x, y) \mid -c < x < c, \ -c < y < c\}$ where $c > \max(r, a, b)$.

A. If we take

$$\gamma(t, u) = [au + r(1 - u)]\cos t + i[bu + r(1 - u)]\sin t,$$

$$0 \le t \le 2\pi, \quad 0 \le u \le 1$$

then:
(i) $\gamma(t, 0) = r \cos t + ir \sin t = re^{it}$, $0 \le t \le 2\pi$, which defines the circle $\partial(0, r)$. For $u = 1$ we have: $\gamma(t, 1) = a \cos t + ib \sin t$, $0 \le t \le 2\pi$, which defines the ellipse with axes a and b.
(ii) $|\operatorname{Re} \gamma(t, u)| = |[au + r(1 - u)]\cos t| \le \max(r, a) < c$ and $|\operatorname{Im} \gamma(t, u)| = |[bu + r(1 - u)]\sin t| \le \max(r, b) < c$, so that $\gamma(t, u)$ maps $[0, 2\pi] \times [0, 1] \to$ open square.
(iii) $\gamma(0, u) = au + r(1 - u) = \gamma(2\pi, u)$.

2.5.: P. With respect to which domain are the circles $\gamma_0(\phi) = \operatorname{cis} \phi$ and $\gamma_1(\phi) = \frac{1}{2} \operatorname{cis} \phi$, $0 \le \phi \le 2\pi$, homotopic?

A. It is easy to see that the two circles are homotopic with respect to any domain which contains the annulus $\frac{1}{2} \le |z| \le 1$.

Indeed, we take

$$\gamma(\phi, u) = (1 - u)\operatorname{cis} \phi + \frac{u}{2} \operatorname{cis} \phi = \left(1 - \frac{u}{2}\right) \operatorname{cis} \phi$$

The function $\gamma(\phi, u)$ maps the Cartesian product $[0, 2\pi] \times [0, 1]$ onto the annulus $\frac{1}{2} \le |z| \le 1$, since $\left|\left(1 - \frac{u}{2}\right) \operatorname{cis} \phi\right| = 1 - \frac{u}{2}$ and, for $0 \le u \le 1$, we have $\frac{1}{2} \le 1 - u/2 \le 1$. Clearly, $\gamma(\phi, u)$ is a p.w.s. closed curve $\forall u, 0 \le u \le 1$.

2.7.: P. Prove

$$\left| \int_\gamma (x^2 - iy^2) \, dz \right| \le 2.5$$

if:
a) γ is the interval $[-i, i]$ on the y-axis
b) γ is the semi circle $z = \operatorname{cis} \phi$, $-\pi/2 \le \phi \le \pi/2$.

A. a) We use the inequality

$$\left| \int_\gamma f(z) \, dz \right| \le \max_{z \in \gamma} |f(z)| \cdot L(\gamma)$$

where $L(\gamma)$ is the length of γ.
On $[-i, i]$ we have $x^2 = 0$ and $|iy^2| \le 1$ so that $|x^2 - iy^2| \le 1$.
Since $L(\gamma) = 2$, it follows that

$$\left| \int_\gamma (x^2 - iy^2) \, dx \right| \le 1 \cdot 2 < 2.5$$

b) We compute the integral. Set $z = \text{cis } \phi = \cos \phi + i \sin \phi$, so that $x = \cos \phi$, $y = \sin \phi$, $dz = (-\sin \phi + i \cos \phi) \, d\phi$.

$$I = \int_\gamma (x^2 - iy^2) \, dz = \int_{-\pi/2}^{+\pi/2} (\cos^2 \phi - i \sin^2 \phi)(-\sin \phi + i \cos \phi) \, d\phi$$

$$= \int_{-\pi/2}^{+\pi/2} (\sin^2 \phi \cos \phi - \cos^2 \phi \sin \phi) \, d\phi + i \int_{-\pi/2}^{+\pi/2} (\cos^3 \phi + \sin^3 \phi) \, d\phi$$

$$= \left[\frac{1}{3} \sin^3 \phi + \frac{1}{3} \cos^3 \phi \right]_{-\pi/2}^{+\pi/2}$$

$$+ i \left[\sin \phi - \frac{1}{3} \sin^3 \phi - \cos \phi + \frac{1}{3} \cos^3 \phi \right]_{-\pi/2}^{+\pi/2}$$

$$= \frac{2}{3} + \frac{4}{3} i, \quad |I|^2 = \frac{4}{9} + \frac{16}{9} = \frac{20}{9}, \quad |I| = \frac{2}{3} \sqrt{5} < 2.5$$

2.9.: P. Compute

$$\int_0^{1+i} (z^2 + z) \, dz$$

Choose two different paths of integration and show that the results are the same.

A. Let us choose the two paths γ_1 and γ_2 where γ_1 consists of the two segments $[0, 1]$ and $[1, 1 + i]$ and γ_2 consists of the segment $[0, 1 + i]$ (see Figure A.8):

$$\int_{\gamma_1} : \int_{\gamma_1} (z^2 + z) \, dz = \int_0^1 (x^2 + x) \, dx + \int_0^1 [(1 + iy)^2 + (1 + iy)] i \, dy$$

$$= \left. \frac{x^3}{3} + \frac{x^2}{2} \right|_0^1 + i \int_0^1 [(2 - y^2) + i \cdot 3y] \, dy$$

$$= \frac{1}{3} + \frac{1}{2} + \left(2 - \frac{1}{3} \right) i - \frac{3}{2} = \frac{5i}{3} - \frac{2}{3}$$

$$\int_{\gamma_2} : \text{Set } z = (1 + i)t, \ 0 \le t \le 1; \text{ then } dz = (1 + i) \, dt \text{ and}$$

$$\int_{\gamma_2} (z^2 + z)\, dz = (1 + i)\int_0^1 [(1 + i)^2 t^2 + (1 + i)t]\, dt$$

$$= (1 + i)^3 \frac{t^3}{3} + (1 + i)^2 \frac{t^2}{2}\bigg|_0^1$$

$$= \frac{(1 + i)^3}{3} + \frac{(1 + i)^2}{2} = \frac{1 + 3i - 3 - i}{3} + i = \frac{5i}{3} - \frac{2}{3}$$

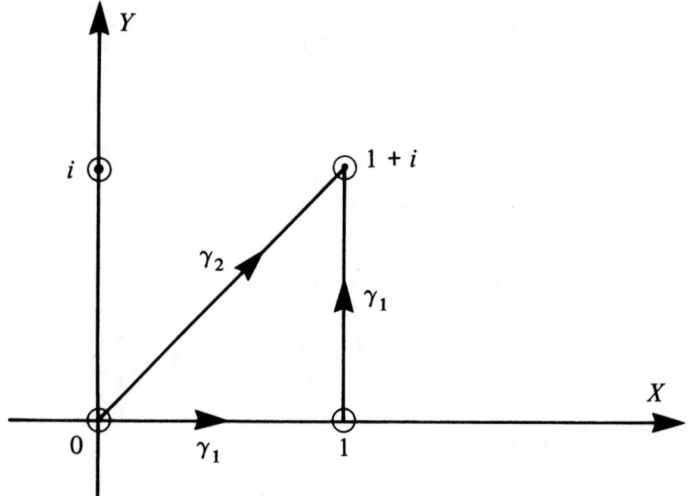

Figure A.8

2.11.: P. Prove that $\left|\int_2^i (z + 1)^2\, dz\right| \leq 9\sqrt{5}$.

A. If we take as path of integration the line $z = i + (2 - i)t$, $0 \leq t \leq 1$, then $\max |(z + 1)^2| = \max |(1 + 2t) + i(1 - t)|^2 = \max |(1 + 2t)^2 + (1 - t)^2| = \max_{0 \leq t \leq 1} |2 + 2t + 5t^2| = 9$, and the length of the path is $\sqrt{1 + 4} = \sqrt{5}$. It follows that

$$\left|\int_2^i (z + 1)^2\, dz\right| \leq 9\sqrt{5}$$

2.13.: P. Let γ be a closed n-sided counterclockwise-oriented polygonal line whose inside is starlike with respect to the point P. Join P to all the vertices of γ and denote by $\Delta_1, \Delta_2, \ldots, \Delta_n$ the counterclockwise oriented triangles.

Show that

$$\int_\gamma f(z)\,dz = \sum_{i=1}^n \int_{\Delta_i} f(z)\,dz$$

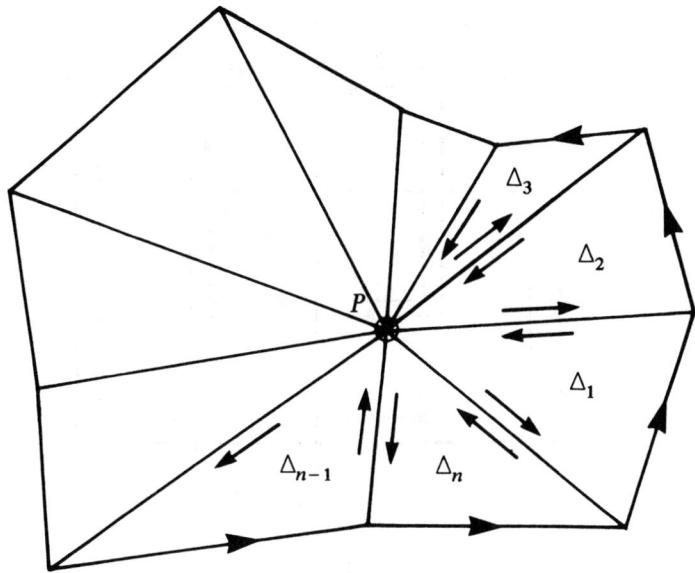

Figure A.9

A. Since the integrals taken over the interior sides vanish, being taken twice but in opposite directions, we find from Figure A.9

$$\int_\gamma f(z)\,dz = \sum_{i=1}^n \int_{\Delta_i} f(z)\,dz$$

2.15.: P. Let D be the domain defined by:

$$D = \{z \mid a < \max(|\operatorname{Re} z|, |\operatorname{Im} z|) < b\}$$

and suppose $f \in H(D)$ and $f \in C(\bar{D})$. State the Cauchy-Goursat theorem for D.

A. Denote by Γ the outer boundary and by γ the inner boundary, counterclockwise oriented. Let $\delta = (b - a)/2$ and consider the closed curves Γ_n and γ_n (Figure A.10), both counterclockwise oriented, and defined by:

$$\Gamma_n = \left\{z \,\middle|\, \max(|\operatorname{Re} z|, |\operatorname{Im} z|) = b - \frac{\delta}{n}\right\}$$

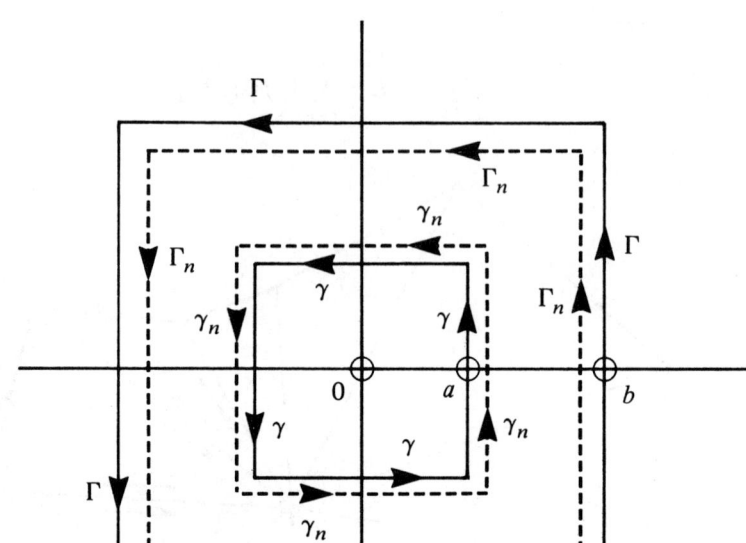

Figure A.10

and

$$\gamma_n = \left\{ z \,\middle|\, \max(|\operatorname{Re} z|, |\operatorname{Im} z|) = a + \frac{\delta}{n} \right\}, \quad n = 1, 2, \ldots$$

Obviously, Γ_n and γ_n are homotopic with respect to D; therefore, by Cauchy's theorem

$$\int_{\Gamma_n} f(z)\,dz = \int_{\gamma_n} f(z)\,dz$$

for $n = 1, 2, \ldots$.

Since $\lim_{n\to\infty} \Gamma_n = \Gamma$, $\lim_{n\to\infty} \gamma_n = \gamma$ and $f \in C(\overline{D})$, it follows that

$$\int_\Gamma f(z)\,dz = \int_\gamma f(z)\,dz$$

or

$$\int_\Gamma f(z)\,dz + \int_{-\gamma} f(z)\,dz = 0$$

Solutions to the Odd-Numbered Problems   237

i.e., the integral taken over the border of D is zero if the orientation of the border is such that the domain lies to the left of the border.

2.17.: P. Prove the Cauchy theorem with the aid of Green's theorem under the assumption that f' is continuous on D.

A. Green's theorem for a triangle Δ: If the functions P and Q possess continuous derivatives P_x, P_y, Q_x, Q_y in D and $\Delta \subset D$, then:

$$\int_{\Delta'} P\,dx + Q\,dy = \iint_\Delta \left(\frac{\partial Q}{\partial x} - \frac{\partial P}{\partial y}\right) dx\,dy$$

If $f = U + iV$ is complex differentiable in D, $\Delta \subset D$ and f' is continuous, then U_x, U_y, V_x and V_y exist and are continuous. It follows that

$$\int_{\Delta'} f(z)\,dz = \int_{\Delta'} [U(x, y) + iV(x, y)][dx + i\,dy]$$

$$= \int_{\Delta'} U\,dx - V\,dy + i\int_{\Delta'} V\,dx + U\,dy$$

$$= \iint_\Delta (-V_x - U_y)\,dx\,dy + i\iint_\Delta (U_x - V_y)\,dx\,dy$$

by Green's theorem. Now, using the Cauchy-Riemann equations, we find:

$$-\iint (V_x + U_y)\,dx\,dy = \iint (U_x - V_y)\,dx\,dy = 0$$

so that $\int_{\Delta'} f(z)\,dz = 0$.

2.19.: P. Compute the value of $f(z) = 2z + 3$ for $z = 0$ by the Cauchy formula, using for γ the circle $\gamma(t) = \operatorname{cis} t$, $0 \le t \le 2\pi$.

A. The Cauchy formula,

$$f(z) = \frac{1}{2\pi i}\int_\gamma \frac{f(u)}{u - z}\,du$$

gives, for $z = 0$ and $f(z) = 2z + 3$,

$$f(0) = \frac{1}{2\pi i}\int_\gamma \frac{2u + 3}{u}\,du = \frac{1}{2\pi i}\int_\gamma \left(2 + \frac{3}{u}\right) du = 3$$

since 2 is holomorphic throughout $\mathbf{C}$.

2.21.: P. Determine the value of

$$\int_\gamma \left(z + \frac{1}{z}\right) dz$$

if γ is the unit circle: $\gamma = \operatorname{cis} t$, $0 \le t \le 2\pi$.

A. Clearly, $\int_\gamma z \, dz = 0$ since z is holomorphic throughout $\mathbf{C}$; there still remains $\int_\gamma \frac{1}{z} dz$. Set $z = \text{cis } t$, $dz = i \text{ cis } t \, dt$. Then

$$\int_\gamma \frac{1}{z} dz = \int_0^{2\pi} \frac{i \text{ cis } t \, dt}{\text{cis } t} = i \int_0^{2\pi} dt = 2\pi i$$

2.23.: P. Compute

$$\int_\gamma \frac{z^2 + 3z + 5}{z + 1} dz$$

where γ is given by $\gamma(t) = a + R \text{ cis } t$, $0 \le t \le 2\pi$, $R > 0$ and $a \in \mathbf{C}$.

A.

$$\frac{z^2 + 3z + 5}{z + 1} = \frac{(z^2 + z) + 2(z + 1) + 3}{z + 1} = z + 2 + \frac{3}{z + 1}$$

Now $\int_\gamma (z + 2) \, dz = 0$ for any closed curve, since $z + 2$ is holomorphic throughout $\mathbf{C}$. Therefore, only $\int_\gamma \frac{3}{z + 1} dz$ remains. We suppose γ does not pass through $z = -1$, since otherwise the integral is not defined.

If $z = -1$ lies outside γ, then $\int_\gamma \frac{3}{z + 1} dz = 0$ since then γ is homotopic to a point with respect to the domain of holomorphicity.

If $z = -1$ lies inside γ, then γ is homotopic with respect to $\mathbf{C}\backslash\{-1\}$ to a circle centered at -1, and

$$\int_\gamma \frac{3}{z + 1} dz = 3 \cdot 2\pi i = 6\pi i$$

2.25.: P. Determine the residues of the functions

a) $f(z) = \dfrac{2z + i}{z^2 - 3z + 2}$

b) $f(z) = \dfrac{z^2 + 1}{z^4 + z^2 + 1}$

at their singularities.

A. a) Since f is a rational function, the singularities are the points where the denominator vanishes:

$$z^2 - 3z + 2 = 0 \quad \text{gives} \quad z_1 = 1, \quad z_2 = 2$$

We now decompose $f(z)$:

$$f(z) = \frac{2z+i}{z^2 - 3z + 2} = \frac{A}{z-1} + \frac{B}{z-2}$$

or $A(z-2) + B(z-1) \equiv 2z + i$. For $z = 1$ we find $-A = 2 + i$, and for $z = 2$ we find $B = 4 + i$. So we have:

$$f(z) = \frac{-2-i}{z-1} + \frac{4+i}{z-2}$$

Now

$$\text{Res}(f, c) = \frac{1}{2\pi i} \int_{\partial(c,r)} f(u)\, du$$

where r is chosen ∋ c is the only singularity inside $\partial(c, r)$. It follows that $\text{Res}(f, 1) = -2 - i$ and $\text{Res}(f, 2) = 4 + i$.

2.27.: P. Let f be a continuous function on $[0, 1]$ and define $F(z)$ by

$$F(z) = \int_0^1 \frac{f(u)}{u - z}\, du$$

Prove that F is holomorphic in $\mathbf{C}\setminus[0, 1]$ and that $\lim_{z \to \infty} zF(z)$ is finite.

A. F is holomorphic by Theorem 2.14.

Now if $M = \max|f(u)|$, $0 \le u \le 1$, then for $|z| \ge 2$ we have

$$|zF(z)| = \left| \int_0^1 \frac{zf(u)}{u - z}\, du \right| \le 2M$$

since $\left| \dfrac{z}{u-z} \right| \le 2$ for $|z| \ge 2$ and $0 \le u \le 1$. It follows that $\lim_{z \to \infty} |zF(z)| \le 2M$.

2.29.: P. Let U and V be conjugate harmonic functions and

$$\frac{V(x, y)}{U(x, y)} = f(x)$$

Show that $f(x) = \tan(\alpha x + \beta)$ with $\alpha, \beta \in \mathbf{R}$, and determine $U(x, y)$ and $V(x, y)$.

A. Since U and V are conjugate harmonic functions, the function $g(z) = U(x, y) + iV(x, y)$ is holomorphic and $\dfrac{V(x, y)}{U(x, y)} = \tan(\arg g(z))$, or $\arg g(z) = \arctan(V/U) = \arctan f(x)$.

Now $\arg g(z) = \text{Im} \log g(z)$. Hence $\dfrac{\partial}{\partial y}[\text{Im} \log g(z)] = 0$, and since log $g(z)$ is complex differentiable whenever $g(z) \neq 0$, we find—using the Cauchy-Riemann equations—that

$$\frac{\partial}{\partial x} \operatorname{Re} \log g(z) = \frac{\partial}{\partial y} \operatorname{Im} \log g(z) = 0$$

It follows that $\operatorname{Re} \log g(z) = \phi(y)$ and

$$\phi'(y) = \frac{\partial}{\partial y} \phi(y) = -\frac{\partial}{\partial x} \operatorname{Im} \log g(z) = -\frac{\partial}{\partial x} \arctan f(x)$$

Since $\phi'(y)$ depends only on y, and $-\frac{\partial}{\partial x} \arctan f(x)$ only on x we have

$$\phi'(y) = -\frac{\partial}{\partial x} \arctan f(x) = -\alpha$$

where $-\alpha$ is a real constant. So $\arctan f(x) = \alpha x + \beta$ and $f(x) = \tan(\alpha x + \beta)$. In the same way, $\phi(y) = -\alpha y + \gamma$, and since $\phi(y) = \operatorname{Re} \log g(z) = \log |g(z)|$, we have

$$|g(z)| = e^{\phi(y)} = e^{-\alpha y + \gamma}$$

Using $g(z) = |g(z)| \operatorname{cis} \arg g(z)$, we find

$$g(z) = e^{-\alpha y + \gamma} \cdot \operatorname{cis}(\alpha x + \beta) = e^{\gamma} \operatorname{cis} \beta \, e^{-\alpha y} \operatorname{cis} \alpha x$$
$$= Ce^{-\alpha y}(\cos \alpha x + i \sin \beta x)$$

where $C = e^{\gamma} \operatorname{cis} \beta$. Hence

$$U(x, y) = Ce^{-\alpha y} \cos \alpha x \quad \text{and} \quad V(x, y) = Ce^{-\alpha y} \sin \alpha x$$

2.31.: P. Find the conjugate harmonic function of

a) $U(x, y) = \dfrac{x}{x^2 + y^2}$

b) $U(x, y) = \sinh x \sin y$

A. a) $$U(x, y) = \frac{x}{x^2 + y^2} = \frac{\operatorname{Re} z}{|z|^2} = \frac{\operatorname{Re} \bar{z}}{z \cdot \bar{z}} = \operatorname{Re} \frac{\bar{z}}{z \cdot \bar{z}} = \operatorname{Re} \frac{1}{z}$$

It follows that

$$U(x, y) + iV(x, y) = \frac{1}{z} = \frac{1}{x + iy} = \frac{x - iy}{x^2 + y^2}, \quad \text{or} \quad V = \frac{-y}{x^2 + y^2}$$

b) Using the Cauchy-Riemann equations, we have:

$$\frac{\partial U}{\partial x} = \frac{\partial V}{\partial y} = \cosh x \sin y \qquad (1)$$

$$-\frac{\partial U}{\partial y} = \frac{\partial V}{\partial x} = -\sinh x \cos y \qquad (2)$$

Integrating (1) with respect to y leads to $V(x, y) = -\cosh x \cos y + g(x)$. It follows, comparing with (2), that

$$\frac{\partial V}{\partial x} = -\sinh x \cos y + g'(x) = -\sinh x \cos y$$

or $g'(x) = 0 \Rightarrow g(x) = C$, where C is an arbitrary real constant. We conclude:

$$V(x, y) = -\cosh x \cos y + C$$

3.1.: P. Prove that the Riemann zeta function $\zeta(z)$ defined by

$$\zeta(z) = \sum_{n=1}^{\infty} n^{-z}$$

converges for Re $z > 1$ and converges uniformly for Re $z \geq 1 + \varepsilon$ where $\varepsilon > 0$ is arbitrarily small. Show that ζ is analytic for Re $z \geq 1 + \varepsilon$.
(Hint: $|n^{-z}| = |e^{-z \log n}| = |e^{-x \log n}| \cdot |e^{-iy \log n}| = e^{-x \log n} = n^{-x}$.)

A. It is well known from real analysis that the numerical series $\sum_{n=1}^{\infty} n^{-(1+\varepsilon)}$ converges for arbitrarily small $\varepsilon > 0$. Now this series is a majorant for the series $\sum_{n=1}^{\infty} n^{-z}$ since $|n^{-z}| = n^{-\text{Re } z}$. The uniform convergence of $\sum_{n=1}^{\infty} n^{-z}$ then follows by the Weierstrass criterion of the numerical majorant (see 3.1.2).

Now $n^{-z} = e^{-z \log n}$ is analytic for any n. So, using Theorem 2.19 we find that ζ is analytic for Re $z \geq 1 + \varepsilon$.

3.3.: P. The Fibonacci numbers are defined by $c_0 = 0$, $c_1 = 1$ and $c_n = c_{n-1} + c_{n-2}$. Prove that $\sum_{n=0}^{\infty} c_n z^n$ is a rational function and determine an expression for c_n. (Hint: Set $\phi(z) = c_2 + c_3 z + \cdots$ and show that $z\phi(z) + 1 + z^2\phi(z) + z = \phi(z)$.)

A.
$$1 + z\phi(z) = 1 + c_2 z + c_3 z^2 + c_4 z^3 + \cdots$$
$$z + z^2\phi(z) = z + c_2 z^2 + c_3 z^3 + \cdots$$

Adding both lines gives:

$$1 + z\phi(z) + z + z^2\phi(z) = 1 + (1 + c_2)z + (c_2 + c_3)z^2 + (c_3 + c_4)z^3$$
$$+ \cdots = c_2 + c_3 z + c_4 z^2 + \cdots = \phi(z)$$

It follows that $\phi(z)(z^2 + z - 1) = -1 - z$ or

$$\phi(z) = \frac{1+z}{1-z-z^2}$$

and $\phi(z)$ is a rational function.

Let α and β be the zeros of $1 - z - z^2$. Then set

$$\frac{1+z}{1-z-z^2} = \frac{A}{1-(z/\alpha)} + \frac{B}{1-(z/\beta)}$$

hence $1 + z \equiv A(1 - (z/\beta)) + B(1 - (z/\alpha))$. Setting $z = \alpha$, we find $1 + \alpha = A(1 - (\alpha/\beta))$, and setting $z = \beta$ we find $1 + \beta = B(1 - (\beta/\alpha))$. So

$$A = \frac{(1+\alpha)\beta}{\beta - \alpha} \quad \text{and} \quad B = \frac{(1+\beta)\alpha}{\alpha - \beta}$$

Since c_n is the coefficient of z^n in

$$\phi(z) = \frac{1+z}{1-z-z^2} = A \sum_{k=0}^{\infty} \left(\frac{z}{\alpha}\right)^k + B \sum_{k=0}^{\infty} \left(\frac{z}{\beta}\right)^k$$

it follows that

$$c_{n+2} = A\left(\frac{1}{\alpha}\right)^n + B\left(\frac{1}{\beta}\right)^n$$

Now by setting $z^2 + z - 1 = 0$, we see that $\alpha\beta = -1$. Hence $c_{n+2} = A(-\beta)^n + B(-\alpha)^n$, with

$$\alpha = \frac{-1+\sqrt{5}}{2} \quad \text{and} \quad \beta = \frac{-1-\sqrt{5}}{2}$$

Finally,

$$c_n = \frac{\sqrt{5}}{5}\left(\frac{1+\sqrt{5}}{2}\right)^n - \frac{\sqrt{5}}{5}\left(\frac{1-\sqrt{5}}{2}\right)^n$$

3.5.: P. Let f be an entire function. Prove the following: If $\exists$ a positive integer n and a positive constant M ∋

$$|f(z)| \leq M|z|^n, \quad \forall z \in \mathbf{C}$$

then f is a polynomial of degree $\leq n$. (Compare with Liouville's theorem.)

A. Develop $f(z)$ in a Taylor series,

$$f(z) = a_0 + a_1 z + \cdots$$

and use Cauchy's inequalities:

$$|a_k| \leq \frac{M(r)}{r^k}, \quad k = 0, 1, \ldots$$

we get

$$|a_k| \leq \frac{Mr^n}{r^k} = \frac{M}{r^{k-n}}$$

This inequality remains true for arbitrarily large r. It follows that $a_k = 0$ for $k > n$, or

$$f(z) = a_0 + a_1 z + \cdots + a_n z^n$$

i.e., $f(z)$ is a polynomial.

3.7.: P. Determine the radius of convergence of

a) $1 + 4z + 9z^2 + \cdots + (n+1)^2 z^n + \cdots$

b) $1 + \alpha z + \binom{\alpha}{2} z^2 + \cdots + \binom{\alpha}{k} z^k + \cdots, \; 0 < \alpha < 1$

A. a) We use the formula $R = 1/(\overline{\lim} \sqrt[n]{|a_n|})$. We have $\overline{\lim} \sqrt[n]{(n+1)^2} = \lim \sqrt[n]{(n+1)^2} = (\lim_{n \to \infty} \sqrt[n]{n+1})^2$. Now $\lim_{n \to \infty} \sqrt[n]{n+1} = 1$; it follows that $R = 1$.

In order to see that $\lim \sqrt[n]{n+1} = 1$, set $\sqrt[n]{n+1} = 1 + h_n$. Obviously, $h_n > 0$. Then, for $n \geq 2$,

$$n + 1 = (1 + h_n)^n > 1 + \binom{n}{2} h_n^2 \quad \text{or} \quad h_n^2 < \frac{n}{\binom{n}{2}} = \frac{2}{n-1}$$

Since $0 < h_n^2 < 2/(n-1)$ with $\lim_{n \to \infty} 2/(n-1) = 0$, it follows that $\lim h_n^2 = \lim h_n = 0$ or

$$\lim \sqrt[n]{n+1} = 1.$$

b) The radius of convergence of the series $\sum_{n=0}^{\infty} a_n z^n$ is given by $R = \lim_{n\to\infty} \left|\dfrac{a_n}{a_{n+1}}\right|$, if that limit exists. Therefore, in the given case, we find

$$R = \lim_{n\to\infty}\left|\binom{\alpha}{n}\Big/\binom{\alpha}{n+1}\right| = \lim_{n\to\infty}\left|\frac{n+1}{\alpha-n}\right| = \lim_{n\to\infty}\frac{n+1}{n-\alpha} = 1.$$

3.9.: P. Expand $f(z) = \log(2+z)$ in a power series of the form $\sum_{n=0}^{\infty} a_n z^n$ and determine the radius of convergence.

A. We first develop $f'(z) = \dfrac{1}{2+z}$ in a power series:

$$f'(z) = \frac{1}{2+z} = \frac{1}{2}\cdot\frac{1}{1+\dfrac{z}{2}} = \frac{1}{2}\left(1 - \frac{z}{2} + \frac{z^2}{4} - \frac{z^3}{8} + \cdots\right)$$

Then we integrate it term by term and obtain

$$f(z) = c + \frac{1}{2}z - \frac{1}{4}\cdot\frac{z^2}{2} + \frac{1}{8}\cdot\frac{z^3}{3} - \cdots.$$

Using $f(0) = \log 2$, we find $c = \log 2$.

The radius of convergence R is given by $1/R = \overline{\lim}\sqrt[n]{|a_n|}$, or

$$\frac{1}{R} = \overline{\lim}\sqrt[n]{\frac{1}{2^n}\cdot\frac{1}{n}} = \frac{1}{2}\lim_{n\to\infty}\sqrt[n]{\frac{1}{n}} \quad \text{or} \quad R = 2\lim_{n\to\infty}\sqrt[n]{n}$$

and, since $\lim_{n\to\infty}\sqrt[n]{n} = 1$, we have $R = 2$.

3.11.: P. Show that the series

$$\sum_{n=0}^{\infty}\left(\frac{z}{3}\right)^n \quad \text{and} \quad \sum_{n=0}^{\infty}\frac{(\frac{1}{3})^n(z-6i)^n}{(1-2i)^{n+1}}$$

are analytic continuations of each other.

A.

$$f_1 = \sum_{n=0}^{\infty}\left(\frac{z}{3}\right)^n = \frac{1}{1-\dfrac{z}{3}} = \frac{3}{3-z}$$

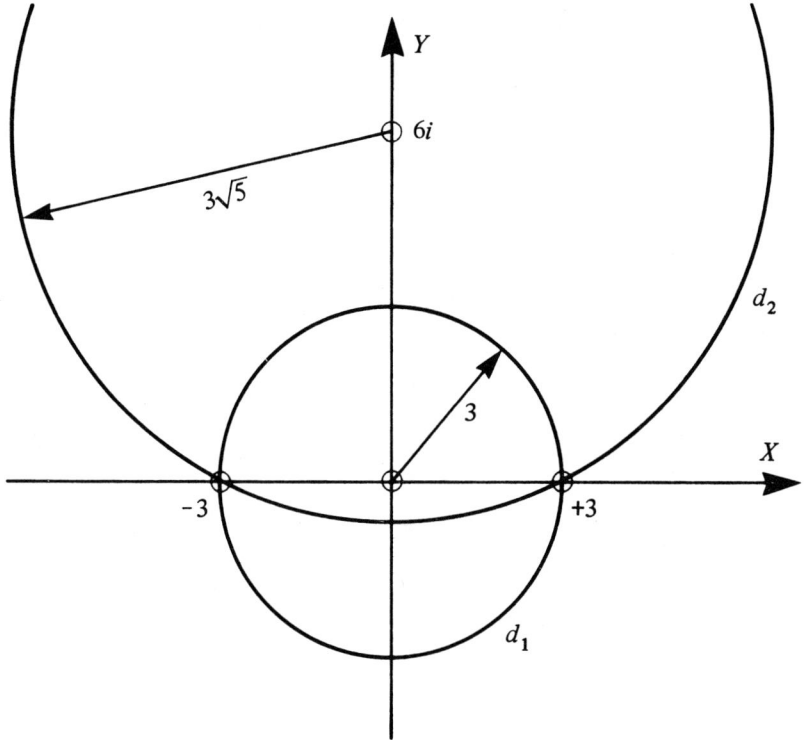

Figure A.11

converges for $|z| < 3$, i.e., in the disc $d_1 = d(0, 3)$, and

$$f_2 = \sum_{n=0}^{\infty} \frac{(\frac{1}{3})^n(z - 6i)^n}{(1 - 2i)^{n+1}} = \frac{1}{1 - 2i} \cdot \frac{1}{1 - \frac{1}{3} \cdot \frac{z - 6i}{1 - 2i}} = \frac{3}{(1 - 2i) \cdot 3 - (z - 6i)}$$

$$= \frac{3}{3 - z}$$

converges for $|z - 6i| < 3|1 - 2i| = 3\sqrt{5}$, i.e., in the disc $d_2 = d(6i, 3\sqrt{5})$.

Since $d_1 \cap d_2 \neq \emptyset$ and, in addition, $f_1 \equiv f_2 = \frac{3}{3 - z}$ on $d_1 \cap d_2$, the two function elements (f_1, d_1) and (f_2, d_2) are both analytic and identical when restricted to $d_1 \cap d_2$, and we say that both elements are analytic continuations of each other.

3.13.: P. Determine the radius of convergence of the following series:

a) $\sum_{n=1}^{\infty} \dfrac{z^n}{n^2}$ b) $\sum_{n=1}^{\infty} \dfrac{nz^n}{2^n}$

c) $\sum_{n=1}^{\infty} n! z^n$ d) $\sum_{n=1}^{\infty} z^{2^n}$

A. a) $a_n = 1/n^2$; hence $1/R = \overline{\lim} \sqrt[n]{1/n^2}$. Now $\lim_{n \to \infty} \sqrt[n]{n} = 1$, (see problem 3.7), so $\lim \sqrt[n]{n^2} = (\lim \sqrt[n]{n})^2 = 1$ and $\lim_{n \to \infty} \sqrt[n]{1/n^2} = 1$. It follows that $R = 1$.

b) $a_n = n/2^n$, $\sqrt[n]{a_n} = \sqrt[n]{n/2^n} = \tfrac{1}{2}\sqrt[n]{n}$. Hence $1/R = \overline{\lim} \sqrt[n]{|a_n|} = \tfrac{1}{2} \lim \sqrt[n]{|a_n|} = \tfrac{1}{2}$, or $R = 2$.

c) $a_n = n!$. Since $\lim \sqrt[n]{n!} = \infty$ (see 3.1.3, example 2), we have $1/R = \lim \sqrt[n]{n!} = \infty$ and $R = 0$.

d) $a_m = 1$ for $m = 2^n$, and $a_m = 0$ for $m \neq 2^n$; thus $\overline{\lim} \sqrt[m]{a_m} = 1$. It follows that $R = 1$.

3.15.: P. At what points of the boundary of the disc of convergence do the following series converge?

a) $\sum_{n=1}^{\infty} z^n$

b) $\sum_{n=1}^{\infty} \dfrac{z^n}{n}$

c) $\sum_{n=1}^{\infty} \dfrac{z^n}{n^2}$

A. a) It is easy to see that $\sum_{n=1}^{\infty} z^n$ does not converge at any point of the boundary. Indeed, a necessary condition for the convergence of the series $\sum a_n$ is $\lim_{n \to \infty} a_n = 0$ (see 4.1.1, Cauchy's criterion). But for $|z| = 1$, $\lim_{n \to \infty} |z^n| = 1$, and $\lim_{n \to \infty} z^n \neq 0$.

b) Obviously, $\sum_{n=1}^{\infty} \dfrac{z^n}{n}$ does not converge for $z = 1$. We want to show that it converges for any other $z \neq 1$ with $|z| = 1$.

Set $z + z^2 + \cdots + z^n = z \dfrac{1 - z^n}{1 - z} = S_n$; then $\sum_{n=1}^{\infty} \dfrac{z^n}{n}$ may be written

$$\sum_{n=1}^{\infty} \dfrac{z^n}{n} = \sum_{n=1}^{\infty} \dfrac{S_n - S_{n-1}}{n} \quad \text{with} \quad S_0 = 0$$

Now

$$\sum_{n=1}^{\infty} \frac{S_n - S_{n-1}}{n} = \sum_{n=1}^{\infty} S_n \left(\frac{1}{n} - \frac{1}{n+1} \right)$$

$$= \frac{z}{1-z} \sum_{n=1}^{\infty} (1 - z^n) \left(\frac{1}{n} - \frac{1}{n+1} \right)$$

Since $|1 - z^n| \leq 2$ for $|z| = 1$, and $\sum_{n=1}^{\infty} \left(\frac{1}{n} - \frac{1}{n+1} \right)$ converges absolutely, we find, using Cauchy's criterion,

$$\left| \sum_{n=N}^{\infty} (1 - z^n) \left(\frac{1}{n} - \frac{1}{n+1} \right) \right| \leq 2 \sum_{n=N}^{\infty} \left(\frac{1}{n} - \frac{1}{n+1} \right) = \frac{2}{N}$$

which proves the convergence of $\sum_{n=1}^{\infty} \frac{z^n}{n}$ for $|z| = 1$, $z \neq 1$.

c) The radius of convergence of $\sum_{n=1}^{\infty} \frac{z^n}{n^2}$ is 1. The series $\sum_{n=1}^{\infty} \frac{1}{n^2}$ converges. Thus $\sum_{n=1}^{\infty} \frac{z^n}{n^2}$ converges absolutely and uniformly for $|z| = 1$.

3.17.: P. Expand the following functions in a Taylor series about $z = 1$:

a) $\dfrac{1}{z}$

b) $\dfrac{1}{z+2}$

c) $\log z$

A. a)
$$\frac{1}{z} = \frac{1}{1 + (z-1)} = 1 - (z-1) + (z-1)^2 - \cdots$$

which converges for $|z - 1| < 1$.

b) $\dfrac{1}{z+2} = \dfrac{1}{3 + (z-1)} = \dfrac{1}{3} \cdot \dfrac{1}{1 + \dfrac{z-1}{3}} = \dfrac{1}{3}\left(1 - \dfrac{z-1}{3} + \left(\dfrac{z-1}{3}\right)^2 - \cdots \right)$

which converges for $|z - 1| < 3$.

c)
$$\log z = \int_1^z \frac{dz}{z} = \int_1^z \frac{dz}{1+(z-1)}$$
$$= \int_1^z [1-(z-1)+(z-1)^2-\cdots]\,dz$$
$$= c + (z-1) - \frac{(z-1)^2}{2} + \frac{(z-1)^3}{3} - \cdots$$

In order to determine c, set $z = 1$; then $\log 1 = 0 = c$. It follows that

$$\log z = (z-1) - \frac{(z-1)^2}{2} + \cdots$$

which converges for $|z-1| < 1$.

3.19.: P. Determine the zeros of the following functions and their order.

a) $e^z - 1$ b) $\sin^2 z$ c) $\sin z^2$ d) $\dfrac{\sin z}{z}$

A. a) $e^z - 1 = 0 \Rightarrow e^z = 1 \Rightarrow e^{x+iy} = 1 \Rightarrow e^x = 1$ and $y = 2k\pi$. It follows that $x = 0$ and $z = 2k\pi i$, $k = 0, \pm 1, \pm 2, \ldots$ Now

$$e^z - 1 = \left(1 + \frac{z}{1!} + \frac{z^2}{2!} + \cdots\right) - 1 = z\left(\frac{1}{1!} + \frac{z}{2!} + \frac{z^2}{3!} + \cdots\right)$$

i.e., $z = 0$ is a simple zero, or zero of order 1, and, since e^z is periodic of period $2\pi i$, all the zeros are simple.

b)
$$\sin z = z - \frac{z^3}{3!} + \frac{z^5}{5!} - \cdots = z\left(1 - \frac{z^2}{3!} + \cdots\right)$$

has a simple zero at $z = k\pi$, $k = 0, \pm 1, \pm 2, \ldots$, since it is periodic with period 2π and $\sin(z+\pi) = -\sin z$. Hence $\sin^2 z$ has a zero of order 2 at $z = k\pi$, $k = 0, \pm 1, \pm 2, \ldots$. These are the only zeros of $\sin^2 z$, since $\sin z = \dfrac{e^{iz} - e^{-iz}}{2i}$ and $\sin^2 z = 0 \Rightarrow e^{iz} - e^{-iz} = 0$, or $e^{2iz} = 1$. In a) we have seen that the zeros of $e^z - 1$ are $2k\pi i$; hence the zeros of $\sin z$ are $\dfrac{2k\pi i}{2i} = k\pi$.

c) In b) we have seen that the zeros of $\sin z$ are $k\pi$, $k = 0, \pm 1, \pm 2, \ldots$. It follows that the zeros of $\sin z^2$ satisfy $z^2 = k\pi$, or $z = \pm\sqrt{k\pi}$, $k = 0, \pm 1, \pm 2, \ldots$. These zeros are of order two: Indeed,

$$\sin z^2 = z^2\left(1 - \frac{z^4}{3!} + \cdots\right)$$

so that $z = 0$ is a zero of order two, and since $\sin(z^2 + k\pi) = (-1)^k \sin z^2$, all the other zeros are of order two.

d) $$\frac{\sin z}{z} = 1 - \frac{z^2}{3!} + \frac{z^4}{5!} - \cdots$$

has no zero at the origin, but has otherwise the same zeros $z = k\pi$, $k = \pm 1, \pm 2, \ldots$, as sin z, and these are simple zeros.

3.21.: P. Determine the Laurent series of the following functions:

a) $\dfrac{e^z - e^{-z}}{z^3}$ about $z = 0$, valid for $|z| > 0$

b) $\dfrac{z^2 + 1}{z - 1}$ about $z = 1$, valid for $|z - 1| > 0$

A. a) We have

$$e^z = 1 + \frac{z}{1!} + \frac{z^2}{2!} + \frac{z^3}{3!} + \cdots$$

$$e^{-z} = 1 - \frac{z}{1!} + \frac{z^2}{2!} - \frac{z^3}{3!} + \cdots$$

$$e^z - e^{-z} = \frac{2z}{1!} + \frac{2z^3}{3!} + \frac{2z^5}{5!} + \cdots$$

and

$$\frac{e^z - e^{-z}}{z^3} = \frac{2}{1! z^2} + \frac{2}{3!} + \frac{2z^2}{5!} + \cdots$$

We see that $z = 0$ is a pole of order 2 of the function $\dfrac{e^z - e^{-z}}{z^3}$.

b) $\dfrac{z^2 + 1}{z - 1} = \dfrac{z^2 - 1 + 2}{z - 1} = z + 1 + \dfrac{2}{z - 1} = (z - 1) + 2 + \dfrac{2}{z - 1}$

We see that $z = 1$ is a simple pole of the function $\dfrac{z^2 + 1}{z - 1}$.

3.23.: P. Find the Laurent series of the following functions:

a) $\dfrac{e^{3z}}{(z - a)^3}$ about $z = a$

b) $(z + 3)\sin \dfrac{1}{z - 3}$ about $z = 3$

A. a)
$$\frac{e^{3z}}{(z-a)^3} = \frac{e^{3(z-a)+3a}}{(z-a)^3} = e^{3a} \cdot \frac{1}{(z-a)^3}\left(1 + \frac{3(z-a)}{1!} + \frac{9(z-a)^2}{2!} + \cdots\right)$$

$$= e^{3a}\left[\frac{1}{(z-a)^3} + \frac{3}{1!(z-a)^2} + \frac{9}{2!(z-a)} + \frac{27}{3!} + \frac{81}{4!}(z-a)\right.$$

$$\left. + \cdots\right]$$

$z = a$ is a pole of order 3.

b) $\quad (z+3)\sin\dfrac{1}{z-3} = (z-3)\sin\dfrac{1}{z-3} + 6\sin\dfrac{1}{z-3}$

Now

$$\sin\frac{1}{z-3} = \frac{1}{z-3} - \frac{1}{3!}\cdot\frac{1}{(z-3)^3} + \frac{1}{5!}\cdot\frac{1}{(z-3)^5} - \cdots$$

It follows that

$$(z+3)\sin\frac{1}{z-3} = 1 - \frac{1}{3!}\cdot\frac{1}{(z-3)^2} + \frac{1}{5!}\cdot\frac{1}{(z-3)^4} - \cdots$$

$$+ \frac{6}{z-3} - \frac{6}{3!(z-3)^3} + \frac{6}{5!(z-3)^5} - \cdots$$

Hence $z = 3$ is an essential singularity.

3.25.: P. Expand in a Laurent series:

a) $\dfrac{1 - e^{2z}}{z^4}$ about the origin

b) $\dfrac{1}{z^2(z-1)}$ about $z = 0$ and $z = 1$

c) $\dfrac{e^{2z}}{(z-1)^2}$ about $z = 1$

d) $\dfrac{1 + e^z}{\sin z + z\cos z}$ about $z = 0$

Compute the residue and determine the order of the poles in each case.

A. a)
$$e^{2z} = 1 + \frac{2z}{1!} + \frac{4z^2}{2!} + \cdots$$

$$\frac{1 - e^{2z}}{z^4} = \frac{1}{z^4}\left(-\frac{2z}{1!} - \frac{4z^2}{2!} - \frac{8z^3}{3!} - \cdots\right)$$

$$= -\frac{2}{z^3} - \frac{4}{2!z^2} - \frac{8}{3!z} - \frac{16}{4!} - \cdots$$

$z = 0$ is a pole of order 3, and the residue at $z = 0$ is $-\frac{8}{3!} = a_{-1}$ since the contribution of all other terms is 0.

b) $\frac{1}{z^2(z-1)}$ about $z = 0$:

$$\frac{1}{z-1} = -\frac{1}{1-z} = -1 - z - z^2 - \cdots$$

and

$$\frac{1}{z^2(z-1)} = -\frac{1}{z^2} - \frac{1}{z} - 1 - z - \cdots$$

It follows that $z = 0$ is a pole of order 2, and that the residue at this point is equal to the coefficient of $1/z$, which is -1.

$$\frac{1}{z^2} = \frac{1}{[1-(1-z)]^2} = [1 + (1-z) + (1-z)^2 + \cdots]^2$$

$$= 1 + 2(1-z) + 3(1-z)^2 + \cdots = 1 - 2(z-1) + 3(z-1)^2 - \cdots$$

and

$$\frac{1}{z^2(z-1)} = \frac{1}{z-1} - 2 + 3(z-1) - 4(z-1)^2 + \cdots$$

It follows that $z = 1$ is a simple pole with residue 1 $\left(1 \text{ is the coefficient of } \frac{1}{z-1}\right)$.

c) We have:

$$e^{2z} = e^2 \cdot e^{2(z-1)} = e^2\left(1 + 2(z-1) + \frac{4(z-1)^2}{2!} + \frac{8(z-1)^3}{3!} + \cdots\right)$$

It follows for the expansion in a Laurent series that

$$\frac{e^{2z}}{(z-1)^2} = e^2 \left[\frac{1}{(z-1)^2} + \frac{2}{z-1} + 2 + \tfrac{4}{3}(z-1) + \cdots \right]$$

Hence the residue is $2e^2$ and $z = 1$ is a pole of order 2.

d) Let

$$\frac{1 + e^z}{\sin z + z \cos z} = f(z)$$

Let $M(z) = 1 + e^z$ and let $N(z) = \sin z + z \cos z$. Then $f(z) = \dfrac{M(z)}{N(z)}$.

$N(z)$ has a zero of order 1 at the origin, so $\dfrac{1}{N(z)}$ has a pole of order 1 (see 3.2.2), and since $M(0) \neq 0$, $f(z)$ also has a pole of order 1 at the origin. The residue of $f(z)$ at the origin can be obtained by

$$\operatorname{Res}(f, 0) = \lim_{z \to 0} z f(z) = \lim_{z \to 0} \frac{1 + e^z}{\dfrac{\sin z}{z} + \cos z} = \frac{2}{2} = 1$$

(See 3.3.1.)

We compute the Laurent series of $f(z)$ using the fact that

$$z f(z) = \frac{1 + e^z}{\dfrac{\sin z}{z} + \cos z}$$

Let $a_0 + a_1 z + a_2 z^2 + \cdots$ be the Taylor series of $zf(z)$. Since $1 + e^z = 2 + z + \tfrac{1}{2}z^2 + \tfrac{1}{6}z^3 + \cdots$, and $\dfrac{\sin z}{z} + \cos z = 2 - \tfrac{2}{3}z^2 + \tfrac{1}{20}z^4 - \cdots$, we must have:

$$2 + z + \tfrac{1}{2}z^2 + \tfrac{1}{6}z^3 + \cdots = (a_0 + a_1 z + a_2 z^2 + \cdots)(2 - \tfrac{2}{3}z^2 + \tfrac{1}{20}z^4 - \cdots)$$

By comparing the coefficients of the same powers right and left, we get:

$$\begin{aligned} 2 &= 2a_0 & &\Rightarrow & a_0 &= 1 \\ 1 &= 2a_1 & & & a_1 &= \tfrac{1}{2} \\ \tfrac{1}{2} &= 2a_2 - \tfrac{2}{3}a_0 & & & a_2 &= \tfrac{7}{12} \\ \tfrac{1}{6} &= 2a_3 - \tfrac{2}{3}a_1 & & & a_3 &= \tfrac{1}{4} \\ &\vdots & & & &\vdots \end{aligned}$$

It follows that

$$z f(z) = 1 + \tfrac{1}{2}z + \tfrac{7}{12}z^2 + \tfrac{1}{4}z^3 + \cdots$$

or
$$f(z) = \frac{1}{z} + \frac{1}{2} + \frac{7}{12}z + \frac{1}{4}z^2 + \cdots$$

which is the Laurent series for $f(z)$.

3.27.: P. Expand the function $e^{1/(z-1)}$ in a Laurent series about $z = 1$.
A. Use the Taylor series for e^z:
$$e^z = 1 + \frac{z}{1!} + \frac{z^2}{2!} + \cdots$$

hence
$$e^{1/(z-1)} = 1 + \frac{1}{1!(z-1)} + \frac{1}{2!(z-1)^2} + \cdots$$

$z = 1$ is an essential singularity of $e^{1/(z-1)}$.

3.29.: P. Prove that $\text{Res}(f + g, c) = \text{Res}(f, c) + \text{Res}(g, c)$.
A. By definition,
$$\text{Res}(f + g, c) = \frac{1}{2\pi i} \int_\partial (f(u) + g(u))\, du$$

where $\partial = \partial(c, r)$ and r is chosen so that c is the only singularity of f and g inside $\partial(c, r)$.
Now
$$\frac{1}{2\pi i} \int_\partial (f(u) + g(u))\, du = \frac{1}{2\pi i} \int_\partial f(u)\, du + \frac{1}{2\pi i} \int_\partial g(u)\, du$$
$$= \text{Res}(f, c) + \text{Res}(g, c).$$

3.31.: P. Let $g(z)$ be analytic at c and let $h(z)$ have a pole of order 2 at c. Show that $\text{Res}(g(z)h(z), c) = g'(c)\text{Res}((z - c)h(z), c) + g(c)\text{Res}(h(z), c)$.
A.
$$g(z) = g_0 + g_1(z - c) + g_2(z - c)^2 + \cdots$$
$$h(z) = \frac{h_{-2}}{(z - c)^2} + \frac{h_{-1}}{z - c} + h_0 + h_1(z - c) + \cdots$$

Hence
$$g(z)h(z) = \frac{g_0 h_{-2}}{(z - c)^2} + \frac{g_0 h_{-1} + g_1 h_{-2}}{z - c} + g_0 h_0 + g_1 h_{-1} + g_2 h_{-2} + \cdots$$

and the coefficient of $1/(z - c)$ is $g_0 h_{-1} + g_1 h_{-2}$.
Now $g_0 = g(c)$ and $g_1 = g'(c)$, $h_{-1} = \text{Res}(h(z), c)$ and $h_{-2} = \text{Res}((z - c)h(z), c)$. Hence
$$\text{Res}[g(z)h(z), c] = g'(c)\text{Res}((z - c)h(z), c) + g(c)\text{Res}(h(z), c)$$

3.33.: P. a) Show that

  i) $\operatorname{Res}\left(\dfrac{z}{\sin z}, 0\right) = 0$

  ii) $\operatorname{Res}\left(\tan z, \dfrac{\pi}{2}\right) = -1$

  iii) $\operatorname{Res}\left(\dfrac{\sin z}{z^2}, 0\right) = 1$

  iv) $\operatorname{Res}\left(\dfrac{1 - \cos z}{z^3}, 0\right) = \dfrac{1}{2}$

b) Determine the residues of $\dfrac{z}{(z-1)(z-2)}$ at $z = 1$ and at $z = 2$.

A. a) i) $\sin z$ has a simple zero at $z = 0$. It follows that $\dfrac{1}{\sin z}$ has a simple pole at $z = 0$, and $\dfrac{z}{\sin z}$ is holomorphic at $z = 0$. Hence

$$\operatorname{Res}\left(\dfrac{z}{\sin z}, 0\right) = 0.$$

ii) We write $\tan z = \dfrac{\sin z}{\cos z} = \dfrac{\cos\left(\dfrac{\pi}{2} - z\right)}{\sin\left(\dfrac{\pi}{2} - z\right)}$

$\sin\left(\dfrac{\pi}{2} - z\right)$ has a simple zero at $z = \pi/2$ and $\cos\left(\dfrac{\pi}{2} - z\right) = 1$ at $z = \pi/2$. Hence $z = \pi/2$ is a simple pole of $\tan z$.
Now $\operatorname{Res}(f, c) = \lim\limits_{z \to c} (z - c) f(z)$ at a simple pole. Thus

$$\operatorname{Res}\left(\tan z, \dfrac{\pi}{2}\right) = \lim_{z \to \pi/2} \left(z - \dfrac{\pi}{2}\right) \cdot \dfrac{\cos\left(\dfrac{\pi}{2} - z\right)}{\sin\left(\dfrac{\pi}{2} - z\right)} = -1$$

since

$$\sin\left(\dfrac{\pi}{2} - z\right) = \left(\dfrac{\pi}{2} - z\right)\left[1 - \dfrac{1}{3!} \cdot \left(\dfrac{\pi}{2} - z\right)^2 + \dfrac{1}{5!} \cdot \left(\dfrac{\pi}{2} - z\right)^4 - \cdots\right]$$

and

$$\lim_{z \to \pi/2} \dfrac{\sin\left(\dfrac{\pi}{2} - z\right)}{z - \dfrac{\pi}{2}} = -1$$

iii)
$$\frac{\sin z}{z^2} = \frac{1}{z} - \frac{1}{3!}z + \frac{1}{5!}z^3 - \cdots$$

Hence
$$a_{-1} = 1 = \text{Res}\left(\frac{\sin z}{z^2}, 0\right)$$

iv)
$$\frac{1 - \cos z}{z^3} = \frac{\frac{z^2}{2!} - \frac{z^4}{4!} + \cdots}{z^3} = \frac{1}{2!z} - \frac{z}{4!} + \cdots$$

Hence
$$a_{-1} = \text{Res}\left(\frac{1 - \cos z}{z^3}, 0\right) = \frac{1}{2}$$

b) The function $\dfrac{z}{(z-1)(z-2)}$ has simple poles at $z = 1$ and at $z = 2$.
It follows that

$$\text{Res}\left(\frac{z}{(z-1)(z-2)}, 1\right) = \lim_{z \to 1} \frac{z}{z-2} = -1$$

and

$$\text{Res}\left(\frac{z}{(z-1)(z-2)}, 2\right) = \lim_{z \to 2} \frac{z}{z-1} = 2$$

3.35.: P. Evaluate:

a) $\displaystyle\int_0^\infty \frac{dx}{x^4 + 1}$

b) $\displaystyle\int_0^\infty \frac{x^2\, dx}{x^4 + 1}$

c) $\displaystyle\int_0^\infty \frac{x^2\, dx}{(x^2 + 1)^2}$

A. a) $\displaystyle\int_0^\infty \frac{dx}{x^4 + 1} = \frac{1}{2}\int_{-\infty}^\infty \frac{dx}{x^4 + 1} = \pi i \sum_{\text{Im } s_j > 0} \text{Res}\left(\frac{1}{z^4 + 1}, s_j\right)$

Now the function $\dfrac{1}{z^4 + 1}$ has simple poles at the roots of $z^4 + 1 = 0$, easily identified as

$$s_1 = \frac{1}{2}\sqrt{2}(1+i), \quad s_2 = \frac{1}{2}\sqrt{2}(-1+i), \quad s_3 = -s_1; \quad s_4 = -s_2$$

and yielding the factorization

$$z^4 + 1 = (z - s_1)(z - s_2)(z - s_3)(z - s_4)$$

Hence

$$\operatorname{Res}\left(\frac{1}{z^4+1}, s_1\right) = \lim_{z \to s_1} \frac{z - s_1}{z^4 + 1} = \frac{1}{(s_1 - s_2)(s_1 - s_3)(s_1 - s_4)}$$
$$= -\frac{\sqrt{2}}{8}(1+i)$$

$$\operatorname{Res}\left(\frac{1}{z^4+1}, s_2\right) = \lim_{z \to s_2} \frac{z - s_2}{z^4 + 1} = \frac{1}{(s_2 - s_1)(s_2 - s_3)(s_2 - s_4)}$$
$$= \frac{\sqrt{2}}{8}(1-i)$$

and consequently

$$\sum_{\operatorname{Im} s_j > 0} \operatorname{Res}\left(\frac{1}{z^4+1}, s_j\right) = -\frac{\sqrt{2}}{4}i, \quad \int_0^\infty \frac{dx}{x^4+1} = \frac{1}{4}\pi\sqrt{2}$$

b) $$\int_0^\infty \frac{x^2\,dx}{x^4+1} = \frac{1}{2}\int_{-\infty}^\infty \frac{x^2\,dx}{x^4+1} = \pi i \sum_{\operatorname{Im} s_j > 0} \operatorname{Res}\left(\frac{z^2}{z^4+1}, s_j\right)$$

The function $\dfrac{z^2}{z^4+1}$ has the same poles as $\dfrac{1}{z^4+1}$ under a). It follows that

$$\operatorname{Res}\left(\frac{z^2}{z^4+1}, s_1\right) = \lim_{z \to s_1}(z - s_1)\frac{z^2}{z^4+1} = \frac{s_1^2}{(s_1 - s_2)(s_1 - s_3)(s_1 - s_4)}$$
$$= \frac{\sqrt{2}}{8}(1-i)$$

$$\operatorname{Res}\left(\frac{z^2}{z^4+1}, s_2\right) = \lim_{z \to s_2} (z-s_2)\frac{z^2}{z^4+1} = \frac{s_2{}^2}{(s_2-s_1)(s_2-s_3)(s_2-s_4)}$$

$$= -\frac{\sqrt{2}}{8}(1+i)$$

and consequently

$$\sum_{\operatorname{Im} s_j > 0} \operatorname{Res}\left(\frac{z^2}{z^4+1}, s_j\right) = -\frac{\sqrt{2}}{4}i, \quad \int_0^\infty \frac{x^2\,dx}{x^4+1} = \frac{1}{4}\pi\sqrt{2}$$

It is not surprising that the two integrals under a) and b) have the same values since each can be transformed into the other by the substitution $x = \frac{1}{t}$.

c) Since $R(x) = \dfrac{x^2}{(1+x^2)^2}$ is an even function, we have:

$$\int_0^\infty \frac{x^2}{(1+x^2)^2}\,dx = \frac{1}{2}\int_{-\infty}^\infty \frac{x^2}{(1+x^2)^2}\,dx$$

The integral $\displaystyle\int_{-\infty}^\infty \frac{x^2}{(1+x^2)^2}\,dx$ is one of category 2 (see 3.3.2). Therefore, we have

$$\int_{-\infty}^\infty R(x)\,dx = 2\pi i \sum_{\operatorname{Im} s_j > 0} \operatorname{Res}(R, s_j)$$

where the s_j are the poles of $R(z)$.

Now, the only pole of $z^2/(1+z^2)^2$ in the upper half plane is of order 2, at $z = i$. Expanding $R(z)$ in a Laurent series about $z = i$ gives us

$$R(z) = \frac{z^2}{(1+z^2)^2} = \frac{a_{-2}}{(z-i)^2} + \frac{a_{-1}}{z-i} + a_0 + a_1(z-i) + \cdots$$

so that $R(z)(z-i)^2 = a_{-2} + a_{-1}(z-i) + a_0(z-i)^2 + a_1(z-i)^3 + \cdots$ is holomorphic in $z = i$ with

$$a_{-1} = \frac{d}{dz}[R(z)(z-i)^2]_{z=i}$$

Since $\text{Res}(R, i) = a_{-1}$ and $R(z)(z - i)^2 = z^2/(z + i)^2$, it follows that

$$\frac{d}{dz}[R(z)(z - i)^2]_{z=i} = \frac{d}{dz}\left[\frac{z^2}{(z + i)^2}\right]_{z=i}$$

$$= \left[\frac{(z + i)^2 \cdot 2z - 2(z + i)z^2}{(z + i)^4}\right]_{z=i}$$

$$= \frac{8i^3 - 4i^3}{16i^4} = \frac{1}{4i} = -\frac{i}{4}$$

Hence

$$\int_{-\infty}^{\infty} R(x)\, dx = -2\pi i \cdot \frac{i}{4} = \frac{\pi}{2} \quad \text{and} \quad \int_0^{\infty} \frac{x^2}{(1 + x^2)^2}\, dx = \frac{\pi}{4}$$

This result can be confirmed in the following elegant way: The substitution $x = \frac{1}{y}$, applied to the integral $I = \int_0^{\infty} \frac{x^2\, dx}{(x^2 + 1)^2}$ yields immediately $I = \int_0^{\infty} \frac{dy}{(y^2 + 1)^2} = \int_0^{\infty} \frac{dx}{(x^2 + 1)^2}$. Adding the two expressions for I, one finds $2I = \int_0^{\infty} \frac{dx}{x^2 + 1} = [\arctan x]_0^{\infty} = \frac{\pi}{2}$ and therefore $I = \frac{\pi}{4}$ as stated above.

3.37.: P. Show that

a) $\int_0^{2\pi} \frac{d\phi}{1 + \sin^2\phi} = \pi\sqrt{2}$

b) $\int_0^{2\pi} \frac{\sin^2\phi\, d\phi}{1 + \frac{1}{2}\cos\phi} = 4\pi(2 - \sqrt{3})$

c) $\int_0^{\pi} \frac{d\phi}{(1 + \frac{1}{2}\cos\phi)^2} = \frac{8\pi}{3\sqrt{3}}$

d) $\int_0^{\pi/2} \frac{d\phi}{1 + \sin^2\phi} = \frac{\pi}{2\sqrt{2}}$

A. a) Set $z = e^{i\phi}$, $dz = iz d\phi$, $\sin \phi = \dfrac{1}{2i}\left(z - \dfrac{1}{z}\right)$. It follows that

$$\int_0^{2\pi} \frac{d\phi}{1 + \sin^2 \phi} = \int_\partial \frac{dz}{iz\left[1 - \dfrac{1}{4}\left(z - \dfrac{1}{z}\right)^2\right]}$$

$$= -i \int_\partial \frac{dz}{z\left[1 - \dfrac{1}{4}\left(z^2 - 2 + \dfrac{1}{z^2}\right)\right]}$$

$$= i \int_\partial \frac{4z\,dz}{z^4 - 6z^2 + 1}$$

where $\partial = \partial(0, 1)$.

The poles of $\dfrac{4z}{z^4 - 6z^2 + 1}$ are simple. They are $z_1 = \sqrt{2} - 1$, $z_2 = 1 - \sqrt{2}$, $z_3 = \sqrt{2} + 1$ and $z_4 = -1 - \sqrt{2}$. Inside ∂ are the two poles, z_1 and z_2. Now

$$\frac{1}{2\pi i}\int_\partial \frac{4z\,dz}{z^4 - 6z^2 + 1} = \sum_{z_j \in d(0,1)} \text{Res}\left(\frac{4z}{z^4 - 6z^2 + 1}, z_j\right)$$

and

$$\text{Res}\left(\frac{4z}{z^4 - 6z^2 + 1}, z_1\right) = \lim_{z \to z_1}(z - z_1)\frac{4z}{z^4 - 6z^2 + 1}$$

$$= \frac{4z_1}{(z_1 - z_2)(z_1 - z_3)(z_1 - z_4)}$$

$$= \frac{4(\sqrt{2} - 1)}{2(\sqrt{2} - 1)(-2)2\sqrt{2}} = -\frac{1}{2\sqrt{2}}$$

$$\text{Res}\left(\frac{4z}{z^4 - 6z^2 + 1}, z_2\right) = \lim_{z \to z_2}(z - z_2)\frac{4z}{z^4 - 6z^2 + 1}$$

$$= \frac{4z_2}{(z_2 - z_1)(z_2 - z_3)(z_2 - z_4)}$$

$$= \frac{4(1 - \sqrt{2})}{2(1 - \sqrt{2})(-2\sqrt{2})2} = -\frac{1}{2\sqrt{2}}$$

$$\int_0^{2\pi} \frac{d\phi}{1 + \sin^2 \phi} = i\int_\partial \frac{4z\,dz}{z^4 - 6z^2 + 1} = i(2\pi i)\left(-\frac{1}{2\sqrt{2}} - \frac{1}{2\sqrt{2}}\right)$$

$$= \frac{2\pi}{\sqrt{2}} = \pi\sqrt{2}$$

b) Setting $z = e^{i\phi}$, we find:

$$I = \int_\partial \frac{-\frac{1}{4}\left(z - \frac{1}{z}\right)^2 dz}{iz\left[1 + \frac{1}{2} \cdot \frac{1}{2}\left(z + \frac{1}{z}\right)\right]} = \frac{i}{4}\int_\partial \frac{(z^4 - 2z^2 + 1)\,dz}{z^2\left[z + \frac{1}{4}z^2 + \frac{1}{4}\right]}$$

$$= i\int_\partial \frac{(z^4 - 2z^2 + 1)\,dz}{z^2(z^2 + 4z + 1)}$$

The function

$$f(z) = \frac{(z^4 - 2z^2 + 1)i}{z^2(z^2 + 4z + 1)}$$

has a pole of order 2 at $z_1 = 0$, and two simple poles at $z_2 = -2 + \sqrt{3}$ and $z_3 = -2 - \sqrt{3}$. The pole at z_3 is outside ∂ and may therefore be neglected.

Using the methods of category 1, we have

$$I = 2\pi i \sum_{s_j \in d(0,\,1)} \text{Res}(f, s_j)$$

In order to determine $\text{Res}(f, 0)$, we use the formula of problem 3.31 with

$$g(z) = \frac{(z^4 - 2z^2 + 1)i}{z^2 + 4z + 1} \quad \text{and} \quad h(z) = \frac{1}{z^2}$$

Then $\text{Res}(f(z), 0) = \text{Res}(g(z)h(z), 0) = g'(0)\text{Res}(1/z, 0) + g(0)\text{Res}(1/z^2, 0) = g'(0)$, since $\text{Res}(1/z, 0) = 1$ and $\text{Res}(1/z^2, 0) = 0$.

Now

$$g'(z) = \frac{(z^2 + 4z + 1)i(4z^3 - 4z) - i(z^4 - 2z^2 + 1)(2z + 4)}{(z^2 + 4z + 1)^2}$$

Hence $g'(0) = -4i$.

$$\text{Res}(f(z), z_2) = \lim_{z \to z_2} (z - z_2) \frac{i(z^4 - 2z^2 + 1)}{z^2(z^2 + 4z + 1)}$$

$$= \lim_{z \to z_2} \frac{i(z^4 - 2z^2 + 1)}{z^2(z - z_3)} = \frac{i(z_2^4 - 2z_2^2 + 1)}{z_2^2(z_2 - z_3)}$$

$$= \frac{i(z_2^2 - 1)^2}{z_2^2(z_2 - z_3)} = \frac{i(6 - 4\sqrt{3})^2}{(\sqrt{3} - 2)^2 2\sqrt{3}} = \frac{12i(7 - 4\sqrt{3})}{(7 - 4\sqrt{3})2\sqrt{3}}$$

$$= \frac{6i}{\sqrt{3}} = 2i\sqrt{3}$$

Hence $I = 2\pi i(-4i + 2i\sqrt{3}) = 8\pi - 4\pi\sqrt{3} = 4\pi(2 - \sqrt{3})$.

c) Since $\cos \phi$ is an even function, we have

$$I = \frac{1}{2}\int_{-\pi}^{\pi} \frac{d\varphi}{(1 + \frac{1}{2}\cos\varphi)^2}$$

Again setting $z = e^{i\phi}$, we find

$$I = \frac{1}{2}\int_{\partial} \frac{dz}{iz\left[1 + \frac{1}{4}\left(z + \frac{1}{z}\right)\right]^2} = -8i\int_{\partial} \frac{z\,dz}{z^4 + 8z^3 + 18z^2 + 8z + 1}$$

$$= -8i \cdot 2\pi i \sum_{s_j \in \mathscr{d}(0,1)} \operatorname{Res}(f, s_j)$$

where

$$f(z) = \frac{z}{z^4 + 8z^3 + 18z^2 + 8z + 1} = \frac{z}{(z^2 + 4z + 1)^2}$$

The poles of f are $s_1 = -2 + \sqrt{3}$ and $s_2 = -2 - \sqrt{3}$, both of order 2. s_1 is inside ∂ and s_2 is outside ∂.

We decompose f into partial fractions—

$$f(z) = \frac{s_1}{12} \cdot \frac{1}{(z - s_1)^2} + \frac{\sqrt{3}}{18} \cdot \frac{1}{z - s_1} + \frac{s_2}{12} \cdot \frac{1}{(z - s_2)^2} - \frac{\sqrt{3}}{18} \cdot \frac{1}{z - s_2}$$

and obtain $\operatorname{Res}(f, s_1) = \sqrt{3}/18$, so that

$$I = 16\pi \cdot \frac{\sqrt{3}}{18} = \frac{8\pi}{3\sqrt{3}}$$

d) $$I = \int_0^{\pi/2} \frac{d\phi}{1 + \sin^2 \phi} = \frac{1}{2}\int_{-\pi/2}^{\pi/2} \frac{d\phi}{1 + \sin^2 \phi}$$

since $1/(1 + \sin^2 \phi)$ is an even function.

$$I = \frac{1}{4}\int_{-\pi}^{\pi} \frac{d\phi}{1 + \sin^2 \phi}$$

since $\sin\left(\frac{\pi}{2} + \phi\right) = \sin\left(\frac{\pi}{2} - \phi\right)$. Finally, $I = \frac{1}{4}\pi\sqrt{2}$ (see a).

3.39.: P. Calculate the residues of

$$z\frac{f'(z)}{f(z)} \quad \text{and} \quad g(z)\frac{f'(z)}{f(z)}$$

at a zero or a pole of $f(z)$ assuming that $g(z)$ is holomorphic at these points.
A. Let c be a zero of f of order α. Then $f(z) = (z - c)^\alpha h(z)$ with $h(c) \neq 0$.
Hence

$$g(z)\frac{f'(z)}{f(z)} = g(z)\left[\frac{\alpha}{z - c} + \frac{h'(z)}{h(z)}\right]$$

It follows that

$$\text{Res}\left(g(z)\frac{f'(z)}{f(z)}, c\right) = \alpha \, \text{Res}\left(\frac{g(z)}{z-c}, c\right) + \text{Res}\left(\frac{g(z)h'(z)}{h(z)}, c\right)$$

$$= \alpha \cdot \frac{1}{2\pi i} \int_{\partial(c,r)} \frac{g(z)}{z-c} \, dz + \frac{1}{2\pi i} \int_{\partial(c,r)} \frac{g(z)h'(z)}{h(z)} \, dz$$

Recalling Cauchy's formula, we have

$$\frac{1}{2\pi i} \int_{\partial(c,r)} \frac{g(z)}{z-c} \, dz = g(c)$$

On the other hand, using Cauchy's theorem,

$$\frac{1}{2\pi i} \int_{\partial(c,r)} \frac{g(z)h'(z)}{h(z)} \, dz = 0$$

since $\dfrac{g(z)h'(z)}{h(z)}$ is holomorphic at c (i.e., in a neighborhood of c).
It follows that

$$\text{Res}\left(g(z)\frac{f'(z)}{f(z)}, c\right) = \alpha g(c)$$

For $g(z) \equiv z$, we find

$$\text{Res}\left(z\frac{f'(z)}{f(z)}, c\right) = \alpha c$$

If c is a pole order α of f, then $f(z) = (z-c)^{-\alpha} h(z)$ with $h(c) \neq 0$, and by the same reasoning we find

$$\text{Res}\left(g(z)\frac{f'(z)}{f(z)}, c\right) = -\alpha g(c)$$

3.41.: P. Show that

$$I = \int_{-\infty}^{\infty} \frac{dx}{(1+x^2)^n} = \frac{\pi}{2^{n-2}} \cdot \frac{(2n-2)!}{[(n-1)!]^2}$$

A. Using the result of category 2,

$$I = 2\pi i \sum_{\text{Im } s_j > 0} \text{Res}\left(\frac{1}{(1+z^2)^n}, s_j\right)$$

The only singularity of $\dfrac{1}{(1+z^2)^n}$ in the upper half plane is $s = i$, which is a pole of order n.

In order to find the residue at i, we expand $\dfrac{1}{(1+z^2)^n}$ in a Laurent series about i. The residue is equal to the coefficient of $\dfrac{1}{z-i}$.

$$\frac{1}{(1+z^2)^n} = \frac{1}{(z-i)^n} \cdot \frac{1}{(z+i)^n} = \frac{1}{(z-i)^n} \cdot \frac{1}{[2i+(z-i)]^n}$$

$$= \frac{1}{(z-i)^n} \cdot \frac{1}{(2i)^n} \cdot \frac{1}{\left(1+\dfrac{z-i}{2i}\right)^n} = \frac{1}{(z-i)^n} \cdot \frac{1}{(2i)^n} \cdot E$$

where

$$E = \left[\binom{n-1}{n-1} - \binom{n}{n-1}\frac{z-i}{2i} + \binom{n+1}{n-1}\left(\frac{z-i}{2i}\right)^2 \cdots\right]$$

Hence the coefficient of $\dfrac{1}{z-i}$ can be written as

$$\operatorname{Res}\left(\frac{1}{(1+z^2)^n}, i\right) = a_{-1} = \frac{1}{(2i)^n} \cdot (-1)^{n-1} \cdot \binom{2n-2}{n-1} \cdot \frac{1}{(2i)^{n-1}}$$

$$= \frac{1}{2^{2n-1}} \cdot \frac{1}{i} \cdot \frac{(2n-2)!}{[(n-1)!]^2}$$

It follows that

$$I = (2\pi i)a_{-1} = \frac{\pi}{2^{2n-2}} \cdot \frac{(2n-2)!}{[(n-1)!]^2}$$

3.43.: P. Show that two roots of the equation

$$z^4 - 3z^2 + 1$$

lie in the annulus $1 < |z| < 2$ and that the other two roots lie in the disc $d(0, 1)$.

A. Obviously, we have on $\partial(0, 2)$:

$$|z^4| > |-3z^2 + 1|$$

since $|z^4| = 2^4 = 16$ and $\max_{z \in \partial(0, 2)} |-3z^2 + 1| = 3 \cdot 2^2 + 1 = 13$.

Now z^4 has four roots in $d(0, 2)$; hence, using Rouché's theorem, $z^4 - 3z^2 + 1$ also has four roots in $d(0, 2)$.

On the other hand, on $\partial(0, 1)$

$$|z^4| < |-3z^2 + 1|$$

since $|z^4| = 1^4 = 1$ and $\min_{z \in \partial(0, 1)} |-3z^2 + 1| = |-3 + 1| = 2$.

Now $-3z^2 + 1$ has two roots in $d(0, 1)$, $z = \pm 1/\sqrt{3}$; it follows from Rouché's theorem that $z^4 - 3z^2 + 1$ has two roots in $d(0, 1)$.

Since there are no roots on $\partial(0, 1)$, it follows that the other two roots of $z^4 - 3z^2 + 1$ lie in the annulus $1 < |z| < 2$.

4.1.: P. Consider the mapping $w = z^2$. Find the image of the following curves:
   a) $z = re^{it}$, $0 \leq t \leq 2\pi$
   b) $y = x + b$
   c) $z = 1 + e^{it}$, $0 \leq t \leq 2\pi$
   d) $x = a$,
   e) $y = b$

A. a) $z = re^{it}$ describes the circle $\partial(0, r)$ if t varies, $0 \leq t \leq 2\pi$, and describes it in the counterclockwise sense. Now $w = z^2$ maps that circle to $w = r^2 e^{2it}$ which if t varies, $0 \leq t \leq 2\pi$, describes the circle $\partial(0, r^2)$ in the counterclockwise sense twice.

b) $w = z^2$ can be split into: $U = x^2 - y^2$ and $V = 2xy$. Hence the image of the straight line $y = x + b$ is given by $U = x^2 - (x + b)^2 = b^2 - 2bx$ and $V = 2x(x + b)$, or, since $x = -(b^2 + U)/2b$, we find

$$V = \frac{b^2 + U}{b} \cdot \frac{U - b^2}{2b}$$

which is a parabola in the (U, V)-plane.

c) $z = 1 + e^{it}$, $0 \leq t \leq 2\pi$, describes the circle $\partial(1, 1)$. $w = z^2 = (1 + e^{it})^2 = 1 + 2e^{it} + e^{2it}$, or

$$U = 1 + 2\cos t + \cos 2t \quad \text{and} \quad V = 2\sin t + \sin 2t$$

which gives us the parametric equations of the image.

d) $x = a$ is mapped into

$$U = a^2 - y^2, \quad V = 2ay$$

or

$$U = a^2 - \left(\frac{V}{2a}\right)^2$$

which is a parabola in the (U, V)-plane.

e) $y = b$ is mapped into

$$\left. \begin{array}{l} U = x^2 - b^2 \\ V = 2xb \end{array} \right\} \text{or } U = \left(\frac{V}{2b}\right)^2 - b^2$$

which again is a parabola in the (U, V)-plane.

4.3.: **P.** Consider the mapping $w = \sin z$. Find the image of the following domains:

a) $-\dfrac{\pi}{2} \leq \operatorname{Re} z \leq \dfrac{\pi}{2}$

b) $-\dfrac{\pi}{2} + k\pi \leq \operatorname{Re} z \leq \dfrac{\pi}{2} + k\pi$

A. We can factorize the mapping $w = \sin z$ into three mappings in the following way:

$$w = \sin z = \frac{e^{iz} - e^{-iz}}{2i} = f \circ g \circ h$$

where $h = iz$, $g = e^h$ and $f = \dfrac{1}{2i}\left(g - \dfrac{1}{g}\right)$.

Clearly a) is a special case of b) ($k = 0$). It is therefore sufficient to treat b).

The image of the strip $St_1 = \{z \mid -\pi/2 + k\pi \leq \operatorname{Re} z \leq \pi/2 + k\pi\}$ by $h = iz$ is again a strip $St_2 = \{z \mid -\pi/2 + k\pi \leq \operatorname{Im} z \leq \pi/2 + k\pi\}$, since $h = iz$ is a rotation by $\pi/2$. Since $|e^z| = e^{\operatorname{Re} z} = e^x$, $\arg e^z = \operatorname{Im} z = y$, we find that the image of a parallel to the x-axis ($y = $ constant) is a ray $\{z \mid \arg z = y, |z| > 0\}$. The image of a segment parallel to the y-axis ($x = $ constant) is the circle $\partial(0, e^x)$. Hence the image of St_2 by g is the half plane H minus the origin, where

$$H = \left\{z \mid -\frac{\pi}{2} + k\pi \leq \arg z \leq \frac{\pi}{2} + k\pi\right\}$$

If k is an even integer, then $H = \{z \mid -\pi/2 \leq \arg z \leq \pi/2\}$; if k is an odd integer, then $H = \{z \mid \pi/2 \leq \arg z \leq (3\pi)/2\}$.

Now let us study the image of $H\setminus\{0\}$ by f: Setting $f = U + iV$ and $z = re^{i\phi}$, we find

$$f(z) = \frac{1}{2i}\left(z - \frac{1}{z}\right) = -\frac{i}{2}\left(re^{i\phi} - \frac{1}{r}e^{-i\phi}\right)$$

$$= -\frac{ri}{2}(\cos\phi + i\sin\phi) + \frac{i}{2r}(\cos\phi - i\sin\phi)$$

Hence

$$U = \frac{\sin\phi}{2}\left(r + \frac{1}{r}\right) \quad \text{and} \quad V = \frac{\cos\phi}{2}\left(\frac{1}{r} - r\right)$$

or

$$\frac{2U}{\sin\phi} = r + \frac{1}{r} \quad \text{and} \quad \frac{2V}{\cos\phi} = \frac{1}{r} - r$$

It follows that $\dfrac{U^2}{\sin^2 \phi} - \dfrac{V^2}{\cos^2 \phi} = 1$; i.e., the image of the ray $\{re^{i\phi} \mid \phi$ is constant$\}$ is the hyperbola whose axes are $\sin \phi$ and $\cos \phi$. On the other hand,

$$\left(\dfrac{2U}{r+\dfrac{1}{r}}\right)^2 + \left(\dfrac{2V}{\dfrac{1}{r}-r}\right)^2 = \sin^2 \phi + \cos^2 \phi = 1$$

which means that the image of a circle $\partial(0, r)$ is the ellipse with axes $\dfrac{1}{2}\left(r + \dfrac{1}{r}\right), \dfrac{1}{2}\left(\dfrac{1}{r} - r\right)$.

In the special case where k is an even integer, we see that for $0 < r < 1$, V is positive since $V = \dfrac{\cos \phi}{2}\left(\dfrac{1}{r} - r\right)$ and $-\pi/2 \leq \phi \leq \pi/2$. However, for $r > 1$, V is negative. U is positive for $0 \leq \phi \leq \pi/2$ and negative for $-\pi/2 \leq \phi < 0$. In conclusion, if k is an even integer, the image of $H \cap d'(0, 1)$ by f is the upper half plane Im $z > 0$ and the image of $H \backslash d'(0, 1)$ is the lower half plane Im $z \leq 0$. Thus the image of the strip St_1 with k even by the mapping $w = \sin z$ is the entire complex plane **C**. But it can be easily checked that the image of St_1 with an arbitrary k is **C**.

4.5.: P. Consider the mapping $w = 1 + z^2$. Find the image C' of the circle C: $|z - 1| = \sqrt{2}$, and show that the angle between the tangent to C at z_0 and the tangent to C' at $1 + z_0^2$ is equal to $\arg(w')_{z_0} = \arg 2z_0 = \arg z_0$.

A. The circle C in parametric equation is $z = 1 + \sqrt{2}e^{it}$, $0 \leq t \leq 2\pi$. The equation for C' is $w = 1 + (1 + \sqrt{2}e^{it})^2 = 2 + 2\sqrt{2}e^{it} + 2e^{2it}$. The direction of the tangent to C is determined by $\arg(z') = \arg(i\sqrt{2}e^{it})$ and the tangent to C' is determined by $\arg(w') = \arg(2\sqrt{2}ie^{it} + 4ie^{2it})$.

$$\arg(w') - \arg(z') = \arg(2\sqrt{2}ie^{it} + 4ie^{2it}) - \arg(i\sqrt{2}e^{it})$$
$$= \arg(2 + 2\sqrt{2}e^{it}) = \arg(1 + \sqrt{2}e^{it}) = \arg z$$

4.7.: P. Determine the locus of points at which the tangent to a curve does not change its direction under the transformation

$$w = az^2 + bz$$

A. Let c be a point where the tangent to a curve does not change its direction under the transformation $w = f(z)$. Then $\arg z'(t) = \arg w'(t)$. Since $\arg w'(t) - \arg z'(t) = \arg f'(z)$, we must have $\arg f'(z) = 0$; i.e., $f'(z)$ must be a real positive number.

If $w = az^2 + bz$ and a and b are real, then $w' = 2az + b$. Hence arg $w' = 0$ for z real and $2az + b > 0$ or $z > b/(2a)$. If a and b are not real, then the locus is the set:

$$\left\{z \mid z = \frac{w' - b}{2a}, \text{ with } w' > 0\right\}$$

which is a ray starting from $-b/(2a)$.

4.9.: P. Show that the map $w = z + 1/z$ maps the circles $|z| = c$ into ellipses and the rays arg $z = \phi$ into hyperbolas. Determine the foci.

A. Set $w = U + iV$ and $z = re^{i\phi}$. Then

$$w = re^{i\phi} + \frac{1}{r}e^{-i\phi}$$

or

$$U = \cos\phi\left(r + \frac{1}{r}\right) \text{ and } V = \sin\phi\left(r - \frac{1}{r}\right)$$

Now if $r = |z| = c$, then

$$\frac{U^2}{\left(r + \frac{1}{r}\right)^2} + \frac{V^2}{\left(r - \frac{1}{r}\right)^2} = \cos^2\phi + \sin^2\phi = 1$$

i.e., the circle $|z| = r$ is mapped into an ellipse whose axes are $r + \frac{1}{r}$ and $r - \frac{1}{r}$.

On the other hand, the ray arg $z = \phi = $ constant is mapped into a hyperbola; indeed, we have

$$\left(\frac{U}{\cos\phi}\right)^2 - \left(\frac{V}{\sin\phi}\right)^2 = \left(r + \frac{1}{r}\right)^2 - \left(r - \frac{1}{r}\right)^2 = 4$$

which is a hyperbola whose axes are $2\cos\phi$ and $2\sin\phi$. The foci lie on $U = \pm 2$.

4.11.: P. Show that the image of the strip $-\pi < y \leq \pi$ under the mapping $w = 1 + e^z$ is $\mathbf{C}\backslash\{1\}$.

A. We shall show that if c is an arbitrary complex number $\neq 1$ then $\exists z \ni c = 1 + e^z$. Indeed: $e^z = c - 1 \Rightarrow |e^z| = e^x = |c - 1|$. For $c \neq 1$, we find $x = \text{Log}|c - 1|$. Also, arg $e^z = \arg e^{iy} = y = \arg(c - 1)$. Choosing $y = \text{Arg}(c - 1)$, we have $c = 1 + e^z$ with $-\pi < \text{Im } z = y \leq \pi$.

4.13.: P. Compute the cross ratio of the fourth roots of i.

A. We have defined the cross ratio of z_1, z_2, z_3, z_4 by:

$$(z_1, z_2, z_3, z_4) = \frac{(z_1 - z_2)(z_3 - z_4)}{(z_1 - z_3)(z_2 - z_4)}$$

Now since $i = e^{i\pi/2}$, its fourth roots are $z_1 = e^{i\pi/8}$, $z_2 = ie^{i\pi/8}$, $z_3 = -e^{i\pi/8}$ and $z_4 = -ie^{i\pi/8}$. Hence

$$(z_1, z_2, z_3, z_4) = \frac{e^{i\pi/8}(1-i)e^{i\pi/8}(-1+i)}{e^{i\pi/8}(1+1)e^{i\pi/8}(i+i)} = -\frac{(1-i)^2}{4i} = \frac{1}{2}$$

4.15.: P. Find the image of a circle and of a straight line by the mapping $w = 1/z$. Which circles are transformed into straight lines and which straight lines remain straight lines? In particular, find the images of—
  a) $|z - 2 - 2i| = \sqrt{5}$
  b) $\operatorname{Re} z = \pm 1, \pm 2$
  c) $\operatorname{Im} z = \pm 1, \pm 2$

A. We use the fact that a straight line contains the point at infinity and that the origin is the only point which is mapped by $w = 1/z$ into the point at infinity. Hence, since the set of circles and straight lines is mapped into itself by $w = 1/z$, it follows that all circles passing through the origin and only these circles are mapped into straight lines and that a straight line remains a straight line only if it passes through the origin.

Particular cases:
a) The circle $|z - 2 - 2i| = \sqrt{5}$ passes through the points 1, 3, i and $3i$ whose images are respectively 1, 1/3, $-i$ and $-i/3$, and does not pass through the origin. It follows by plotting the points that the image of the circle $|z - 2 - 2i| = \sqrt{5}$ is again a circle whose center is at

$$\frac{1 + \frac{1}{3}}{2} - i\frac{1 + \frac{1}{3}}{2} = \frac{2}{3}(1 - i)$$

and whose radius is $\sqrt{(\frac{1}{3})^2 + (\frac{2}{3})^2} = \frac{1}{3}\sqrt{5}$.
b) The images of these four straight lines are circles passing through the origin which have their centers respectively at $\pm\frac{1}{2}, \pm\frac{1}{4}$.
c) The images of these four straight lines are circles passing through the origin which have their centers respectively at $\mp i/2, \mp i/4$.

4.17.: P. Find a homography which maps:
  a) the points a, b, c into $0, 1, \infty$
  b) the points $0, 1, \infty$ into a, b, c
  c) the points $-1, 0, 2$ into $0, 3, 6$.

A. The homography $w = \dfrac{z - a}{\gamma z + \delta}$ maps a into 0, the homography $w = \dfrac{\alpha z + \beta}{z - c}$ maps c into ∞, and $w = k\dfrac{z - a}{z - c}$ maps a into 0 and c into ∞.

In order to have $b \to 1$, we determine k by

$$1 = k\frac{b - a}{b - c} \quad \text{or} \quad k = \frac{b - c}{b - a}$$

which finally gives

$$w = \frac{z-a}{z-c} \cdot \frac{b-c}{b-a}$$

for the homography which transforms the points a, b, c into $0, 1, \infty$.
b) Take the inverse of the homography in a) which may be written:

$$\frac{w-a}{w-c} \cdot \frac{b-c}{b-a} = z \quad \text{or} \quad w = \cdots$$

c) If we want to determine a homography mapping a, b, c into d, e, f, we factor it by first mapping $a, b, c \to 0, 1, \infty$ and then $0, 1, \infty \to d, e, f$. $a, b, c \to 0, 1, \infty$ is given by a),

$$w = \frac{z-a}{z-c} \cdot \frac{b-c}{b-a}$$

$0, 1, \infty \to d, e, f$ is given by b),

$$\frac{w-d}{w-f} \cdot \frac{e-f}{e-d} = z$$

Hence for $a, b, c \to d, e, f$, we find

$$\frac{w-d}{w-f} \cdot \frac{e-f}{e-d} = \frac{z-a}{z-c} \cdot \frac{b-c}{b-a}$$

In particular, $-1, 0, 2 \to 0, 3, 6$ gives

$$\frac{w}{w-6} \cdot \frac{-3}{3} = \frac{z+1}{z-2} \cdot \frac{-2}{1} \quad \text{or} \quad \frac{w}{6-w} = \frac{2(z+1)}{2-z}$$

Then

$$\frac{w}{6} = \frac{2(z+1)}{4+z}$$

and, finally,

$$w = \frac{12(z+1)}{4+z}$$

4.19.: P. Let $a, b, a \neq b$, be the two fixed points of a homography h. Show that h may be written as:

$$\frac{w-a}{w-b} = k \frac{z-a}{z-b}$$

where k is a complex number.

Show that the cross ratio of a, b, z, w is constant. A particular case: Determine a, b and k for the homography

$$w = \frac{5z - 2}{z + 3}$$

Show that w maps the upper half plane Im $z \geq 0$ into itself.

A. The fixed points of a homography h are the points for which $z = h(z)$, i.e., the points which remain unchanged by h.

Now suppose a and b are the fixed points of the homography; then $a = h(a)$ and $b = h(b)$. Let $h_1 = \dfrac{z - a}{z - b}$. h_1 maps $a \to 0$ and $b \to \infty$ and $h_1 \circ h \circ h_1^{-1}$ maps $0 \to 0$ and $\infty \to \infty$. $h_1 \circ h \circ h_1^{-1}$ is therefore given by:

$$h_1 \circ h \circ h_1^{-1}(z') = kz'$$

It follows that $h = h_1^{-1} \circ kz' \circ h_1$, or

$$\frac{w - a}{w - b} = k \frac{z - a}{z - b}$$

defines h.

Now consider $w = \dfrac{5z - 2}{z + 3}$; its fixed points are found by setting

$$z = \frac{5z - 2}{z + 3}$$

or $z^2 + 3z = 5z - 2$. Hence $z^2 - 2z + 2 = 0$ or $z = 1 \pm \sqrt{-1} = 1 \pm i$.

$$\frac{w - (1 + i)}{w - (1 - i)} = \frac{\dfrac{5z - 2}{z + 3} - (1 + i)}{\dfrac{5z - 2}{z + 3} - (1 - i)} = \frac{5z - 2 - z(1 + i) - 3(1 + i)}{5z - 2 - z(1 - i) - 3(1 - i)}$$

$$= \frac{z(4 - i) - (5 + 3i)}{z(4 + i) - (5 - 3i)} = \frac{4 - i}{4 + i} \cdot \frac{z - \dfrac{5 + 3i}{4 - i}}{z - \dfrac{5 - 3i}{4 + i}}$$

$$= \frac{15 - 8i}{17} \cdot \frac{z - (1 + i)}{z - (1 - i)}$$

Thus $k = (15 - 8i)/17$.

Since all the coefficients of $w = \dfrac{5z - 2}{z + 3}$ are real, the real axis is mapped into itself. Now w could map the upper half plane onto itself or onto the lower half plane. It is then sufficient to consider

$$w(z = i) = \frac{5i - 2}{i + 3} = \frac{(5i - 2)(3 - i)}{(i + 3)(3 - i)} = \frac{17i - 1}{10}$$

which is mapped into the upper half plane. Hence the upper half plane is mapped onto the upper half plane.

More generally, if $w = \dfrac{az + b}{cz + d}$ where a, b, c and d are real, then w maps the real axis onto the real axis and maps the upper half plane onto the upper half plane if the determinant

$$\begin{vmatrix} a & b \\ c & d \end{vmatrix} > 0$$

and maps the upper half plane into the lower half plane if

$$\begin{vmatrix} a & b \\ c & d \end{vmatrix} < 0$$

Indeed

$$w(z = i) = \frac{ai + b}{ci + d} = \frac{(ai + b)(d - ci)}{c^2 + d^2}$$

and

$$\operatorname{Im} w(z = i) = \begin{vmatrix} a & b \\ c & d \end{vmatrix}$$

In order to see that the fact that one point of the upper half plane maps into a point of the upper half plane is sufficient to prove that the whole upper half plane is mapped onto the upper half plane, let us use the connectedness principle. Denote by U the set $\{z \mid \operatorname{Im} z > 0\}$ and by E the set $\{z \mid z \in U,\ h(z) \in U\}$. E is not empty since $i \in E$. E is *open in the relative topology* since $z \in E \Rightarrow \exists\, \varepsilon \ni d(z, \varepsilon) \subset E$ by virtue of the continuity of h. E is *closed in the relative topology*: indeed, if $\{z_n\} \to z$ with $z_n \in U$, and $z \in U$, and if $\operatorname{Im}(h(z_n)) > 0$ then $\operatorname{Im}(h(z)) \geq 0$ by virtue of the continuity of h. Now $\operatorname{Im}(h(z)) \neq 0$ since $z \notin$ real axis. Hence $\operatorname{Im} h(z) > 0$ and $E = U$.

4.21.: P. Find the homographies which carry the circle $|z - c| = r$ onto the circle $|z| = 1$.

A. We construct first an affine transformation which maps $c \to 0$ and reduces the radius to 1:

$$z' = \frac{z - c}{r}$$

We continue with a homography which maps the circle $|z| = 1$ onto itself.

Such a homography maps a point $\alpha \to 0 (|\alpha| \neq 1)$ and $1/\bar{\alpha} \to \infty$. It follows that

$$w = k \cdot \frac{z' - \alpha}{z' - \dfrac{1}{\bar{\alpha}}} \quad \text{or} \quad w = k' \cdot \frac{z' - \alpha}{1 - \bar{\alpha} z'}$$

Now for $z' = 1$ we must have $|w| = 1$; hence

$$\left| k' \frac{1-\alpha}{1-\bar{\alpha}} \right| = 1, \quad \text{or since} \quad \left| \frac{1-\alpha}{1-\bar{\alpha}} \right| = 1$$

we must have $|k'| = 1$, i.e., $k' = e^{it}$. Finally,

$$w = e^{it} \frac{z' - \alpha}{1 - \bar{\alpha} z'}$$

is the most general homography which leaves the circle $|z| = 1$ unchanged.

It follows that the most general homography which carries the circle $|z - c| = r$ onto the circle $|z| = 1$ is given by:

$$w = e^{it} \cdot \frac{\frac{z-c}{r} - \alpha}{1 - \bar{\alpha} \cdot \frac{z-c}{r}} = e^{it} \cdot \frac{z - c - \alpha r}{r - \bar{\alpha} z + \bar{\alpha} c}$$

4.23.: P. Show that the six homographies

$$z, \quad \frac{1}{z}, \quad 1 - z, \quad \frac{1}{1-z}, \quad \frac{z-1}{z}, \quad \frac{z}{z-1}$$

form a subgroup of H.

A. Denote respectively by A, B, C, D, E, F the six homographies. Then we have the following multiplication table

		second factor					
		A	B	C	D	E	F
first factor	A	A	B	C	D	E	F
	B	B	A	D	C	F	E
	C	C	E	A	F	B	D
	D	D	F	B	E	A	C
	E	E	C	F	A	D	B
	F	F	D	E	B	C	A

which shows us that the six homographies form a group.

Example: $E \cdot B$ means: Table B, $z' = \frac{1}{z}$, then

$$E: z'' = \frac{z' - 1}{z'} = \frac{\frac{1}{z} - 1}{\frac{1}{z}} = 1 - z,$$

which is C. The elements involved in this operation are underlined in the table.

4.25.: P. Find a mapping which transforms the domain D defined by $0 < \arg z < \pi/4$ onto the unit disc.

A. We first use the transformation $z' = z^4$. Since $\arg z' = 4 \arg z$, the transformation $z' = z^4$ maps the sector $0 < \arg z < \pi/4$ onto the upper half plane $\operatorname{Im} z > 0$ or $0 < \arg z < \pi$. Then we use a homography h which maps the upper half plane onto $d(0, 1)$. Let $\alpha(\operatorname{Im} \alpha > 0)$ be mapped into the origin; then $\bar{\alpha}$ is mapped into ∞. Hence

$$h(z') = k \frac{z' - \alpha}{z' - \bar{\alpha}}$$

Furthermore, we want a point on the real axis to be mapped into the circle $|z'| = 1$. For real values of z' we have $\left|\frac{z' - \alpha}{z' - \bar{\alpha}}\right| = 1$. Therefore $|k| = 1$ or $k = e^{i\phi}$. Hence

$$h(z') = e^{i\phi} \frac{z' - \alpha}{z' - \bar{\alpha}}$$

and the mapping which transforms the angular region $D = \{z \mid 0 < \arg z < \pi/4\}$ onto the unit disc is

$$w = e^{i\phi} \frac{z^4 - \alpha}{z^4 - \bar{\alpha}}$$

4.27.: P. Show that $\operatorname{Log} \dfrac{1 + z}{1 - z}$ is an isomorphism of the disc $d(0, 1)$ with the strip $-\pi/2 < \operatorname{Im} z < \pi/2$.

A. Following the hint: $z_1 = 1 - z$ is an isomorphism $d(0, 1) \to d(1, 1)$; $z_2 = 1/z_1$ is an isomorphism $d(1, 1) \to \operatorname{Re}(z_2) > \frac{1}{2}$, $z_3 = -1 + 2z_2$ is an isomorphism $\operatorname{Re}(z_2) > \frac{1}{2} \to \operatorname{Re}(z_3) > 0$; and $z_4 = \operatorname{Log} z_3 = \log |z_3| + i$. $\operatorname{Arg} z_3$ defines an isomorphism $\operatorname{Re}(z_3) > 0 \to -\pi/2 < \operatorname{Im}(z_4) < \pi/2$. Hence $z_4 = \operatorname{Log} z_3 = \operatorname{Log}(1 + 2z_2) = \operatorname{Log}(1 + 2/z_1) = \operatorname{Log}\left(1 + \dfrac{2}{1 - z}\right)$

$= \operatorname{Log} \dfrac{1 + z}{1 - z}$ defines an isomorphism $d(0, 1) \to \{z \mid -\pi/2 < \operatorname{Im} z < \pi/2\}$.

4.29.: P. Determine the image of the given domain D by the following conformal mappings:
  a) $w = e^z$; D: rectangle with vertices at a, $a + ib$, c, $c + ib$ (a, b, c real and $b < 2\pi$)

  b) $w = \dfrac{1}{z}$; D: $\operatorname{Re} z > 0$

  c) $w = \dfrac{i - z}{i + z}$; D: $\operatorname{Im} z > 0$

d) $w = \dfrac{z-1}{z+1}$; $\quad D: \operatorname{Re} z > 0$

e) $w = z + \dfrac{1}{z}$; $\quad D: |z| < 1, \quad \operatorname{Im} z > 0$

f) $w = \dfrac{1 - \cos z}{1 + \cos z}$; $\quad D: 0 < \operatorname{Re} z < \dfrac{\pi}{2}$

g) $w = \left(\dfrac{z+1}{z-1}\right)^2$; $\quad D: |z| < 1, \quad \operatorname{Im} z > 0$

h) $w = \sqrt{z}$; $\quad D: y^2 > 4ax$

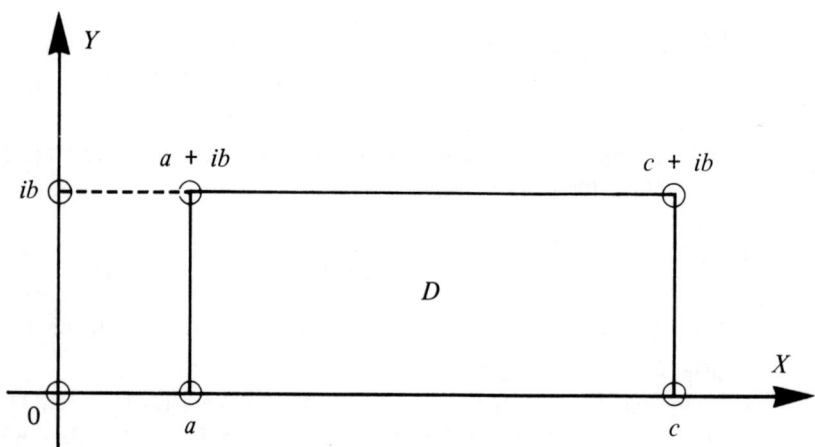

Figure A.12

A. a) Using $|e^z| = |e^{x+iy}| = e^x$ and $\arg e^z = \arg e^{x+iy} = y$, we have: $a \le x \le c \Leftrightarrow e^a \le e^x \le e^c \Leftrightarrow e^a \le |e^z| \le e^c$ and $0 \le y \le b \Leftrightarrow 0 \le \arg e^z \le b$. Hence the image of D by $w = e^z$ covers the angle $0 \le \arg w \le b$, between the circles $\partial(0, e^a)$ and $\partial(0, e^c)$.

b) Using $|1/z| = 1/|z|$ and $\arg(1/z) = -\arg z$, we see that $w = 1/z$ is an isomorphism of $\operatorname{Re} z > 0$ onto $\operatorname{Re} z > 0$. Indeed: $-\pi/2 < \operatorname{Arg} z < \pi/2 \Leftrightarrow \pi/2 > \operatorname{Arg} w > -\pi/2$ and $0 < |z| < \infty \Leftrightarrow \infty > |w| > 0$.

c) $|w| = \left|\dfrac{i-z}{i+z}\right| = 1$ for real z, i.e., the real axis is mapped onto the unit circle $|w| = 1$. Since $z = i \to w = 0$, the upper half plane $\operatorname{Im} z > 0$ is mapped onto the unit disc $\mathscr{d}(0, 1)$.

d) $|w| = \left|\dfrac{z-1}{z+1}\right| = 1$ for imaginary z, i.e., the imaginary axis is mapped onto the unit circle $|w| = 1$. Since $z = 1 \to w = 0$, we conclude that $\operatorname{Re} z > 0$ is mapped onto $d(0, 1)$.

e) Setting $w = U + iV$ and $z = re^{i\phi}$, we have

$$U = \cos\phi\left(r + \dfrac{1}{r}\right) \quad \text{and} \quad V = \sin\phi\left(r - \dfrac{1}{r}\right)$$

Now $\operatorname{Im} z > 0 \Leftrightarrow \pi > \phi > 0$ and $|z| < 1 \Leftrightarrow r < 1$. Hence $\sin\phi > 0$ and $r - 1/r < 0 \Rightarrow V < 0$ with V taking all negative values. $U \geq 0$ for $0 < \phi \leq \pi/2$ and $U \leq 0$ for $\pi/2 \leq \phi < \pi$, U taking all real values.

In conclusion, $w = z + \dfrac{1}{z}$ maps the semidisc $(|z| < 1) \cap (\operatorname{Im} z > 0)$ onto the lower half plane $\operatorname{Im} w < 0$ (Figure A.13).

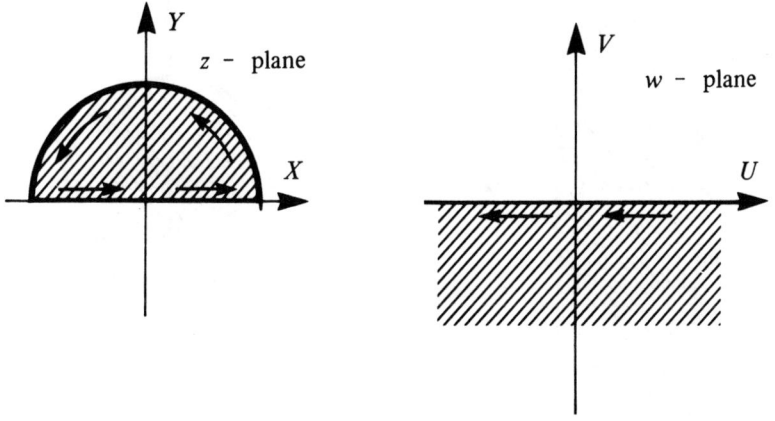

Figure A.13

f) $$w = \dfrac{1 - \cos z}{1 + \cos z} = \dfrac{1 - \dfrac{e^{iz} + e^{-iz}}{2}}{1 + \dfrac{e^{iz} + e^{-iz}}{2}} = -\dfrac{e^{2iz} - 2e^{iz} + 1}{e^{2iz} + 2e^{iz} + 1}$$

$$= -\left(\dfrac{e^{iz} - 1}{e^{iz} + 1}\right)^2$$

Let us factor w in the following way: $z_1 = e^{iz}$, $z_2 = \dfrac{z_1 - 1}{z_1 + 1}$ and $z_3 = -z_2^2 = w$. Now $\arg z_1 = \arg e^{iz} = \arg e^{ix-y} = x = \operatorname{Re} z$. Hence $0 < \operatorname{Re} z < \pi/2 \Rightarrow 0 < \arg z_1 < \pi/2$. (Figure A.14).

From part d) we know that $\operatorname{Re} z_1 > 0$ is mapped onto $|z_2| < 1$ by $z_2 = \dfrac{z_1 - 1}{z_1 + 1}$. Since all the coefficients in $z_2 = \dfrac{z_1 - 1}{z_1 + 1}$ are real, the real axis is mapped onto itself. Since $i - 1$ is mapped into $\dfrac{i-2}{i} = 1 + 2i$, the upper half plane is mapped onto itself by z_2. So $z_2 = \dfrac{z_1 - 1}{z_1 + 1}$ is an isomorphism of $0 < \arg z_1 < \pi/2$ onto the semidisc $(|z_2| < 1) \cap (\operatorname{Im} z_2 > 0)$, or $(0 < \arg z_2 < \pi) \cap (|z_2| < 1)$.

Now since $\arg z_3 = \pi + 2 \arg z_2$ and $|z_3| = |z_2|^2$, it follows that $0 < \arg z_2 < \pi \Leftrightarrow \pi < \arg z_3 < 3\pi$ and $|z_2| < 1 \Leftrightarrow |z_3| < 1$.

In conclusion, we find that the strip $0 < \operatorname{Re} z < \pi/2$ is mapped by $w = \dfrac{1 - \cos z}{1 + \cos z}$ onto $(|w| < 1) \cap (-\pi < \arg w < \pi)$.

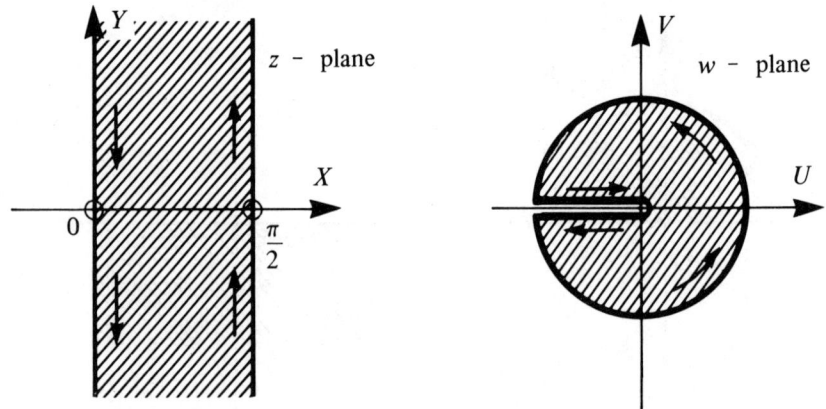

Figure A.14

g) We factor again by $z_1 = \dfrac{z+1}{z-1}$ and $w = z_1^2$. Since all the coefficients of $z_1 = \dfrac{z+1}{z-1}$ are real, it follows that z_1 maps the real axis onto itself. Since i is mapped into $\dfrac{1+i}{i-1} = -\dfrac{(1+i)^2}{2} = -i$, the upper half plane is mapped by z_1 onto the lower half plane. On the other hand, if z describes the circle $|z| = 1$, $\dfrac{z+1}{z-1}$ describes the imaginary axis $\operatorname{Re} z = x = 0$; indeed, $z = -1 \to z_1 = 0$, $z = 1 \to z_1 = \infty$ and $z = i \to z_1 = \dfrac{i+1}{i-1} = -\dfrac{(1+i)^2}{2} = -i$. Since $z = 0 \to z_1 = -1$, the disc $d(0, 1)$ is mapped by z_1 onto $\operatorname{Re} z_1 < 0$.

We see thus that the semidisc $(|z| < 1) \cap (\operatorname{Im} z > 0)$ is mapped by $z_1 = \dfrac{z+1}{z-1}$ onto $(\operatorname{Re} z_1 < 0) \cap (\operatorname{Im} z_1 < 0)$, or $-\pi < \arg z_1 < -\pi/2$.
Now since $\arg w = 2 \arg z$, the set $-\pi < \arg z_1 < -\pi/2$ is mapped onto $-2\pi < \arg w < -\pi$ or $0 < \arg w < \pi$, which is the upper half plane $\operatorname{Im} z > 0$. In conclusion, $w = \left(\dfrac{z+1}{z-1}\right)^2$ maps the set $D = (|z| < 1) \cap (\operatorname{Im} z > 0)$ onto the upper half plane $\operatorname{Im} z > 0$.

h) Setting $w = U + iV$ and $z = x + iy$ the mapping $w = \sqrt{z}$ or $w^2 = z$ may be written as

$$U^2 - V^2 = x$$
$$2UV = y$$

It follows that $y^2 > 4ax \Leftrightarrow 4U^2V^2 > 4a(U^2 - V^2)$ or

$$\frac{1}{V^2} - \frac{1}{U^2} < \frac{1}{a}$$

which is then the image of $y^2 > 4ax$ by $w = \sqrt{z}$.

4.31.: P. Determine the function which is harmonic in the annulus $r \leq |z - c| \leq R$ and which attains the value a on the circle $\partial(c, r)$ and the value A on the circle $\partial(c, R)$. (*Hint*: Consider $\operatorname{Re} \log(z - c)$).

A. The function $\operatorname{Re} \log(z - c) = \log|z - c|$ as the real part of an analytic function is harmonic throughout $\mathbf{C}$ except at $z = c$. Clearly $\alpha \log|z - c| + \beta$ is also harmonic and we can determine α and $\beta \ni \alpha \log r + \beta = A$ and $\ni \alpha \log R + \beta = B$; hence

$$\alpha = \frac{B - A}{\log R - \log r} \quad \text{and} \quad \beta = \frac{A \log R - B \log r}{\log R - \log r}$$

It follows that the harmonic function h we are looking for can be written as

$$h(x, y) = \frac{B - A}{\log R - \log r} \log|z - c| + \frac{A \log R - B \log r}{\log R - \log r}$$

Index

Abelian group, 9
absolute value, 40
affine transformation, 200
algebra, 11
analytic continuation, 149
    uniqueness of, 150
analyticity, 37
argument, 40
automorphism, 73

Banach space, 27
boundary of a set, 14
boundedness, 29
    total, 30

Cartesian product, 4
Cauchy-Goursat theorem, 95
Cauchy-Riemann equations, 68
Cauchy's criterion, 135
Cauchy sequence, 25
Cauchy's formula, 106, 116
Cauchy's inequalities, 145
Cauchy's theorem, 99
chain rule, 65
closed disc, 51
closed set, 51
compact exhaustion, 121
compactness, 20
complete metric space, 26
complex differentiability, 37
conformality, 38, 192
conjugate numbers, 41

connectedness, 17
    polygonal, 55
continuity, sequential, 28
    uniform, 32
continuous function, 15
convex set, 57
countable set, 7
cross ratio, invariance of, 204

DeMoivre's relations, 3, 13
DeMoivre's rule, 44
dense sets, 30
differentiability, 63

Euler's formula, 149
equivalence relation, 4

field, 10
functions, equibounded, 34
    continuous, 15
    equicontinuous, 34
    meromorphic, 164

Gauss-d'Alembert theorem, 145
Gaussian plane, 40
greatest lower bound, 18
group, 9

harmonic conjugate, 127
harmonic function, 126
Hausdorff space, 20
holomorphicity, 37, 93

Index

homeomorphic sets, 16
homeomorphism, 15
homography—Moebius transformation, 199
homotopy of closed curves, 78

index of closed curve, 110
interior point, 14, 51
inversion, 202
isomorphicity, 10
isomorphism, 73

Jordan curve, 58, 59

Laurent series, 156
    principle part, 157
least upper bound, 18
Lebesgue number, 31
Liouville's theorem, 145
logarithmic residue, 183
lower bound, 18

meromorphic function, 164
metric space, 22
Moebius transformation, 199
Morera's theorem, 117

neighborhood of a point, 13
normed vector space, 24

open disc, 51
open set, 51

periodic functions, 165
permanence of functional relations, 153
poles of order p, 161
precompact set, 30
punctured disc, 57

radius of convergence, 138
relative topology, 16
residues, 103, 169
residue theorem, 171
Riemann relation, 63
Riemann's mapping theorem, 208
Riemann sphere, 47
ring, 10
Rouché's theorem, 185

Schwarz's lemma, 146
singularities, isolated, 160
    essential, 161
starlike domain, 56
symmetry, principle of, 205

Taylor series, 143
    uniqueness of, 144
topological space, 13
topology, 12
triangle inequality, 42

upper bound, 18

vector space, 11

Weierstrass' criterion, 137
Weierstrass' theorem, 163